Proceedings
of the
STEKLOV INSTITUTE
OF MATHEMATICS

1984, ISSUE 3

Theory and Applications of Differentiable Functions of Several Variables. IX

Edited by

S. M. Nikol'skiĭ

Translation of
ТРУДЫ
ордена Ленина
МАТЕМАТИЧЕСКОГО ИНСТИТУТА
имени В. А. СТЕКЛОВА

Том 161 (1983)

АКАДЕМИЯ НАУК

СОЮЗА СОВЕТСКИХ СОЦИАЛИСТИЧЕСКИХ РЕСПУБЛИК

ТРУДЫ

ордена Ленина

МАТЕМАТИЧЕСКОГО ИНСТИТУТА

имени В. А. СТЕКЛОВА

CLXI

ИССЛЕДОВАНИЯ ПО ТЕОРИИ ДИФФЕРЕНЦИРУЕМЫХ ФУНКЦИЙ МНОГИХ ПЕРЕМЕННЫХ И ЕЕ ПРИЛОЖЕНИЯМ. IX

Заместитель ответственного редактора (Assistant to the editor-in-chief)
профессор Е. А. ВОЛКОВ (E. A. Volkov)

издательство "наука"
МОСКВА 1983

Library of Congress Cataloging in Publication Data

Issledovaniia po teorii differentsiruemykh funktsiĭ mnogikh peremennykh i ee prilozheniiam.
IX. English. Theory and applications of differentiable functions of several variables. IX.
(Proceedings of the Steklov Institute of Mathematics, ISSN 0081-5438; 1984, issue 3)
Translation of: Issledovaniia po teorii differentsiruemykh funktsiĭ mnogikh peremennykh i ee prilozheniiam. IX.
Number in Russian series statement: tom 161 (1983).
Includes bibliographical references.
1. Differentiable functions–Addresses, essays, lectures. 2. Functions of several real variables–Addresses, essays, lectures. 3. Differential equations, Partial–Addresses, essays, lectures. I. Nikol′skiĭ, S. M. (Sergeĭ Mikhaĭlovich), 1905– . II. Series: Trudy Ordena Lenina Mathematicheskogo instituta imeni V. A. Steklova. English; 1984, issue 3.
QA1.A413 no. 161 [QA331.5] 510s 84-24501
ISBN 0-8218-3083-X [515.8′4]

September 1984

Translation authorized by the All-Union Agency for Author's Rights, Moscow

The paper used in this journal is acid-free and falls within the guidelines established to ensure permanence and durability.

Reprinted 1986

PROCEEDINGS OF THE STEKLOV INSTITUTE OF MATHEMATICS IN THE ACADEMY OF SCIENCES OF THE USSR

TABLE OF CONTENTS

BERKOLAĬKO, M. Z. and OVCHINNIKOV, V. I. Inequalities for entire functions of exponential type in symmetric spaces 1

BERKOLAĬKO, M. Z. Imbedding theorems with different metrics and different dimensions for generalized Besov spaces 19

BESOV, O. V. On the denseness of the compactly supported functions in a weighted Sobolev space 33

BURENKOV, V. I. and GOL′DMAN, M. L. On the interconnection of norms of operators in periodic and nonperiodic function spaces 53

DOMYSHEVA, L. N. On the denseness of the compactly supported functions in weighted spaces 113

KALYABIN, G. A. Estimates of the capacity of sets with respect to generalized Lizorkin-Treibel classes and weighted Sobolev classes 119

KOKILASHVILI, V. M. On weighted Lizorkin-Triebel spaces. Singular integrals, multipliers, imbedding theorems 135

KUDRYAVTSEV, L. D. On the question of polynomial traces 163

LIZORKIN, P. I. and NIKOL′SKIĬ, S. M. Coercive properties of an elliptic equation with degeneracy and a generalized right-hand side 171

MAGARIL-IL′YAEV, G. G. Inequalities for derivatives, and duality 199

OTELBAEV, M. Coercive estimates and separation theorems for elliptic equations in R^n 213

STEPANOV, V. D. Characteristic properties of convolution integral operators 241

KALYABIN, G. A. Correction to the paper "Descriptions of functions from classes of Besov-Lizorkin-Triebel type 251

ABSTRACT. This collection is devoted to the study of various problems in the theory of differentiable functions of several real variables and its applications to partial differential equations. The papers herein 1) establish imbedding theorems and the boundedness of certain operators in new function spaces, 2) obtain coercive estimates and separability theorems for elliptic operators, 3) find a connection between operator norms in periodic and nonperiodic function spaces, 4) study questions of the denseness of compactly supported functions in weighted spaces, 5) investigate properties of polynomial traces, 6) establish estimates of the capacity of sets in function spaces, 7) study singular integral operators and multipliers in weighted spaces.

The preceding parts of this series on differentiable functions have appeared in Trudy Mat. Inst. Steklov. **77** (1965), **89** (1967), **105** (1969), **117** (1972), **131** (1974), **140** (1976), **150** (1979), **156** (1980) = Proc. Steklov Inst. Math. **77** (1967), **89** (1968), **105** (1971), **117** (1974), **131** (1975), **1979**; issue 1; **1981**, issue 4; **1983**, issue 2.

1980 *Mathematics Subject Classification*. Primary 26C99, 26D10, 30D05, 30D15, 31C15, 35B45, 35J30, 35J50, 35J55, 41A25, 41A30, 41A44, 41A50, 42B15, 42B20, 44A35, 46E30, 46E35, 47F05; Secondary 30D15, 35B65, 42A85, 42B15, 42B25, 44A15, 46E30, 46E35, 46F05.

RUSSIAN TABLE OF CONTENTS*

Берколайко М. З., Овчинников В. И. Неравенства для целых функций экспоненциального типа в симметричных пространствах 3

Берколайко М. З. Теоремы вложения разных метрик и измерений обобщенных пространств Бесова 18

Бесов О. В. О плотности финитных функций в весовом пространстве С. Л. Соболева 29

Буренков В. И., Гольдман М. Л. О взаимосвязи норм операторов в периодических и непериодических функциональных пространствах 47

Домышева Л. И. О плотности финитных функций в весовых пространствах 106

Калябин Г. А. Оценки емкости множеств относительно обобщенных классов Лизоркина—Трибеля и весовых классов Соболева 111

Кокилашвили В. М. Весовые пространства Лизоркина—Трибеля. Сингулярные интегралы, мультипликаторы, теоремы вложения 125

Кудрявцев Л. Д. К вопросу о полиномиальных следах 149

Лизоркин П. И., Никольский С. М. Коэрцитивные свойства эллиптического уравнения с вырождением и обобщенной правой частью 157

Магарил-Ильяев Г. Г. Неравенства для производных и двойственность . . 183

Отелбаев М. Коэрцитивные оценки и теоремы разделимости для эллиптических уравнений в R^n 195

Степанов В. Д. Характеристические свойства интегральных операторов свертки 217

Калябин Г. А. Уточнение к работе «Описания функций из классов типа Бесова—Лизоркина—Трибеля» 226

---

* The American Mathematical Society scheme for transliteration of Cyrillic may be found at the end of index issues of *Mathematical Reviews*.

Correspondence between *Trudy Mat. Inst. Steklov.* and *Proc. Steklov Inst. Math.*

Russian imprint	Vol.	English imprint
1966	74	1967
1965	75	1967
1965	76	1967
1965	77	1967
1965	78	1967
1965	79	1966
1965	80	1968
1966	81	1968
1966	82	1967
1965	83	1967
1965	84	1968
1966	85	1967
1965	86	1967
1966	87	1967
1967	88	1969
1967	89	1968
1967	90	1969
1967	91	1969
1966	92	1968
1967	93	1970
1968	94	1969
1968	95	1971
1968	96	1970
1968	97	1969
1968	98	1971
1967	99	1968
1971	100	1974
1972	101	1975
1967	102	1970
1968	103	1970
1968	104	1971
1969	105	1971
1969	106	1972

Russian imprint	Vol.	English imprint
1971	107	1976
1968	108	1971
1971	109	1974
1970	110	1972
1970	111	1972
1971	112	1973
1970	113	1972
1970	114	1974
1971	115	1974
1971	116	1973
1972	117	1974
1972	118	1976
1973	119	1976
1974	120	1976
1972	121	1974
1973	122	1975
1973	123	1975
1976	124	1978, Issue 3
1973	125	1975
1973	126	1975
1975	127	1977
1972	128	1974
1973	129	1976
1978	130	1979, Issue 4
1974	131	1975
1973	132	1975
1973	133	1977
1975	134	1977
1975	135	1978, Issue 1
1975	136	1978, Issue 2
1976	137	1978, Issue 4
1975	138	1977

Russian imprint	Vol.	English imprint Year	Issue
1980	139	1982	1
1976	140	1979	1
1976	141	1979	2
1976	142	1979	3
1977	143	1980	1
1979	144	1980	2
1980	145	1981	1
1978	146	1980	3
1980	147	1981	2
1978	148	1980	4
1978	149	1981	3
1979	150	1981	4
1980	151	1982	2
1980	152	1982	3
1981	153	1982	4
1983	154	1984	4
1981	155	1983	1
1980	156	1983	2
1981	157	1983	3
1981	158	1983	4
1983	159	1984	2
1982	160	1984	1
1983	161	1984	3

Труды Матем. Инст. Стеклов.
161(1983)

Proceedings of the Steklov
Institute of Mathematics
1984, Issue 3

INEQUALITIES FOR ENTIRE FUNCTIONS OF EXPONENTIAL TYPE IN SYMMETRIC SPACES

UDC 517.83

M. Z. BERKOLAĬKO AND V. I. OVCHINNIKOV

ABSTRACT. Sharp estimates are given for the norms of operators of the imbeddings $\mathfrak{M}_{\bar{\mu},E_1} \hookrightarrow \mathfrak{M}_{\bar{\mu},E_2}$ (inequality with different metrics) and $\mathfrak{M}_{\bar{\mu},E_1(R^n)} \hookrightarrow (L_\infty(\mathfrak{M}_{\bar{\mu},E_2(R^m)}))$ (inequality with different metrics and dimensions; inequality with different dimensions when $n \neq m$). Here the E_i are arbitrary symmetric spaces on R^n, $\bar{\mu} = (\mu_1,\dots,\mu_n)$, $\mu_j \geqslant 1$, $1 \leqslant j \leqslant n$, and $\mathfrak{M}_{\bar{\mu},E_i}$ is the subspace of the entire functions of exponential type $\bar{\mu}$ whose restrictions to R^n belong to E_i. Classical inequalities of Jackson and Nikol'skiĭ are obtained from these results when $E_i = L_{p_i}$ $(1 \leqslant p_1 \leqslant p_2 \leqslant \infty)$.

Bibliography: 10 titles.

Let $\mathfrak{M}_{\bar{\mu},p}(R^n)$ $(1 \leqslant p \leqslant \infty)$ be the subspace of $L_p(R^n)$ consisting of the entire functions of exponential type $\bar{\mu}$ (EFET $\bar{\mu}$), i.e., the subspace of all entire functions which admit for any $\varepsilon > 0$ an estimate

$$|g(z_1,\dots,z_n)| \leqslant A_\varepsilon \exp\left(\sum_{j=1}^{n} (\mu_j + \varepsilon)|z_j|\right) \qquad (z_j \in C)$$

and whose restrictions to R^n belong to $L_p(R^n)$ (see [1]).

A well-known role is played by the inequalities of Nikol'skiĭ [1], which give upper estimates of the norms of operators effecting the imbedding $\mathfrak{M}_{\bar{\mu},p_1}(R^n) \hookrightarrow \mathfrak{M}_{\bar{\mu},p_2}(R^n)$ for $1 \leqslant p_1 \leqslant p_2 \leqslant \infty$ (inequality with different metrics) and of the imbedding $\mathfrak{M}_{\bar{\mu},p}(R^n) \hookrightarrow L_\infty(\mathfrak{M}_{\bar{\mu},p}(R^m))$ for $1 \leqslant p \leqslant \infty$, $R^n = R^m \times R^{n-m}$ (inequality with different dimensions) that are sharp with respect to $\bar{\mu}$ and do not depend on p_i $(i = 1, 2)$. Generalization of these inequalities to the case where $L_p(R^n)$ is replaced by an arbitrary symmetric space $E(R^n)$(1) is not only of natural interest from the point of view of function theory, but also can be used to obtain sharp imbedding theorems for generalized Besov spaces [4].

In [5] Rodin obtained a sharp inequality with different metrics for trigonometric polynomials of a single variable in the case of the two types of imbeddings $\mathfrak{T}_{n,E} \hookrightarrow \mathfrak{T}_{n,\infty}$ and $\mathfrak{T}_{n,\mathscr{M}_{\psi_1}} \hookrightarrow \mathfrak{T}_{n,\Lambda_{\psi_2}}$, where $\mathfrak{T}_n$ is the subspace of the corresponding symmetric space of measurable functions on $[0, 2\pi]$ which consists of trigonometric polynomials of degree at most n, and $\mathscr{M}_\psi$ and Λ_ψ are Marcinkiewicz and Lorentz spaces, respectively. In [6] Ganzburg carried over these results to subspaces of EFET

1980 *Mathematics Subject Classification*. Primary 30D15, 46E30, 46E35.

(1) See, for example, [2] for all the definitions and facts relating to symmetric spaces.

(isotropic case), and used the second result to solve the problem of the imbedding $\mathfrak{M}_{\mu,E_1} \hookrightarrow \mathfrak{M}_{\mu,E_2}$ and to give an upper estimate for the norm of the operator of this imbedding. But, aside from the fact that this estimate is not sharp in a number of cases, it does not yet give us Nikol'skiĭ's inequality in full scope: for $E_i = L_{p_i}$ $(1 \leqslant p_1 < p_2 \leqslant \infty)$ there appears a factor on the right-hand side that grows unboundedly as $p_1 \to p_2$ and thus does not allow us to consider the trivial case $p_1 = p_2$ (or, more generally, $E_1 = E_2$).

We mention also the article [7] of Nessel and Wilmes, in which the inequality with different metrics is extended from spaces forming Banach couples to spaces constructed by the K-method from these couples.

In this article we give a sharp inequality with different metrics for arbitrary symmetric spaces and an inequality with different dimensions that cannot be improved in the whole class of symmetric spaces. There are four sections: the necessary notation is introduced in §0; an inequality with different metrics is given in §1; in §2 we discuss situations in which the norm of the operator of the imbedding $\mathfrak{M}_{\bar{\mu},E_1} \hookrightarrow \mathfrak{M}_{\bar{\mu},E_2}$ is described in terms of the fundamental functions of the spaces E_i (such a description is most convenient in applications); finally, in §3 we prove a sharp inequality with different dimensions.

§0. Some notation and definitions

Let $E(0,\infty)$ be a symmetric space of measurable functions on $(0,\infty)$. By $E(R^n)$ we denote the corresponding space of functions on the n-dimensional Euclidean space R^n, i.e., the collection of measurable (with respect to n-dimensional Lebesgue measure) functions $f(x)$, $x = (x_1, \dots, x_n)$, such that the nonincreasing rearrangement $f^*(t)$ (with respect to n-dimensional Lebesgue measure) belongs to $E(0,\infty)$; then $\|f\|_{E(R^n)} = \|f^*\|_{E(0,\infty)}$. It is not hard to verify that this scheme encompasses all the examples of symmetric spaces, in particular, L_p, the Lorentz spaces $L_{p,q}$ and Λ_ψ, the Marcinkiewicz spaces $\mathscr{M}_\psi$, the Orlicz spaces L_N^*, and so on. In what follows we omit the symbol R^n in the notation for spaces wherever the dimension of the domain does not play a role. The characteristic function of a set $e \subset R^n$ will be denoted by κ_e.

Let E_d^t denote the sequence space generated as follows by E for any $t > 0$: $\{a_j\}_1^\infty \in E_d^t$, if

$$\sum_{j=1}^{\infty} a_j \kappa_{((j-1)/t,\, j/t)} \in E, \qquad \|\{a_j\}\|_{E_d^t} = \left\| \sum_{j=1}^{\infty} a_j \kappa_{((j-1)/t,\, j/t)} \right\|_E; \quad E_d = E_d^1.$$

It is well known that the dilation operator $(\sigma_\tau f)(s) = f(s/\tau)$ is bounded on any symmetric space $E(0,\infty)$. Therefore, the sequence spaces $E_d^{t_1}$ and $E_d^{t_2}$ coincide and the norms are equivalent for different t_1 and t_2.

The fundamental function of the space E is the quasiconcave function $\psi_E(t)$ ($\psi_E(t) \nearrow$, $\psi_E(t)/t \searrow$) defined as $\psi_E(t) = \|\kappa_e\|_E = \|\kappa_{(0,t)}(s)\|_{E(0,\infty)}$, where $\operatorname{meas} e = t$.

We define the involution

$$\psi(t) \mapsto \psi^*(t) = 1/\psi(1/t)$$

on the collection of quasiconcave functions.

As usual, E' denotes the symmetric space dual to E.

Suppose that the spaces E_i $(i = 1, 2)$ are such that $E_1 \cap L_\infty \hookrightarrow E_2$. For $0 < s$, $t < \infty$ we consider the quantity (see [8])

$$\varphi(s,t) = \sup_{\|f\|_{E_1} \leqslant s,\, \|f\|_{L_\infty} \leqslant t} \|f\|_{E_2} \qquad (f \in E_1 \cap L_\infty). \tag{1}$$

The function $\varphi(s, t)$ obviously is nondecreasing in each argument and is 1-homogeneous, and $\varphi(1, t)$ is quasiconcave. The main property of this function (which follows easily from the definition) is that

$$\|f\|_{E_2} \leqslant \varphi\left(\|f\|_{E_2}, \|f\|_{L_\infty}\right) \qquad (f \in E_1 \cap L_\infty). \tag{2}$$

In particular, if $E_1 = L_{p_1}$ and $E_2 = L_{p_2}$, then $\varphi(s,t) = s^{1-\theta} \cdot t^\theta$, where $\theta = 1 - p_1/p_2$.

Further, let $\bar{\mu} = (\mu_1, \ldots, \mu_n)$ and $|\bar{\mu}| = \prod_1^n \mu_j$ $(\mu_j > 0)$. Denote by $\mathfrak{M}_{\bar{\mu},E}$ the (linear subspace of EFET $\bar{\mu}$ whose restrictions to R^n are in E. It will be shown below that $\mathfrak{M}_{\bar{\mu},E}$ is closed in E. When $E = L_p$ $(1 \leqslant p \leqslant \infty)$, the classical notation $\mathfrak{M}_{\bar{\mu},p}$ is preserved for $\mathfrak{M}_{\bar{\mu},E}$.

By $\mathscr{I}_{E,F}$ we denote the norm of the operator of an imbedding $E \hookrightarrow F$ of two Banach spaces E and F. For brevity, the norm of the operator of the imbedding $\mathfrak{M}_{\bar{\mu},E_1} \hookrightarrow \mathfrak{M}_{\bar{\mu},E_2}$ is denoted by $\mathscr{I}_{E_1,E_2}(\bar{\mu})$.

Following [1], we use the sign $\ll$ in cases where an inequality holds with a universal constant depending at most on n and whose exact value is not of interest to us. Two quantities A and B are said to be *equivalent* $\asymp$ if $A \ll B \ll A$ (with different constants).

§1. An inequality with different metrics

THEOREM 1. *For any symmetric space E*

$$\|g\|_{L_\infty} \ll \psi_E^*(|\bar{\mu}|)\|g\|_E \qquad (g \in \mathfrak{M}_{\bar{\mu},E}), \tag{3}$$

$$\|g\|_E \ll |\bar{\mu}|\psi_E(1/|\bar{\mu}|)\|g\|_{L_1} \qquad (g \in \mathfrak{M}_{\bar{\mu},1}). \tag{4}$$

PROOF. Consider the anisotropic Vallée-Poussin kernel

$$V_{\bar{\mu}}(x) = \frac{2^n}{|\bar{\mu}|} \prod_{j=1}^n \frac{\sin(\mu_j x_j/2)\sin(3\mu_j x_j/2)}{x_j^2}.$$

The imbedding $E \hookrightarrow L_1 + L_\infty$ is valid for any symmetric space E, so it follows from [3] (§84) that $\mathfrak{M}_{\bar{\mu},E} \hookrightarrow \mathfrak{M}_{\bar{\mu},\infty}$. Therefore, by [1],

$$g(x) = \frac{1}{\pi^n} \int_{R^n} V_{\bar{\mu}}(x-u) g(u)\, du. \tag{5}$$

It is easily seen that the Vallée-Poussin kernel belongs to the intersection of the spaces L_1 and L_∞ on R^n, and

$$\|V_{\bar{\mu}}\|_{L_\infty} \ll |\bar{\mu}|, \qquad \|V_{\bar{\mu}}\|_{L_1} \ll 1.$$

We denote by $F(q)$ the linear functional generated by the kernel $(1/\pi^n)V_{\bar{\mu}}(x-u)$ and defined on the sum $L_1 + L_\infty$, and we consider the one-dimensional operator $T(g) = F(g)\kappa_{(0,1/|\bar{\mu}|)}$.

It is known [2] that any symmetric space is a normalized interpolation space for one-dimensional operators bounded on the spaces L_1 and L_∞. Therefore,

$$\|T\|_{E\to E} \leqslant \max\left(\|T\|_{L_1}, \|T\|_{L_\infty}\right),$$

and, moreover,

$$\|T\|_E = \|V_{\bar\mu}\|_{E'}\|\kappa_{(0,1/|\bar\mu|)}\|_E,$$

$$\|T\|_{L_1} = \|V_{\bar\mu}\|_{L_\infty}\|\kappa_{(0,1/|\bar\mu|)}\|_{L_1} \ll 1,$$

$$\|T\|_{L_\infty} = \|V_{\bar\mu}\|_{L_1}\|\kappa_{(0,1/|\bar\mu|)}\|_{L_\infty} \ll 1.$$

Consequently, $\|V_{\bar\mu}\|_{E'} \leqslant \psi_E^*(|\bar\mu|)$. By (5),

$$|g(x)| \leqslant \|V_{\bar\mu}\|_{E'}\|g\|_E \ll \psi_E^*(|\bar\mu|)\|g\|_E \qquad (x \in R^n).$$

To prove inequality (4), we use (2) with $E_2 = E$ and $E_1 = L_1$; then

$$\|g\|_E \leqslant \varphi\left(\|g\|_{L_1}, \|g\|_{L_\infty}\right) \ll \varphi\left(\|g\|_{L_1}, |\bar\mu|\|g\|_{L_1}\right) = \varphi(1, |\bar\mu|)\|g\|_{L_1}.$$

It is known [8] that (cf. Lemma 2 in §2)

$$\psi_E(t) = \varphi(\psi_{L_1}(t), \psi_{L_\infty}(t)) = \varphi(t, 1).$$

Therefore, $\|g\|_E \ll |\bar\mu|\psi_E(1/|\bar\mu|)$. The theorem is proved.

In the isotropic case ($\mu_1 = \mu_2 = \cdots = \mu_n = \mu$) inequality (3) was proved in another way in [6]. Moreover, the proof was carried out for spaces E whose elements satisfy the condition that $\|f\kappa_D\|_E \to 0$ as meas $D \to 0$ for all $f \in E$, and was then carried over to arbitrary E with the help of (3), which is not completely obvious and was presented without proof. Here we have made practical use of the approach of V. A. Rodin, who proved (3) for trigonometric polynomials of a single variable.

Let us consider "model" EFET μ in one variable, namely,

$$Q_\mu(t) = \frac{1}{\mu}\frac{\sin^2(\mu t/2)}{t^2}$$

($\mu > 0$), and the EFET $\bar\mu = (\mu_1, \ldots, \mu_n)$: $Q_{\bar\mu}(x) = (Q_{\mu_1} \otimes \cdots \otimes Q_{\mu_n})(x) = Q_{\mu_1}(x_1) \cdots Q_{\mu_n}(x_n)$.

It is known that $\|Q_{\bar\mu}\|_{L_1} \asymp 1$ and $\|Q_{\bar\mu}\|_{L_\infty} \asymp |\bar\mu|$; therefore (3) and (4) give us that

$$\|Q_{\bar\mu}\|_E \asymp |\bar\mu|\psi_E(1/|\bar\mu|).$$

From this we get

$$\mathscr{I}_{E_1,E_2}(\bar\mu) \gg \psi_{E_1}^*(|\bar\mu|)\psi_{E_2}(1/|\bar\mu|). \tag{6}$$

Thus, if the right-hand side of (6) appears in an inequality with different metrics, then such an inequality is certainly sharp, and the norm of the imbedding operator is attained on sufficiently simple functions. This explains the fact that in the next section we look for situations in which the inequality with different metrics has the form

$$\mathscr{I}_{E_1,E_2}(\bar\mu) \ll c(E_1, E_2)\psi_{E_1}^*(|\bar\mu|)\psi_{E_2}(1/|\bar\mu|), \tag{7}$$

where $c(E_1, E_2)$ is a constant depending on the spaces E_i ($i = 1, 2$).

PROPOSITION 1. *In any symmetric space E the linear subspace $\mathfrak{M}_{\bar\mu,E}$ is closed.*

The proof follows in a standard way from inequality (3), which means that $\mathfrak{M}_{\bar\mu,E}$ is continuously imbedded in $\mathfrak{M}_{\bar\mu,\infty}$.

Let $\{a_j\}_1^\infty$ be a nonincreasing nonnegative sequence. Denote by $a(s)$ the function

$$\sum_{j=1}^\infty a_j\kappa_{((j-1)/\mu_1, j/\mu_1)}(s),$$

let

$$\bar{g}(x) = \left(a \otimes \kappa_{(0,\mu_2^{-1})} \otimes \kappa_{(0,\mu_3^{-1})} \otimes \cdots \otimes \kappa_{(0,\mu_n^{-1})}\right)(x)$$
$$= a(x_1)\kappa_{(0,\mu_2^{-1})}(x_2)\kappa_{(0,\mu_3^{-1})}(x_3) \cdots \kappa_{(0,\mu_n^{-1})}(x_n),$$

and let $g_{\bar{\mu}}(x)$ be the function $Q_{\bar{\mu}} * \bar{g}$, where $\bar{\mu} = (\mu_1, \dots, \mu_n)$.

LEMMA 1. *The sequence* $\{a_j\}$ *belongs to* $E_d^{|\bar{\mu}|}$ *if and only if* $g_{\bar{\mu}}$ *belongs to* $\mathfrak{M}_{\bar{\mu},E}$; *in this connection*

$$\|g_{\bar{\mu}}\|_E \asymp \|\{a_j\}\|_{E_d^{|\bar{\mu}|}}. \tag{8}$$

PROOF. The function $g_{\bar{\mu}}$ is an EFET $\bar{\mu}$; therefore, it suffices to prove that $\{a_j\} \in E_d^{|\bar{\mu}|} \Leftrightarrow g_{\bar{\mu}} \in E$, which is equivalent to (8).

For simplicity we prove (8) for $n = 2$; the general case is settled similarly.

We first show that $g_{\bar{\mu}}(x) \gg \bar{g}(x)$. By the definition of the function $\bar{g}(x)$, it suffices to check this only on the cube $T_k = \{(k-1)/\mu_1 < x_1 \leqslant k/\mu_1,\ 0 < x_2 < 1/\mu_2\}$ for any integer k.

If $x \in T_k$, then $\bar{g}(x) = a_k$. Let us estimate $g_{\bar{\mu}}(x)$ for $x \in T_k$. By definition,

$$g_{\bar{\mu}} = \left(Q_{\mu_1} * a\right) \otimes \left(Q_{\mu_2} * \kappa_{(0,\mu_2^{-1})}\right) = \sum_{j=1}^{\infty} a_j\left(Q_{\mu_2} * \kappa_{((j-1)/\mu_1,\, j/\mu_1)}\right) \otimes \left(Q_{\mu_2} * \kappa_{(0,\mu_2^{-1})}\right).$$

Observe that if $(k-1)/\mu < s < k/\mu$ for some integer k and positive μ, then

$$Q_\mu * \kappa_{((k-1)/\mu,\, k/\mu)}(s) = \frac{1}{\mu}\int_{(k-1)/\mu}^{k/\mu} \frac{\sin(\mu(s-u)/2)}{(s-u)^2}\,du \geqslant \int_0^{1/2} \frac{\sin^2(t/2)}{t^2}\,dt \gg 1;$$

therefore, if $x \in T_k$, then

$$g_{\bar{\mu}}(x) \geqslant a_k\left(Q_{\mu_1} * \kappa_{((k-1)/\mu_1,\, k/\mu_1)}\right)(x_1)\left(Q_{\mu_2} * \kappa_{(0,\mu_2^{-1})}\right)(x_2) \gg a_k.$$

Thus, $g_{\bar{\mu}}(x) \gg \bar{g}(x)$ and

$$\|g_{\bar{\mu}}\|_E \gg \|\bar{g}\|_E = \|\bar{g}^*\|_{E(0,\infty)} = \|\{a_j\}\|_{E_d^{|\bar{\mu}|}},$$

since $\bar{g}^*(s) = \sum_{j=1}^{\infty} a_j \kappa_{(j-1/\mu_1,\, j/\mu_1)}(s)$.

To verify the reverse inequality, note that for any positive μ

$$Q_\mu \ll \mu \sum_{k=-\infty}^{\infty} \beta_k \kappa_{((k-1)/\mu,\, k/\mu)},$$

where $\beta_k = 1/k^2$ if $k \neq 0$, and $\beta_0 = 1$. From this estimate we get

$$g_{\bar{\mu}} \ll \sum_{k=-\infty}^{\infty} \sum_{l=-\infty}^{\infty} \beta_k \beta_l \mu_1 \mu_2 \left(a * \kappa_{((k-1)/\mu_1,\, k/\mu_1)}\right) \otimes \left(\kappa_{(0,\mu_2^{-1})} * \kappa_{((l-1)/\mu_2,\, l/\mu_2)}\right)$$
$$\ll \sum_{-\infty}^{\infty} \sum_{-\infty}^{\infty} \beta_k \beta_l \mu_1 \left(a * \kappa_{((k-1)/\mu_1,\, k/\mu_1)}\right) \otimes \kappa_{((l-1)/\mu_2,\, (l+1)/\mu_2)}.$$

We estimate $\mu_1 a * \kappa_{(0,\mu_1^{-1})}$:

$$\mu_1 a * \kappa_{(0,\mu_1^{-1})}(x_1) = \sum_{j=1}^{\infty} a_j \kappa_{((j-1)/\mu_1,\, j/\mu_1)} * \kappa_{(0,\mu_1^{-1})}(x_1)$$
$$\leqslant \sum_{j=1}^{\infty} a_j \kappa_{((j-1)/\mu_1,\, (j+1)/\mu_1)}(x_1) = a(x_1) + a(x_1 - \mu_1^{-1}).$$

Since the function $a * \kappa_{((k-1)/\mu_1, k/\mu_1)}$ is obtained from $a * \kappa_{(0,\mu_1^{-1})}$ under a translation by $(k-1)/\mu_1$, it follows that

$$\mu_1 a * \kappa_{((k-1)/\mu_1, k/\mu_1)}(x_1) \leqslant a\left(x_1 - \frac{k-1}{\mu_1}\right) + a\left(x_1 - \frac{k}{\mu_1}\right).$$

Therefore

$$g_{\bar\mu}(x_1, x_2) \ll \sum_{-\infty}^{\infty} \sum_{-\infty}^{\infty} \beta_k \beta_l \left(a\left(x_1 - \frac{k-1}{\mu_1}\right) + a\left(x_1 - \frac{k}{\mu_1}\right)\right) \kappa_{((l-1)/\mu_2, l/\mu_2)}(x_2).$$

Thus, the right-hand side of the preceding inequality is a sum of terms of the form

$$\sum_{-\infty}^{\infty} \sum_{-\infty}^{\infty} \beta_k \beta_l a(x_1 - \alpha_1) \kappa_{(0,\mu_2^{-1})}(x_2 - \alpha_2),$$

where α_1 and α_2 are real numbers.

The rearrangement of the function $a(x_1 - \alpha_1)\kappa_{(0,\mu_2^{-1})}(x_2 - \alpha_2)$ is equal to

$$\sum_{j=1}^{\infty} a_j \kappa_{((j-1)/|\bar\mu|, j/|\bar\mu|)}(s) \qquad (s > 0),$$

since the rearrangement of $a(x_1 - \alpha_1)$ is equal to

$$\sum_{j=1}^{\infty} a_j \kappa_{((j-1)/\mu_1, j/\mu_1)}(s) \qquad (s > 0).$$

Consequently, by the fact that $\sum_{-\infty}^{\infty} \beta_k < \infty$ and the completeness of the space E, we have that $g_{\bar\mu} \in E$, and

$$\|g_{\bar\mu}\|_E \ll \left\| \sum_{j=1}^{\infty} a_j \kappa_{((j-1)/|\bar\mu|, j/|\bar\mu|)} \right\|_E = \|\{a_j\}\|_{E_d^{|\mu|}}.$$

The lemma is proved.

Let E^t denote the symmetric space of functions on $(0, \infty)$ coinciding with $E \cap L_\infty$, where E is a symmetric space, and let the norm in it be

$$\|f\|_{E^t} = \max\left(\psi_E(1/t)\|f\|_{L_\infty}, \|f\|_E\right).$$

The space E^t is related to the space E_d^t.

Proposition 2. *E_1^t is imbedded in E_2 if and only if $E_{1d}^t \hookrightarrow E_{2d}^t$; in this connection*

$$\mathscr{I}_{E_1^t, E_2} \asymp \mathscr{I}_{E_{1d}^t, E_{2d}^t}.$$

Proof. Suppose that E_1^t is imbedded in E_2, and consider a sequence $\{a_j\}_1^\infty \in E_{1d}^t$. Since E_{1d}^t is symmetric, it can be assumed that $\{a_j\}$ is a nonnegative nonincreasing sequence. By definition,

$$a = \sum_{j=1}^{\infty} a_j \kappa_{((j-1)/t, j/t)} \in E_1,$$

and $\|\{a_j\}\|_{E_{1d}^t} = \|a_1\|_{E_1}$. It is easy to see that

$$\psi_{E_1}(1/t) a_1 \leqslant \left\| \sum_{j=1}^{\infty} a_j \kappa_{((j-1)/t, j/t)} \right\|_{E_1} = \|a_1\|_{E_1}.$$

Therefore, $\|a\|_{E_1^t} = \|a\|_{E_1} = \|\{a_j\}\|_{E_{1d}^t}$, because $\|a\|_{L_\infty} = a_1$.

Consequently,

$$\|\{a_j\}\|_{E_{2d}^t} = \|a\|_{E_2} \leqslant \mathscr{I}_{E_1^t E_2} \|a\|_{E_1^t} = \mathscr{I}_{E_1^t E_2} \|\{a_j\}\|_{E_{1d}^t}.$$

Hence, $\mathscr{I}_{E_{1d}^t, E_{2d}^t} \leqslant \mathscr{I}_{E_1^t, E_2}$.

Suppose now that E_{1d}^t is imbedded in E_{2d}^t, and consider $g(s) \in E_1^t$. Then it is easy to see that $g \in E_2$, and

$$\begin{aligned}
\|g\|_{E_2} = \|g^*\|_{E_2} &\leqslant \left\| \sum_{j=1}^{\infty} g^*\left(\frac{j-1}{t}\right) \kappa_{((j-1)/t, j/t)} \right\|_{E_2} \\
&= \left\| \left\{ g^*\left(\frac{j-1}{t}\right) \right\} \right\|_{E_{2d}^t} \leqslant \mathscr{I}_{E_{1d}^t, E_{2d}^t} \left\| \left\{ g^*\left(\frac{j-1}{t}\right) \right\} \right\|_{E_{1d}^t} \\
&\leqslant \mathscr{I}_{E_{1d}^t, E_{2d}^t} \left(\left\| \sum_{j=2}^{\infty} g^*\left(\frac{j-1}{t}\right) \kappa_{((j-1)/t, j/t)} \right\|_{E_1} + \|g^*(0) \kappa_{(0, t^{-1})}\|_{E_1} \right) \\
&\leqslant \mathscr{I}_{E_{1d}^t, E_{2d}^t} \left(\|g\|_{L_\infty} \psi_{E_1}(1/t) + \left\| \sum_{j=1}^{\infty} g^*\left(\frac{j}{t}\right) \kappa_{((j-1)/t, j/t)} \right\|_{E_1} \right) \\
&\leqslant 2\mathscr{I}_{E_{1d}^t, E_2} \max\left(\|g\|_{L_\infty} \psi_{E_1}(1/t), \|g^*\|_{E_1} \right) = 2\mathscr{I}_{E_{1d}^t, E_{2d}^t} \|g\|_{E_1^t}.
\end{aligned}$$

Thus, $\mathscr{I}_{E_1^t, E_2} \leqslant 2\mathscr{I}_{E_{1d}^t, E_{2d}^t}$. The proposition is proved.

THEOREM 2. *$\mathfrak{M}_{\bar{\mu}, E_1}$ is continuously imbedded in $\mathfrak{M}_{\bar{\mu}, E_2}$ if and only if $E_{1d}^{|\bar{\mu}_0|} \hookrightarrow E_{2d}^{|\bar{\mu}_0|}$ for at least one $\bar{\mu}_0$, and in this case*

$$\mathscr{I}_{E_2, E_2}(\bar{\mu}) \asymp \mathscr{I}_{E_{1d}^{|\bar{\mu}|}, E_{2d}^{|\bar{\mu}|}} \asymp \mathscr{I}_{E_1^{|\bar{\mu}|}, E_2}. \tag{9}$$

PROOF. As already noted, the space $E_d^{|\bar{\mu}|}$ does not depend on the number $|\bar{\mu}|$ so far as the set of its elements goes. Therefore, for at least one $\bar{\mu}_0$ the imbedding $E_{1d}^{|\bar{\mu}_0|} \hookrightarrow E_{2d}^{|\bar{\mu}_0|}$ is equivalent to the analogous imbedding for any $\bar{\mu}$.

By Propsosition 2, the imbedding $E_{1d}^{|\bar{\mu}|} \hookrightarrow E_{2d}^{|\bar{\mu}|}$ implies the imbedding $E_1^{|\bar{\mu}|} \hookrightarrow E_2$; therefore, if $g \in \mathfrak{M}_{\bar{\mu}, E_1}$, then $\psi_{E_1}(1/|\bar{\mu}|)\|g\|_{L_\infty} \ll \|g\|_{E_1}$ by Theorem 1, i.e., $\|g\|_{E_1^{|\bar{\mu}|}} \ll \|g\|_{E_1}$ and $g \in E_1^{|\bar{\mu}|}$, thus, $g \in E_2$ and

$$\|g\|_{E_2} \leqslant \mathscr{I}_{E_1^{|\bar{\mu}|}, E_2} \|g\|_{E_1^{|\bar{\mu}|}} \ll \mathscr{I}_{E_1^{|\bar{\mu}|}, E_2} \|g\|_{E_1},$$

or

$$\mathscr{I}_{E_1, E_2}(\bar{\mu}) \ll \mathscr{I}_{E_1^{|\bar{\mu}|}, E_2} \asymp \mathscr{I}_{E_{1d}^{|\bar{\mu}|}, E_{2d}^{|\bar{\mu}|}}.$$

If now $\{a_j\} \in E_{1d}^{|\bar{\mu}|}$, then $g_{\bar{\mu}} \in \mathfrak{M}_{\bar{\mu}, E_1}$ and $\|\{a_j\}\|_{E_{1d}^{|\bar{\mu}|}} \asymp \|g_{\bar{\mu}}\|_{E_1}$ by Lemma 1. By the same lemma, $\|\{a_j\}\|_{E_{2d}^{|\bar{\mu}|}} \asymp \|g_{\bar{\mu}}\|_{E_2}$, and, by the imbedding $\mathfrak{M}_{\bar{\mu}, E_1} \hookrightarrow \mathfrak{M}_{\bar{\mu}, E_2}$, we have that

$$\|g_{\bar{\mu}}\|_{E_2} \ll \mathscr{I}_{E_1, E_2}(\bar{\mu}) \|g_{\bar{\mu}}\|_{E_1} \asymp \mathscr{I}_{E_1, E_2}(\bar{\mu}) \|\{a_j\}\|_{E_{1d}^{|\bar{\mu}|}}.$$

Thus, $\{a_j\} \in E_{2d}^{|\bar{\mu}|}$ and

$$\|\{a_j\}\|_{E_{2d}^{|\bar{\mu}|}} \ll \mathscr{I}_{E_1, E_2}(\bar{\mu}) \|\{a_j\}\|_{E_{1d}^{|\bar{\mu}|}},$$

i.e., $\mathscr{I}_{E_{1d}^{|\bar{\mu}|}, E_{2d}^{|\bar{\mu}|}} \ll \mathscr{I}_{E_1, E_2}(\bar{\mu})$. The theorem is proved.

The quantity $\mathscr{I}_{E_1^t, E_2}$ can be computed, for example, as follows.

By definition,

$$\mathscr{I}_{E_1', E_2} = \sup_{\|g\|_{E_1'} \leqslant 1} \|g\|_{E_2},$$

i.e., the supremum is over all g such that $\|g\|_{E_1} \leqslant 1$, and $\psi_{E_1}(1/t)\|g\|_{L_\infty} \leqslant 1$ or $\|g\|_{L_\infty} \leqslant \psi^*_{E_1}(t)$. Thus, by (1),

$$\mathscr{I}_{E_1', E_2} = \varphi\big(1, \psi^*_{E_1}(t)\big). \tag{10}$$

Proposition 3. *If $E_1 \cap L_\infty \hookrightarrow E_2$, then*

$$\varphi\big(1, \psi^*_{E_1}(t)\big) \asymp \theta_{E_1, E_2}(t), \tag{11}$$

where

$$\Theta_{E_1, E_2}(t) = \sup_{\|g\|_{E_1} \leqslant 1} \|g^* \kappa_{(t^{-1}, \infty)}\|_{E_2}.$$

Proof. We estimate $\theta_{E_1, E_2}(t)$ above: if $\|g\|_{E_1} \leqslant 1$, then

$$\begin{aligned} \|g^*(s)\kappa_{(t^{-1},\infty)}(s)\|_{E_2} &= \|g^*(s + t^{-1})\|_{E_2} \leqslant \mathscr{I}_{E_1', E_2}\|g^*(s + t^{-1})\|_{E_1'} \\ &\leqslant \mathscr{I}_{E_1', E_2}\max\big(g^*(t^{-1})\psi_{E_1}(t^{-1}), \|g^*(s + t^{-1})\|_{E_1}\big) \\ &\leqslant \mathscr{I}_{E_1', E_2}\big(g^*(t^{-1})\psi_{E_1}(t^{-1}) + \|g^*(s + t^{-1})\|_{E_1} \\ &\leqslant 2\mathscr{I}_{E_1', E_2}\|g\|_{E_1} \leqslant 2\mathscr{I}_{E_1', E_2}. \end{aligned}$$

By definition,

$$\mathscr{I}_{E_1', E_2} = \sup_{\|g^*\|_{E_1'} \leqslant 1} \|g^*\|_{E_2}.$$

Further,

$$\|g^*\|_{E_2} \leqslant \|g^*\kappa_{(0,t^{-1})}\|_{E_2} + \|g^*\kappa_{(t^{-1},\infty)}\|_{E_2} \leqslant \psi^*_{E_1}(t)\psi_{E_2}(1/t) + \theta_{E_1, E_2}(t).$$

Consider the function $g_0(s) = \frac{1}{2}\psi^*_{E_1}(t)\kappa_{(0,2t^{-1})}(s)$. Obviously, $\|g_0\|_{E_1'} = 1$ and

$$\psi_{E_1}(t)\psi_{E_2}(1/t) = 2\|g_0^*\kappa_{(t^{-1},\infty)}\|_{E_2}.$$

Finally,

$$\mathscr{I}_{E_1', E_2} \leqslant 3\theta_{E_1, E_2}(t).$$

The proposition is proved.

We have thus proved the main assertion in this section.

Theorem 3 (inequality with different metrics). *The imbeddings* ($E_{id} = E^1_{id}$; $i = 1, 2$)

a) $E_{1d} \hookrightarrow E_{2d}$,

b) $E_1 \cap L_\infty \hookrightarrow E_2$,

c) $\mathfrak{M}_{\bar{\mu}, E_1} \hookrightarrow \mathfrak{M}_{\bar{\mu}, E_2}$ *are equivalent, and the inequality*

$$\|g\|_{E_2} \ll \theta_{E_1, E_2}(|\bar{\mu}|)\|g\|_{E_1} \qquad \big(g \in \mathfrak{M}_{\bar{\mu}, E_1}\big)$$

is sharp.

REMARK 1. It is easy to see that $\mathscr{I}_{E_1,E_2}(\bar\mu) \gg \mathscr{I}_{E_1,E_2}(\bar 1)$ for $\mu_j \geqslant 1$, $j = 1,\dots,n$. Therefore, it follows from Theorem 2 that $\mathscr{I}_{E_1,E_2}(\bar\mu) \gg \mathscr{I}_{E_{1d},E_{2d}}$. This means that the "coalescence" of the spaces E_i $(i = 1,2)$ for a fixed $\bar\mu$ $(\mu_j \geqslant 1)$ can lead to unbounded growth of the constant in the inequality with different metrics, which happens if the constant of the imbedding $E_{1d} \hookrightarrow E_{2d}$ increases, for example, when $E_1(0,\infty) = \mathscr{M}_{1-1/p_1}$ and $E_2 = \Lambda_{1/p_2}$, $p_1 < p_2$. Therefore, it is impossible to obtain Nikol′skiĭ's inequality from the formula given in [6]; the fact is that this formula was established by replacing the imbedding $\mathfrak{M}_{\bar\mu,E_1} \hookrightarrow \mathfrak{M}_{\bar\mu,E_2}$ by the coarser imbedding $\mathfrak{M}_{\bar\mu,\mathscr{M}_{t/\psi_{E_1}(t)}} \hookrightarrow \mathfrak{M}_{\bar\mu,\Lambda_{\psi_{E_2}}}$, which, first, does not necessarily follow from the former, and, second, leads to the phenomenon mentioned above in the case of power fundamental functions.

REMARK 2. Formulas (9)–(11) enable us, in particular, to show that an inequality with different metrics does not always hold in the form (7). Let $E_1 = \mathscr{M}_{1/2}$ be the Marcinkiewicz space constructed from the function $t^{1/2}$, and let $E_2 = \Lambda_{\varphi_2}$, where $\varphi_2(t) = \min(t^{1/2}, 1)$. Then

$$\theta_{E_1,E_2}(t) \asymp \left\| s^{-1/2}\kappa_{(t^{-1},\infty)}(s)\right\|_{\Lambda_{\varphi_2}} \asymp 1 + \int_{t^{-1}}^{1} s^{-1/2}\, ds^{1/2} = 1 + \frac{1}{2}\ln t,$$

for $t > 1$, i.e., $\mathscr{I}_{E_1,E_2}(\bar\mu) \asymp 1 + \ln|\bar\mu|$ for $|\bar\mu| > 1$. At the same time, $\psi^*_{E_1}(|\bar\mu|)\psi_{E_2}(|\bar\mu|^{-1}) \equiv 1$ for $|\bar\mu| > 1$.

§2. Cases where $\theta(t) \asymp \psi^*_{E_1}(t)\psi_{E_2}(t^{-1})$

Suppose that $E_1 \cap L_\infty \hookrightarrow E_2$. It follows from Proposition 3 that E_2 can be regarded as intermediate between E_1 and L_∞, i.e., the imbedding $E_2 \hookrightarrow E_1 + L_\infty$ can be assumed. Indeed, $\varphi(1,u) \geqslant \varphi(1,1)\min(1,u)$. Therefore, for the space $E_3 = E_2 \cap (E_1 + L_\infty)$, which is intermediate between E_1 and L_∞,

$$\begin{aligned}\sup_{\|f\|_{E_1}\leqslant 1,\ \|f\|_{L_\infty}\leqslant u} \|f\|_{E_3} &= \sup_{\|f\|_{E_1}\leqslant 1,\ \|f\|_{L_\infty}\leqslant u} \max\left\{\|f\|_{E_2}, \varphi(1,1)\|f\|_{E_1+L_\infty}\right\} \\ &= \max\left\{\varphi(1,u), \varphi(1,1)\sup_{\|f\|_{E_1}\leqslant 1,\ \|f\|_{L_\infty}\leqslant u}\|f\|_{E_1+L_\infty}\right\} = \varphi(1,u).\end{aligned}$$

Thus, for computing $\theta(t)$ the space E_2 can always be replaced by E_3.

We construct one more space in terms of whose fundamental function the quantity $\theta(t)$ can be expressed.

By construction, E_4 is the collection of $f \in E_3$ such that $Tf \in E_3$ for any linear operator T acting continuously from E_1 to E_1 and from L_∞ to L_∞ and such that the norm

$$\|f\|_{E_4} = \sup_{\|T\|_{E_1\to E_1}\leqslant 1,\ \|T\|_{L_\infty\to L_\infty}} \|Tf\|_{E_3}$$

is finite.

Obviously, E_4 is a normalized interpolation space between E_1 and L_∞. Moreover, E_4 is normally imbedded in E_3, and

$$\varphi(1,u) \leqslant \varphi_4(1,u) = \sup_{\substack{\|f\|_{E_1}\leqslant 1\\ \|f\|_{L_\infty}\leqslant u}} \ \sup_{\substack{\|T\|_{E_1\to E_1}\leqslant 1\\ \|T\|_{L_\infty\to L_\infty}\leqslant 1}} \|Tf\|_{E_3} \leqslant \sup_{\substack{\|Tf\|_{E_1}\leqslant 1\\ \|Tf\|_{L_\infty}\leqslant u}} \|Tf\|_{E_3} \leqslant \varphi(1,u).$$

Thus, $\varphi(1,u) = \varphi_4(1,u)$.

THEOREM 4. $\theta_{E_1,E_2}(t) \asymp \psi^*_{E_1}(t)\psi_{E_4}(t^{-1})$.

The proof of this theorem rests on a lemma which is apparently of independent interest.

LEMMA 2. *If E, F, and G are symmetric spaces and G is a normalized interpolation space between E and F, then*

$$\psi_G(u) = \varphi(\psi_E(u), \psi_F(u)),$$

where

$$\varphi(s,t) = \sup_{\|g\|_E \leqslant s,\, \|g\|_F \leqslant t} \|g\|_G \qquad (g \in E \cap F).$$

PROOF. Let us show that $\psi_G(u) \geqslant \varphi(\psi_E(u), \psi_F(u))$; the reverse inequality is obvious. Consider the one-dimensional operator $Tf = \mathscr{F}(f)g$, where $g \in E \cap F$, and the functional $\mathscr{F}$ is defined on $E + F$ and is generated by the function $\kappa_{(0,u)}(t)$, i.e., $\mathscr{F}(f) = \int_0^u f(t)\,dt$. Then

$$\|T\|_{E\to E} = \psi_{E'}(u)\|g\|_E, \quad \|T\|_{F\to F} = \psi_{F'}(u)\|g\|_F, \quad \|T\|_{G\to G} = \psi_{G'}(u)\|g\|_G.$$

By the condition,

$$\psi_{G'}(u)\|g\|_G \leqslant \max\{\psi_{E'}(u)\|g\|_E, \psi_{F'}(u)\|g\|_F\},$$

i.e., $\|g\|_G \leqslant (\psi_{G'}(u))^{-1}$ for $\|g\|_E \leqslant (\psi_{E'}(u))^{-1}$ and $\|g\|_F \leqslant (\psi_{F'}(u))^{-1}$ $(g \in E \cap F)$. This means that

$$\varphi\left((\psi_{E'}(u))^{-1}, (\psi_{F'}(u))^{-1}\right) \leqslant (\psi_{G'}(u))^{-1},$$

which implies the required result, since $\psi_{E'}(u) = u/\psi_E(u)$.

PROOF OF THEOREM 4. Applying Lemma 2, we get that

$$\theta_{E_1,E_2}(t) \asymp \varphi(1, \psi^*_{E_1}(t)) = \psi^*_{E_1}(t)\varphi(\psi_{E_1}(t^{-1}), 1) = \psi^*_{E_1}(t)\psi_{E_4}(t^{-1}).$$

THEOREM 5. *If E_2 is an interpolation space between E_1 and L_∞ with the interpolation constant C, then*

$$\theta_{E_1,E_2}(t) \ll C\psi_{E_2}(t^{-1})\psi^*_1(t). \tag{12}$$

This follows from the fact that $E_2 = E_4$ as sets and $\|f\|_{E_4} \leqslant C\|f\|_{E_2}$.

COROLLARY 1. (12) *and the Riesz interpolation theorem yield the Nikol'skiĭ inequality when $E_i = L_{p_i}$ $(1 \leqslant p_1 \leqslant p_2 \leqslant \infty)$.*

The Kreĭn-Semenov interpolation theorem ([2], Chapter II, §6.4) gives us

COROLLARY 2. *Suppose that the spaces E_i $(i = 1, 2)$ are such that*

$$\int_0^1 \|\sigma_{1/\tau}\|_{E_2}\, d\bar{\psi}_{E_1}(\tau) < \infty,$$

where

$$\bar{\psi}_{E_1}(\tau) = \sup_{t>0} \frac{\psi_{E_1}(\tau t)}{\psi_{E_1}(t)} \qquad (0 < \tau < 1).$$

Then (12) *holds.*

COROLLARY. 3. *Suppose that $E_5 \hookrightarrow E_2$ and E_5 is interpolated between E_1 and L_∞ with interpolation constant C'. Then*

$$\theta_{E_1,E_2}(t) \ll C'\psi_{E_5}(t^{-1})\psi^*_{E_1}(t).$$

We now consider the case (important in applications) where the spaces E_i are "of the same nature": for example, the $E_i = L^*_{N_i}$ are Orlicz spaces, the $E_i = \Lambda_{\psi_i}$ are Lorentz spaces, or the $E_i = \mathcal{M}_{\psi_i}$ are Marcinkiewicz spaces.

THEOREM 6. *Under the assumptions made above*

$$\theta_{E_1,E_2}(t) \asymp \widetilde{\psi^*_{E_1}(t)\psi_{E_2}(t^{-1})},$$

where the tilde denotes the smallest quasiconcave majorant of a function.(²)

PROOF. We consider the case where $E_1 = L^*_{N_1}$ and $E_2 = L^*_{N_2}$. The cases $E_1 = \mathcal{M}_{\psi_1}$, λ_{ψ_2} and $E_2 = \mathcal{M}_{\psi_2}$, Λ_{ψ_2} are handled similarly.

For brevity let $\alpha(t)$ denote the ratio $\psi_{E_2}(t^{-1})/\psi_{E_1}(t^{-1})$, and let $\hat{\psi}_{E_2}(t)$ denote the function obtained from the equality $\hat{\psi}_{E_2}(t) = \tilde{\alpha}(t)\psi_{E_1}(t^{-1})$. Clearly, $\hat{\psi}_{E_2}(t) \geqslant \psi_{E_2}(t)$. We show that $\hat{\psi}_{E_2}(t)$ is quasiconcave and $\hat{\psi}_{E_2}(t) = \lambda(\lambda_{E_1}(t))$, where $\lambda(v)$ is a quasiconcave function.

The smallest quasiconcave majorant can be constructed according to the formula

$$\tilde{\alpha}(t) = \inf_{\xi,\eta} \max(\xi, \eta t),$$

where the infimum is over all ξ and η such that $\alpha(t) \leqslant \max(\xi, \eta t)$ for all t.

We consider another form of majorants

$$\tilde{\tilde{\alpha}}(t) = \inf_{\xi,\eta} \max\left(\xi, \eta\psi^*_{E_1}(t)\right),$$

where the infimum is over all ξ and η such that $\alpha(t) \leqslant \max(\xi, \eta\psi^*_{E_1}(t))$ for all t.

Let us show that $\tilde{\tilde{\alpha}}(t) = \tilde{\alpha}(t)$. Obviously, $\tilde{\tilde{\alpha}}(t) \geqslant \tilde{\alpha}(t)$, since the latter function is the smallest quasiconcave majorant.

Suppose now that ξ_0 and η_0 are such that for all $\tau > 0$

$$\alpha(t) \leqslant \max(\xi_0, \eta_0 t); \tag{13}$$

then

$$\alpha(t) \leqslant \max\left(\xi_0, \xi_0\psi_{E_1}(\eta_0/\xi_0)\psi^*_{E_1}(t)\right),$$

since this inequality coincides with (13) for $t \leqslant \xi_0/\eta_0$, while for $t > \xi_0/\eta_0$

$$\alpha(t) = \psi_{E_2}(t^{-1})\psi^*_{E_1}(t) \leqslant \psi_{E_2}(\eta_0/\xi_0)\psi^*_{E_1}(t)$$

and $\psi_{E_2}(\eta_0/\xi_0) \leqslant \xi_0\psi_{E_1}(\eta_0/\xi_0)$ for $t = \xi_0/\eta_0$, by (13). Therefore,

$$\alpha(t) \leqslant \xi_0\psi_{E_1}(\eta_0/\xi_0)\psi^*_{E_1}(t) \leqslant \max\left(\xi_0, \xi_0\psi_{E_1}(\eta_0/\xi_0)\psi^*_{E_1}(t)\right).$$

By the quasiconcavity of $\psi^*_{E_1}(t)$,

$$\max\left(\xi_0, \xi_0\psi_{E_1}(\eta_0/\xi_0)\psi^*_{E_1}(t)\right) \leqslant \max(\xi_0, \eta_0 t),$$

(²) It is not hard to see that the smallest quasiconcave majorant of this function coincides with its smallest increasing majorant.

and so $\tilde{\tilde{\alpha}}(t) \leqslant \tilde{\alpha}(t)$. Hence,

$$\hat{\psi}_{E_2}(t^{-1}) = \tilde{\tilde{\alpha}}(t)\psi_{E_1}(t^{-1}) = \inf_{\eta,\xi} \max\left(\xi\psi_{E_1}(t^{-1}), \eta\psi^*_{E_1}(t)\psi_{E_1}(t^{-1})\right)$$
$$= \inf_{\eta,\xi} \max\left(\xi\psi_{E_1}(t^{-1}), \eta\right),$$

i.e., $\hat{\psi}_{E_2}(u) = \inf_{\xi,\eta} \max(\xi\psi_{E_1}(u), \eta)$ is a quasiconcave function.

If we now set

$$\lambda(v) = \inf_{\xi,\eta} \max(\xi v, \eta),$$

where the infimum is over all ξ and η such that $\alpha(t) \leqslant \max(\xi, \eta\psi^*_{E_1}(t))$ for all $t > 0$, then $\hat{\psi}_{E_2}(t) = \lambda(\psi_{E_1}(t))$.

Consider now the Orlicz space L^*_N with fundamental function $\hat{\psi}_{E_2}(t)$, i.e., the Orlicz space for the function $N(t) = 1/\hat{\psi}^{-1}_{E_2}(1/t)$. By construction, $L^*_{N_2} \hookleftarrow L^*_N$ and $N^{-1}(t) = \lambda^*(N_1^{-1}(t))$. Consequently, by [10] L_N is interpolated between L_{N_1} and L_∞. Therefore, by Corollary 3,

$$\theta_{E_1,E_2}(t) \ll \hat{\psi}_{E_2}(t^{-1})\psi^*_{E_1}(t) = \tilde{\alpha}(t).$$

But, on the other hand,

$$\psi_{E_1}(t^{-1})\psi^*_{E_1}(t) \ll \theta_{E_1,E_2}(t)$$

and $\theta_{E_1,E_2}(t)$ is quasiconcave. Therefore,

$$\tilde{\alpha}(t) \ll \theta_{E_1,E_2}(t).$$

This proves the theorem.

REMARK 3. To consider the cases of Lorentz spaces or Marcinkiewicz spaces we should use the interpolation theorems for these spaces: $\mathcal{M}_{\varphi(\varphi_1,\varphi_2)}$ ($\Lambda_{\varphi(\varphi_1,\varphi_2)}$) is interpolated between $\mathcal{M}_{\varphi_1}$ (Λ_{φ_1}) and $\mathcal{M}_{\varphi_2}$ (Λ_{φ_2}) for any first-order homogeneous function $\varphi(\cdot,\cdot)$ that is increasing in each variable.

Finally, to conclude this section we given an example of one more situation where an inequality with different metrics in the form (7) holds, but with a less precise constant $c(E_1, E_2)$ (see [4]).

THEOREM 7. *Suppose that there is a $\delta > 0$ such that the function $t^\delta\psi_{E_2}(t)/\psi_{E_1}(t)$ is decreasing. Then*

$$\theta_{E_1,E_2}(t) \ll \left(1 + \frac{1}{\delta}\right)\psi_{E_2}(t^{-1})\psi^*_{E_1}(t).$$

PROOF. For any function such that $\|g\|_{E_1} \leqslant 1$ it is true that $g^*(t) \leqslant 1/\psi_{E_1}(t)$. Therefore,

$$\theta_{E_1,E_2}(t) \leqslant \left\|\frac{1}{\psi_{E_1}}\kappa_{(t^{-1},\infty)}\right\|_{E_2} \leqslant \left\|\frac{1}{\psi_{E_1}}\kappa_{(t^{-1},\infty)}\right\|_{\Lambda_{\psi_2}} \leqslant \frac{\psi_{E_2}(t^{-1})}{\psi_{E_1}(t^{-1})} + \int_{t^{-1}}^\infty \frac{1}{\psi_{E_1}(s)}d\psi_{E_2}(s)$$
$$\leqslant \frac{\psi_{E_2}(t^{-1})}{\psi_{E_1}(t^{-1})} + \int_{t^{-1}}^\infty \frac{\psi_{E_2}(s)}{\psi_{E_1}(s)}\frac{ds}{s} \leqslant \frac{\psi_{E_2}(t^{-1})}{\psi_{E_1}(t^{-1})} + \frac{\psi_{E_2}(t^{-1})}{\psi_{E_1}(t^{-1})}t^{-\delta}\int_{t^{-1}}^\infty \frac{ds}{s^{1+\delta}},$$

from which we get what is required.

The theorem is proved.

§3. An inequality with different dimensions

Denote by $\mathfrak{M}^m_{\bar{\mu}'',E}$ the subspace of measurable functions on R^n belonging to a symmetric space E and such that for almost all values of the first m variables $x' = (x_1,\dots,x_m)$ these functions are EFET $\bar{\mu}'' = (\mu_{m+1},\dots,\mu_n)$ with respect to the last $n-m$ variables $x'' = (x_{m+1},\dots,x_n)$.

THEOREM 8. *Suppose that E is interpolated between L_1 and L_∞. If $g \in \mathfrak{M}^m_{\bar{\mu}'',E}$, then $g(x', x'') \in E(R^m)$ and*

$$\left\| g\left(|\bar{\mu}''|^{1/m} x', x''\right)\right\|_{E(R^m)} \ll \|g\|_{E(R^n)}$$

for any x''.

PROOF. For simplicity it is assumed that $m = 1$, $n = 2$, and $x'' = 0$. As already mentioned, for $g \in \mathfrak{M}^1_{\mu'',E}$

$$g(x', x'') = \frac{1}{\pi}\int_{-\infty}^{\infty} V_{\mu''}(x'' - u) g(x', u)\, du$$

for almost all x', where

$$V_{\mu''}(u) = \frac{2}{\mu''}\,\frac{\sin(\mu'' u/2)\sin(3\mu'' u/2)}{u^2}.$$

Consider the operator K: $(L_1 + L_\infty)(R) \to (L_1 + L_\infty)(R^2)$ acting according to the formula $K\colon f \to \pi^{-1} V_{\mu''}(x'') f(x')$. Obviously,

$$\|Kf\|_{L_1} \ll \|f\|_{L_1}, \qquad \|Kf\|_{L_\infty} \ll \mu''\|f\|_{L_\infty}.$$

For the dual operator K', which acts from $(L_1 + L_\infty)(R^2)$ to $(L_1 + L_\infty)(R)$, we get, respectively, that

$$\|K'g\|_{L_1} \ll \mu''\|g\|_{L_1}, \qquad \|K'g\|_{L_\infty} \ll \|g\|_{L_\infty}.$$

Let us now show that for $g \in \mathfrak{M}^1_{\mu'',E}$

$$(K'g)(x') = g(x', 0).$$

By definition, for $f \in (L_1 \cap L_\infty)(R)$

$$\langle f, g(\cdot, 0)\rangle = \int_{-\infty}^{\infty} f(x') g(x', 0)\, dx' = \frac{1}{\pi}\int_{-\infty}^{\infty} f(x')\left(\int_{-\infty}^{\infty} V_{\mu''}(x'') g(x', x'')\, dx''\right) dx'$$

$$= \frac{1}{\pi}\int_{-\infty}^{\infty}\int_{-\infty}^{\infty} f(x') V_{\mu''}(x'') g(x', x'')\, dx''\, dx'.$$

The integrand in this iterated integral is summable, since

$$f(x')V_{\mu''}(x'') \in (L_1 \cap L_\infty)(R^2) \quad \text{and} \quad g(x', x'') \in E \subset (L_1 + L_\infty)(R^2);$$

consequently, this integral is a Lebesgue integral on the plane, and

$$\langle f, g(\cdot, 0)\rangle = \int_{-\infty}^{\infty}\int_{-\infty}^{\infty} (Kf)(x', x'') g(x', x'')\, dx'\, dx'' = \langle K(f), g\rangle,$$

i.e., $(K'g)(x') = g(x', 0)$.

We now introduce the operator T: $(L_1 + L_\infty)(R^2) \to (L_1 + L_\infty)(R^2)$, acting according to the formula

$$(Tg)(x', x'') = \kappa_{(0,1/\mu'')}(x'')(K'g)(x').$$

It is easy to see that $\|T\|_{L_1} \ll 1$ and $\|T\|_{L_\infty} \ll 1$. Therefore, since E is interpolated between L_1 and L_∞,

$$\|(Tg)(x', x'')\|_{E(R^2)} \ll \|g(x', x'')\|_{E(R^2)}.$$

Consider $g \in \mathfrak{M}^1_{\mu'',E}$. The rearrangement of the function $(Tg)(x', x'') = g(x', 0)\kappa_{(0,1/\mu'')}(x'')$ is obviously equal to

$$g(\cdot, 0)^*(\mu'' t) = g(\mu'' x', 0)^*(t).$$

Therefore,

$$\|g(\mu'', x', 0)\|_{E(R)} \ll \|g\|_{E(R^2)}. \tag{14}$$

The theorem is proved.

COROLLARY 1. *If* $g \in \mathfrak{M}^m_{\bar{\mu}'',E}$, *then*

$$\|g(\cdot, x'')\|_{E(R^m)} \ll \|\sigma_{|\bar{\mu}''|}\|_{E\to E}\|g\|_{E(R^n)}. \tag{15}$$

We now consider the question of sharpness for the estimate (15). Let $\chi_{\mu''}(t)$ denote the function

$$\frac{A}{\mu''}\int_t^{t+1/\mu''} \frac{\sin^2(\mu'' s/2)}{s^2}\, ds,$$

where A is chosen so that $\chi_{\mu''}(0) = 1$.

It is easy to see that this function has type μ''. Moreover,

$$\kappa_{(0,1/\mu'')}(t) \ll \chi_{\mu''}(t) \quad \text{and} \quad \chi_{\mu''}(t) \ll \chi_{(0,1/\mu'')}(t) + \sum_{|k|=1}^{\infty} \frac{1}{k^2}\kappa_{(k/\mu'',(k+1)/\mu'')}(t).$$

Denote by $(g \otimes \chi_{\mu''})(x', x'')$ the function $g(x')\chi_{\mu''}(x'')$. We estimate the rearrangement of this function with respect to Lebesgue measure on the plane.

The inequality $\chi_{(0,1/\mu'')} \ll \chi_{\mu''}$ implies that

$$[g \otimes \chi_{(0,1/\mu'')}]^*(t) \ll [g \otimes \chi_{\mu''}]^*(t),$$

from which we obtain

$$g^*(t\mu'') \ll [g \otimes \chi_{\mu''}]^*(t). \tag{16}$$

On the other hand, let us use the inequality

$$\int_0^t (F_1 + F_2)^*(S)\, ds \leqslant \int_0^1 F_1^*(s)\, ds + \int_0^t F_2^*(s)\, ds$$

to get an upper estimate of $[g \otimes \chi_{\mu''}]^*(t)$. Since

$$g(x')\chi_{\mu''}(x'') \ll x_{(0,1/\mu'')}(x'')g(x') + \sum_{|k|=1}^{\infty} \frac{1}{k^2}\kappa_{(k/\mu'',(k+1)/\mu'')}(x'')g(x')$$

and $\sum_{|k|=1}^{\infty} 1/k^2 < \infty$, it follows that

$$\begin{aligned}\int_0^1 (g \otimes \chi_{\mu''})^*(s)\, ds &\ll \int_0^t (g \otimes \chi_{(0,1/\mu'')})^*(s)\, ds \\ &\quad + \sum_{|k|=1}^{\infty} \frac{1}{k^2}\int_0^t (g \otimes \kappa_{(k/\mu'',(k+1)/\mu'')})^*(s)\, ds \\ &\ll \int_0^t g^*(\mu'' s)\, ds.\end{aligned} \tag{17}$$

From (16) and (17) we get

$$\int_0^t (g \otimes \chi_{\mu''})^*(s)\, ds \asymp \int_0^t g^*(\mu'' s)\, ds.$$

Because E is interpolated between L_1 and L_∞, the Calderón theorem (see, for example, [2]) gives us that

$$\|g \otimes \chi_{\mu''}\|_{E(R^2)} \asymp \|g^*(\mu'' s)\|_{E(0,\infty)} = \|g(\mu'' x')\|_{(R)E}. \tag{18}$$

Since $\chi_{\mu''}(0) = 1$, it follows that $(g \otimes \chi_{\mu''})(x', 0) = g(x')$, and $g \otimes \chi_{\mu''}$ is of exponential type μ'' with respect to the second variable. Therefore, (18) means that the estimate (14) is sharp, along with (15). The remaining part of this section deals with the inequality with different dimensions in the case where the function is of finite type with respect to all the variables.

THEOREM 9. *If a function $g(x_1, \ldots, x_n)$ has exponential type $\bar{\mu} = (\mu_1, \ldots, \mu_n)$ and belongs to a symmetric space E which is interpolated between L_1 and L_∞, then for any $x'' = (x_{m+1}, \ldots, x_n)$*

$$\|g(x', x'')\|_{E_2(R^m)} \ll \|\sigma_{|\bar{\mu}''|}\|_{E_1^{|\bar{\mu}|} \to E_2} \|g\|_{E_1(R^n)}, \tag{19}$$

where $\bar{\mu}'' = (\mu_{m+1}, \ldots, \mu_n)$.

PROOF. As shown in Theorem 8,

$$\left\|g\left(|\mu''|^{1/m} x', x''\right)\right\|_{E_1^{|\bar{\mu}|}(R^m)} \ll \|g\|_{E_1^{|\bar{\mu}|}(R^n)}.$$

By inequality (3), $g \in \mathfrak{M}_{\bar{\mu}, E_1}$, $\psi_{E_1}(|\bar{\mu}|^{-1})\|g\|_{E_1} \ll \|g\|_{E_1}$ for $g \in \mathfrak{M}_{\bar{\mu}, E_1}$. Therefore,

$$\begin{aligned} \|g(x', x'')\|_{E_2(R^m)} &\leqslant \|\sigma_{|\bar{\mu}''|}\|_{E_1^{|\bar{\mu}|} \to E_2} \left\|g\left(|\bar{\mu}''|^{1/m} x', x''\right)\right\|_{E_1^{|\bar{\mu}|}(R^m)} \\ &\ll \|\sigma_{|\bar{\mu}''|}\|_{E_1^{|\bar{\mu}|} \to E_2} \|g(x', x'')\|_{E_1^{|\bar{\mu}|}(R^m)} \ll \|\sigma_{|\bar{\mu}''|}\|_{E_1^{|\bar{\mu}|} \to E_2} \|g\|_{E_1(R^n)}. \end{aligned}$$

The theorem is proved.

We now show that the estimate (19) is sharp. The detailed arguments are given for the case $n = 2$ and $m = 1$. The general case is settled similarly, as is clear from what follows. For convenience it is assumed that the space under consideration is the space of functions of exponential type μ' with respect to the first variable x' and of type μ'' with respect to the second variable x''.

By (18), for any $g \in E(R)$ it is possible to choose $\chi_{\mu''}$ of type μ'' so that

$$\|g \otimes \chi_{\mu''}\|_{E(R^2)} \asymp \|g(\mu'' x')\|_{E(R)} \tag{20}$$

in any symmetric space E interpolated between L_1 and L_∞.

We break up the semiaxis $(0, \infty)$ into intervals of length $1/\mu'$ by the points j/μ', where the j are positive integers and $\mu' > 0$, and denote by $D_{\mu'}$ the collection of measurable functions on $(0, \infty)$ that are constant on these intervals.

LEMMA 3. *For any $\mu' > 0$ and $\mu'' > 0$*

$$\sup_{g \in D_{\mu'} \cap E_1} \frac{\|g(t)\|_{E_1}}{\|g(\mu'' t)\|_{E_1^{\mu'\mu''}}} \leqslant \|\sigma_{\mu''}\|_{E_1^{\mu'\mu''} \to E_2} < 2 \sup_{g \in D_{\mu'} \cap E_1} \frac{\|g(t)\|_{E_2}}{\|g(\mu'')\|_{E_1^{\mu'\mu''}}}. \tag{21}$$

PROOF. Since the left-hand inequality is obvious, we consider the right-hand one. If $\|g(\mu''t)\|_{E_1^{\mu'\mu''}} \leqslant 1$, then the function $\bar{g}$ equal to $g^*(k/\mu')$ on $(k/\mu', (k+1)/\mu']$ $(k = 0, 1, \ldots)$ satisfies $\|\bar{g}(\mu't)\|_{E_1^{\mu'\mu''}} \leqslant 2$. We have that $\bar{g} \geqslant g^*$, and the operator $\sigma_{\mu''}$ is monotone; hence, the lemma is proved.

By (21), $\bar{g}$ can be chosen so that

$$\|\sigma_{\mu''}\|_{E_1^{\mu'\mu''} \to E_2} \|\bar{g}(\mu''t)\|_{E_1^{\mu'\mu''}} \ll \|\bar{g}(t)\|_{E_2}$$

and $\bar{g}(t)$ is nonnegative and constant on the intervals $((k-1)/\mu', k/\mu')$. We extend $\bar{g}$ by zero to the negative semiaxis and consider the convolution of $\bar{g}$ with the function

$$Q_{\mu'}(t) = \frac{1}{\mu'} \frac{\sin^2(\mu't/2)}{t^2}.$$

As in §1, denote it by $g_{\mu'}(x')$.

The function $g_{\mu'}(x')$ has exponential type μ', and thus $g \otimes \chi_{\mu''}$ has exponential type (μ', μ'').

By (20), $g_{\mu'} \otimes \chi_{\mu''} \in E_1^{\mu'\mu''}$ and

$$\begin{aligned}\|g_{\mu'} \otimes \chi_{\mu''}\|_{E_1(R^2)} \|g_{\mu'} \otimes \chi_{\mu''}\|_{E_1^{\mu'\mu''}(R^2)} &\asymp \|g_{\mu'}(\mu''x')\|_{E_1^{\mu'\mu''}(R)} \\ &= c\left|Q_{\mu'\mu''} * \bar{g}(\mu''x')\right|_{E_1^{\mu'\mu''}(R)} \ll \|\bar{g}(\mu''x')\|_{E_1^{\mu'\mu''}(R)}.\end{aligned}$$

By the choice of $\bar{g}(x')$,

$$\|\sigma_{\mu''}\|_{E_1^{\mu'\mu''} E_2} \|\bar{g}(\mu''x')\|_{E_1^{\mu'\mu''}} \ll \|\bar{g}(x')\|_{E_2(R)}.$$

Therefore

$$\|\sigma_{\mu''}\|_{E_1^{\mu'\mu''} E_2} \|g_{\mu'} \otimes \chi_{\mu''}\|_{E_1(R^2)} \ll \|\bar{g}\|_{E_2(R)}.$$

As already noted earlier (see §2), $\bar{g} \ll g_{\mu'}$; consequently,

$$\|\sigma_{\mu''}\|_{E_1^{\mu'\mu''} \to E_2} \|g_{\mu'} \otimes \chi_{\mu''}\|_{E_1(R^2)} \ll \|g_{\mu'}\|_{E_2(R)}.$$

Thus, (19) is a sharp estimate.

Denote by $\mathfrak{m}_{\tau, E}(R)$ the subspace of EFET τ that depend on a single variable and belong to $E(R)$.

Arguing as in the proof of Theorem 2, we can show that the quantity $\|\sigma_{\mu''}\|_{E_1^{\mu'\mu''} \to E_2}$ in (19) can be replaced by $\|\sigma_{\mu''}\|_{\mathfrak{m}_{\mu'\mu'', E_1} \to E_2}$. Then the inequality with different metrics and different dimensions takes the form

$$\|g(x', x'')\|_{E_2(R^m)} \ll \|\sigma_{|\bar{\mu}''|}\|_{\mathfrak{m}_{|\mu|, E_1} \to E_2} \|g\|_{E_1(R^n)}. \tag{22}$$

From (19) we can get (9) when $m = n$ (it is true that (19) assumes that E_1 is interpolated between L_1 and L_∞, while (9) is true for any symmetric spaces), and the "pure" inequality with different dimensions when $E_1 = E_2$. Moreover, it is easy to see that (22) becomes Nikol'skiĭ's inequality with different metrics and different dimensions when $E_i = L_{p_i}$ $(1 \leqslant p_1 \leqslant p_2 \leqslant \infty)$.

The authors thank the participants of the Kreĭn-Semenov seminar for their attention to the results presented here, and also the referee for important remarks leading to an improvement of the paper.

BIBLIOGRAPHY

1. S. M. Nikol'skiĭ, *Approximation of functions of several variables and embedding theorems*, 2nd rev. aug. ed., "Nauka", Moscow, 1977; English transl. of 1st ed., Springer-Verlag, 1974.

2. S. G. Kreĭn, Yu. I. Petunin and E. M. Semenov, *Interpolation of linear operators*, "Nauka", Moscow, 1978; English transl., Amer. Math. Soc., Providence, R. I., 1982.

3. N. I. Akheizer, *Lectures on the theory of approximation*, 2nd rev. ed., "Nauka", Moscow, 1965; English transl. of 1st ed., Ungar, New York, 1956.

4. M. Z. Berkolaĭko, *Imbedding theorems for generalized Besov and Höder spaces*, Dokl. Akad. Nauk SSSR **251** (1980), 521–525; English transl. in Soviet Math. Dokl. **21** (1980).

5. V. A. Rodin, *The inequalities of Jackson and Nikol'skiĭ for trigonometric polynomials in a symmetric space*, Mathematical Programming and Related Questions (Proc. Seventh Winter School, Drogobych, 1974), Theory of Functions and Functional Analysis, Tsentral. Èkonom.-Mat. Inst. Akad. Nauk SSSR, Moscow, 1976, pp. 133–139. (Russian)

6. M. I. Ganzburg, *Inequalities for entire functions of exponential type in symmetric spaces*, Studies in the Theory of Functions of Several Real Variables (Yu. A. Brudnyĭ, editor), Yaroslav. Gos. Univ., Yaroslavl, 1976, pp. 22–33.

7. R. J. Nessel and G. Wilmes, *Nikok'skiĭ-type inequalities in connection with spectral measures*, Acta Math. Acad. Sci. Hungar. **33** (1979), 169–182.

8. V. I. Dmitriev, S. G. Kreĭn and V. I. Ovchinnikov, *Foundations of the theory of interpolation of linear operators*, The Geometry of Linear Spaces and Operator Theory (B. S. Mityagin, editor), Yaroslav. Gos. Univ., Yaroslav, 1977, pp. 31–74. (Russian)

9. Einar Hille and Ralph S. Phillips, *Functional analysis and semi-groups*, rev. ed., Amer. Math. Soc., Providence, R. I., 1957.

10. V. I. Ovchinnikov, *Interpolation theorems which follow from Grothendieck's inequality*, Funktsional. Anal. i Prilozhen. **10** (1976), no. 4, 45–54; English transl. in Functional Anal. Appl. **10** (1976).

Translated by H. H. McFADEN

Труды Матем. Инст. Стеклов.
161(1983)

Proceedings of the Steklov
Institute of Mathematics
1984, Issue 3

IMBEDDING THEOREMS WITH DIFFERENT METRICS AND DIFFERENT DIMENSIONS FOR GENERALIZED BESOV SPACES

UDC 517.5 + 517.83

M. Z. BERKOLAĬKO

ABSTRACT. Necessary and sufficient conditions are given for imbedding generalized Besov spaces

$$B^{a,\varphi_1}_{E_1,F_{1*}}(R^n) \hookrightarrow B^{b,\varphi_2}_{E_2,F_{2*}}(R^m) \qquad (n \geqslant m)$$

with different metrics and with different dimensions, where the E_i are symmetric spaces on R^n, the F_{i*} are symmetric spaces on $(0,1]$ with the Haar differential measure dt/t, and the $\varphi_i(t)$ are smoothness functions satisfying the Lozinskiĭ-Stechkin conditions. Some known and some new results on imbeddings of the Besov spaces $B^r_{p\theta}$ are obtained for $E_i = L_{p_i}$ $(1 \leqslant p_1 \leqslant p_2 \leqslant \infty)$, $F_{i*} = L_{\theta*}$, and $\varphi = t^{r-a}$ (if the second-order modulus of continuity is taken in the definition of the spaces).

Bibliography: 14 titles.

Generalization of the classical Besov spaces $B^r_{p\theta}$ [1] went in two directions. The first was generalization of the smoothness, when power smoothness functions were initially replaced by functions that were not powers but satisfied the Lozinskiĭ-Stechkin conditions (see [2] and [3]), and later replaced by arbitrary functions that even were not monotone (see [4] and [5]). The imbedding theorems obtained here bear a character of sharpness.

The second direction involves general concepts of Lipschitz spaces (see [6]–[9]) and amounts, roughly speaking, to replacement of the Lebesque spaces $L_p(R^n)$ and $L_\theta((0,1],\ dt/t)$ in the definition of $B^r_{p\theta}$ by more general structures. Here only sufficient conditions for imbedding were obtained. We mention also [10], in which imbeddings of generalized Sobolev spaces were studied.

In this article we consider the generalized Besov spaces $B^{a,\varphi}_{E,F_*}$, where φ is a smoothness function satisfying the Lozinskiĭ-Stechkin conditions, the E are symmetric spaces([1]) with respect to n-dimensional Lebesgue measure, and the F_* are symmetric spaces on $(0,1]$ with the Haar differential measure, and we give precise conditions for the imbedding

$$B^{a,\varphi_1}_{E_1,F_{1*}}(R^n) \hookrightarrow B^{b,\varphi_2}_{E_2,F_{2*}}(R^m) \qquad (n \geqslant m).$$

1980 *Mathematics Subject Classification*. Primary 46E30, 46E35; Secondary 30D15.

([1]) All terms and facts relating to symmetric spaces and to interpolation of linear operators can be found, for example, in [11].

The proofs of the main theorems are based on Nikol'skiĭ's approach, which is connected with the idea of approximating functions in the given classes by entire functions of exponential type (EFET), but the classes themselves, unlike in [4] and [5], are defined in the traditional way in terms of moduli of continuity. It is for this reason that we impose the Lozinskiĭ-Stechkin conditions on the functions φ: as shown in [3], this is necessary even in the case of L_p and $L_{\theta,*}$ in order that the structural and constructive definitions of the Besov spaces coincide.

Some of the results of this article, often in lesser generality, were announced in [12].

§1. Some definitions and notation. Auxiliary statements

Let $E(0, \infty)$ be a symmetric space on $(0, \infty)$. Denote by $E(R^n)$ the Banach space of measurable (with respect to n-dimensional Lebesgue measure) functions $f(x)$ whose rearrangements $f^*(t)$ ($f^*(t)) = \inf\{h: \lambda_f(h) < t\}$, $0 < t < \infty$, where $\lambda_f(h) = \operatorname{meas}_n(x: |f(x)| > h)$ is the distribution function of $f(x)$), belong to $E(0, \infty)$, and let $\|f\|_{E(R^n)} = \|f^*\|_{E(0,\infty)}$. The symbols R^n and $(0, \infty)$ will be omitted in the notation for spaces wherever the domain of a function is clear from the context.

Further, let F_* denote the Banach space of functions $u(t)$ on $(0, 1)$ that are measurable with respect to the Haar measure dt/t and whose rearrangements $u^*_H(t)$ with respect to this measure belong to the symmetric space $F(0, \infty)$, and let $\|u\|_{F_*} = \|u^*_H\|_F$. The subspace of F consisting of the functions of the form $\sum_{s=1}^{\infty} a_s \kappa_{(s,s+1]}(t)$ is denoted by F^d. Here and below, κ_e denotes the characteristic function of a set e.

The fundamental function $\psi_E(t)$ $(0 < t < \infty)$ of the space E is the quasiconcave $(\psi_E(t) \nearrow, \psi_E(t)/t \searrow)$ function defined by $\psi_E(t) = \|\kappa_{(0,t)}\|_E$. Recall that the dilation operator $(\sigma_\tau f)(t) = f(t/\tau)$ acts in a symmetric space and is continuous. As usual, E' denotes the space associated (dual) to E, and E'' denotes the second associated space.

Unless the contrary is stated by the words "for any symmetric space", the space E below is assumed to be an interpolation space between L_1 and L_∞. The space F_* is always assumed to be maximal: $F_*'' \cong F_*$.

Let F_1^d and F_2^d be Banach ideal spaces of sequences. Denote by $F_2^d \ominus F_1^d$ the space of sequences for which

$$\|y\|_{F_2^d \ominus F_1^d} = \sup_{\|x\|_{F_1^d} \leqslant 1} \|xy\|_{F_2^d} < \infty.$$

The symbol Φ_k (k a positive integer) stands for the collection of continuous functions $\varphi(t)$ on $[0, 1]$ with $\varphi(0) = 0$ such that the function $\varphi(t)t^{-k}$ is almost decreasing.

We say that a function $\varphi \in \Phi_k$ satisfies the Lozinskiĭ-Stechkin conditions ($\varphi \in (s) \cap (S_k)$) if: a) there exists a number $\alpha > 0$ such that $\varphi(t)t^{-\alpha}$ is almost increasing (condition (S)), and b) there exists a number β, $0 < \beta < k$, such that $\varphi(t)t^{-k+\beta}$ is almost decreasing (condition (S_k)).

Let $x = (x_1, \dots, x_n) \in R^n$, $x = (y, z)$, where $y = (x_1, \dots, x_m)$ and $z = (x_{m+1}, \dots, x_n)$, $m \leqslant n$, and let $l = (l_1, \dots, l_n)$ be a multi-index, with $|l| = l_1 + \cdots + l_n$, and $f^{(l)}(x)$ the Sobolev generalized partial derivative of order l, i.e., of order l_j with respect to x_j.

Denote by $\omega_k(f, t)_E$ the modulus of continuity of order k in the norm of E.

DEFINITION 1. $B_{E,F_*}^{a,\varphi}$ ($a \geqslant 0$ an integer $\varphi \in \Phi_k$) is the Banach space of functions $f \in E(R^n)$ having all derivatives of order l with $|l| = a$ such that

$$\|f\|_B = \|f\|_E + \sum_{|l|=a} \left\| \frac{\omega_k(f^{(l)}, \cdot)}{\varphi(\cdot)} \right\|_{F_*} < \infty.$$

For $E = L_p$ and $F_* = L_{\theta,*}$ we get the spaces $B_{p,\theta}^{a,\varphi}$ in [3], and for $k = 2$ and $\varphi = t^{r-a}$ the Besov spaces $B_{p\theta}^r$.

Let $\mathfrak{M}_{\mu,E}$ ($\mathfrak{M}_{\mu,p}$ for $E = L_p$) denote the subspace of E consisting of entire functions of exponential type μ (EFET μ) whose restrictions to R^n belong to E. What follows will be based essentially on assertions generalizing known inequalities of Bernstein, Nikol'skiĭ, and Jackson.

PROPOSITION 1 [14]. *For any symmetric space* E([2])

$$\|g_\mu^{(l)}\|_E \ll \mu^{|l|}\|g_\mu\|_E \qquad (g_\mu \in \mathfrak{M}_{\mu,E}). \tag{1}$$

PROPOSITION 2 [13]. *For any symmetric space* E

$$\|g_\mu\|_{L_\infty} \ll \frac{1}{\psi_E(\mu^{-n})}\|g_\mu\|_E \qquad (g_\mu \in \mathfrak{M}_{\mu,E}),$$

$$\|g_\mu\|_E \ll \mu^n \psi_E(\mu^{-n})\|g_\mu\|_{L_1} \qquad (g_\mu \in \mathfrak{M}_{\mu,1}).$$

Consider the EFET

$$Q_\mu(x) = \prod_{j=1}^{n} \frac{\sin^2(\mu/2)x_j}{x_j^2};$$

it follows from Proposition 2 that([3])

$$\|Q_\mu\|_E \asymp \mu^{2n}\psi_E(\mu^{-n}). \tag{2}$$

In considering arbitrary symmetric spaces E_i ($i = 1, 2$) below, we assume that $E_1 \cap L_\infty \hookrightarrow E_2$ (see [13]).

PROPOSITION 3 (inequality with different metrics [13]). *For any symmetric spaces* E_i *the norm of the imbedding operator* $\mathfrak{M}_{\mu,E_1} \hookrightarrow \mathfrak{M}_{\mu,E_2}$ *is equivalent to the quantity*

$$\theta(\mu) = \sup_{\|g\|_{E_1(0,\infty)} \leqslant 1} \|g^* \kappa_{(\mu^{-n},\infty)}\|_{E_2(0,\infty)}.$$

We give examples of situations where the inequality with different metrics takes a form more convenient for applications.

PROPOSITION 4 [13]. *Suppose that one of the following conditions holds* ($\psi_i = \psi_{E_i}$):

a) *The function* $\psi_2(t)/\psi_1(t)$ *is decreasing, and the spaces* E_i *are of a "single type", i.e.,* $E_i = L_{N_i}^*$ *are Orlicz spaces, or* $E_i = \Lambda_{\psi_i}$, $L_{p_i q_i}$ *are Lorentz spaces* ($L_{pp} = L_p$), *or* $E_i = \mathcal{M}_{\psi_i}$ *are Marcinkiewicz spaces.*

b) *There exists a* $\delta > 0$ *such that the function* $\psi_2(t)t^\delta/\psi_1(t)$ *is decreasing.*

c) E_2 *is an interpolation space between* E_1 *and* L_∞.

([2]) $\ll$ stands for $\leqslant$ with a constant that is possibly not sharp.

([3]) Two quantities A and B are taken to be *equivalent* ($\asymp$) if $A \ll B \ll A$.

Then the inequality

$$\|g_\mu\|_{E_2} \ll \frac{\psi_2(\mu^{-n})}{\psi_1(\mu^{-n})}\|g_\mu\|_{E_1} \qquad (g_\mu \in \mathfrak{M}_{\mu,E_1}) \tag{3}$$

holds and is sharp with respect to order.

The sharpness of (3) can be checked with the functions Q_μ and follows directly from (2).

PROPOSITION 5 (inequality with different metrics and different dimensions). *The norm of the imbedding operator $\mathfrak{M}_{\mu,E_1}(R^n) \hookrightarrow \mathfrak{M}_{\mu,E_2}(R^m)$ is equivalent for any $z \in R^{n-m}$ to the quantity*

$$T_\mu = \|\sigma_{\mu^{n-m}}\|_{\mathfrak{M}_{\mu^n,E_1} \to E_2}, \tag{4}$$

where $\mathfrak{M}_{\mu^n,E_1}$ is the subspace of entire functions of a single variable having exponential type μ^n, and the equivalence constants do not depend on z.

For $m = n$ we get Proposition 3 (but under the superfluous assumption that the E_i are interpolation spaces), and for $E_1 = E_2$ we get the inequality with different dimensions.

PROPOSITION 6. *For any symmetric space E and any function $f \in E$ having an lth derivative* ($|l| = 0, 1, \ldots$)

$$\mathscr{E}_\mu(f)_{E''} \ll \mu^{-|l|}\omega_k(f^{(l)}, 1/\mu)_E, \tag{5}$$

where $\mathscr{E}_\mu(f)_{E''}$ is the best approximation of f by functions in $\mathfrak{M}_{\mu,E}$ in the norm of the space E''. If E is maximal, then

$$\mathscr{E}_\mu(f)_E \ll \mu^{-|l|}\omega_k(f^{(l)}, 1/\mu)_E. \tag{6}$$

The proof literally repeats that of the corresponding assertion in [1] (§5.2.1), except that instead of Minkowski's inequality for L_p the generalization of it in [11] should be used.

§2. Equivalent norms in $B^{a,\varphi}_{E,F_*}$

Taking into account that F_* is maximal, we can formulate the condition that a function $f \in E(R^n)$ belongs to the space $B^{a,\varphi}_{E,F_*}$ as follows:

$$f \in B^{a,\varphi}_{E,F_*} \Leftrightarrow \|f\|_E + \max_{|l|=a} \sup_{\substack{\|g\|_{F'_*} \leqslant 1 \\ g \geqslant 0}} \int_0^1 \frac{\omega_k(f^{(l)}, t)_E}{\varphi(t)} g(t)\,\frac{dt}{t} < \infty,$$

and the corresponding norms are equivalent.

It will be convenient for us to use another equivalent norm

$$\|f\|_B = \|f\|_E + \max_{|l|=a} \sup_{\substack{\|g\|_{F'_*} \leqslant 1 \\ g \geqslant 0}} \sum_{s=0}^{\infty} \frac{\omega_k(f^{(l)}, 2^{-s})_E}{\varphi(2^{-s})} b_s, \tag{7}$$

where

$$b_s = \int_{2^{-s}}^{2^{-s+1}} g(t)\,\frac{dt}{t},$$

$g(t) \equiv 0$ for $t \in [1,2]$; this is obtained if we break up the integral over binary intervals and take into account the inequalities $\omega_k(f, 2t)_E \ll \omega_k(f, t)_E$ and $\varphi(2t) \ll \varphi(t)$.

THEOREM 1. *The second lower index F_* in the notation for the Besov space can be replaced by F^d, with the norm defined as*

$$\|f\|_B = \|f\|_E + \sum_{|l|=a} \left\| \left\{ \frac{\omega_k(f^{(l)}, 2^{-s})_E}{\varphi(2^{-s})} \right\}_1^\infty \right\|_{F_d}.$$

PROOF. The function $\omega_k(f^{(l)}, t)_E / \varphi(t)$ is equivalent to the step function

$$a(t) = \sum_{s=1}^{\infty} \frac{\omega_k(f^{(l)}, 2^{-s})_E}{\varphi(2^{-s})} \kappa_{(2^{-s}, 2^{-s+1}]}(t),$$

whose distribution function with respect to Haar measure is equivalent to the distribution function of the step function

$$\bar{a}(t) = \sum_{s=0}^{\infty} \frac{\omega_k(f^{(l)}, 2^{-s})_E}{\varphi(2^{-s})} \kappa_{(s, s+1]}(t)$$

with respect to Lebesgue measure, and this gives us the theorem.

COROLLARY.

$$B^{a,\varphi}_{E, L_{1*}} \hookrightarrow B^{a,\varphi}_{E, F_*} \hookrightarrow B^{a,\varphi}_{E, L_{\infty*}}.$$

We consider the expansion $f = \sum_0^\infty q_s$ of a function $f(x)$ in a series of Vallée-Poussin type (the q_s are EFET 2^{s+1}, and $\|q_s\|_E \ll \mathscr{E}_{2^s}(f)_E$) that converges weakly in S', and we introduce the norm

$$^{(1)}\|f\|_B = \sup_{\substack{g \geqslant 0 \\ \|g\|_{F'_*} \leqslant 1}} \sum_{s=0}^{\infty} \frac{2^{as}\|q_s\|_E}{\varphi(2^{-s})} b_s, \tag{8}$$

where b_s is defined as in (7).

Let $^{(1)}B^{a,\varphi}_{E,F_*}$ be the Banach space of functions $f \in L_1 + L_\infty$ which can be expanded in series $f = \sum_0^\infty q_s$ of Vallée-Poussin type converging weakly in S' and which have finite norm $^{(1)}\|f\|_B$.

THEOREM 2. *If the space E is maximal, then $B \hookrightarrow {}^{(1)}B$. If $\varphi \in (S) \cap (S_k)$, then the imbedding $^{(1)}B \hookrightarrow B$ is valid for any symmetric space E.*

PROOF. If we observe that $\sigma_{2^s}(f, x) \to f(x)$ $(s \to \infty)$ weakly in S', where $f \in L_1 + L_\infty$, and

$$\sigma_{2^s}(f, x) = (2/\pi)^{n/2}(V_{2^s} * f),$$

where V_{2^s} is the corresponding Vallée-Poussin kernel (this follows from the validity of this limit relation in L_1 and in L_∞; [1], §8.6), then the first assertion of the theorem can easily be proved by contradiction when Proposition 6 and the expression (7) for the norm in B are used.

We prove the second assertion. The finiteness of the norm $^{(1)}\|f\|_B$ gives us that $\sum_0^\infty q_s$ converges absolutely in E. First of all we have that

$$2^{as}\|q_s\|_E(\varphi(2^{-s}))^{-1} = O(1).$$

Indeed, if this is not so, then let $\{s_k\} \uparrow \infty$ be a subsequence such that

$$2^{as_k}\|q_{s_k}\|_E(\varphi(2^{-s_k}))^{-1} \geqslant 2^{2k}$$

and consider the function

$$g_0(t) = \begin{cases} 2^{-k} & \text{for } 2^{-s_k} \leqslant t \leqslant 2^{-s_k+1}, \\ 0 & \text{elsewhere.} \end{cases}$$

Obviously, $g_0 \in L_{1*} \cap L_{\infty *}$, but

$$\sum_{k=0}^{\infty} \frac{2^{as_k}\|q_{s_k}\|_E}{\varphi(2^{-s_k})} \int_{2^{-s_k}}^{2^{-s_k+1}} g_0(t)\,\frac{dt}{t} = \infty.$$

Thus,

$$\sum_{s=0}^{\infty} \|q_s\|_E \ll \sum_{s=0}^{\infty} 2^{-as}\varphi(2^{-s}) \asymp \int_0^1 t^a\varphi(t)\,\frac{dt}{t} < \infty$$

for any $\varphi \in \Phi_k$ if $a \geqslant 1$, and for $\varphi \in (S)$ if $a = 0$.

We estimate the second part of the norm. It follows from (1) that

$$\omega_k(f^{(l)}, 2^{-s})_E \ll 2^{-sk} \sum_{N=0}^{s-1} 2^{N(a+k)}\|q_N\|_E + \sum_{N=s}^{\infty} 2^{aN}\|q_N\|_E \qquad (|l| = a),$$

and the series on the right-hand side converges because of the arguments above. For any $g \geqslant 0$ with $\|g\|_{F'_*} \leqslant 1$

$$\begin{aligned}\sum_{s=0}^{\infty} \frac{\omega_k(f^{(1)}, 2^{-s})_E}{\varphi(2^{-s})} b_s &\ll \sum_{s=0}^{\infty} b_s(\varphi(2^{-s}))^{-1} \sum_{N=s}^{\infty} 2^{aN}\|q_N\|_E \\ &\quad + \sum_{s=0}^{\infty} b_s 2^{-sk}(\varphi(2^{-s}))^{-1} \sum_{N=0}^{s-1} 2^{N(a+k)}\|q_N\|_E \\ &= A_1 + A_2.\end{aligned}$$

Further, denoting by α the number in condition (S), we get that

$$\begin{aligned}A_1 &= \sum_{s=0}^{\infty} b_s(\varphi(2^{-s}))^{-1} \sum_{N=s}^{\infty} 2^{aN}\|q_N\|_E(\varphi(2^{-N}))^{-1}\varphi(2^{-N})2^{N\alpha}2^{-N\alpha} \\ &\ll \sum_{s=0}^{\infty} 2^{s\alpha}b_s \sum_{N=s}^{\infty} 2^{aN}\|q_N\|_E(\varphi(2^{-N}))^{-1}2^{-N\alpha}.\end{aligned} \tag{9}$$

Consider the function $Q_f(t) = 2^{aN}\|q_N\|_E(\varphi(2^{-N}))^{-1}$ for $t \in (2^{-N}, 2^{-N+1}]$, $N = 1, 2, \ldots$. Obviously, ${}^{(1)}\|f\|_B < \infty \Leftrightarrow Q_f(t) \in F_*$; this follows from properties of the function $t^{-1}(\varphi(t))^{-1}$. We show that $s^{-\alpha}\int_0^s Q_f S(t)t^{\alpha-1}\,dt \in F_*$. For any $g \geqslant 0$ with $\|g\|_{F'_*} \leqslant 1$ the fact that $\|\sigma_\tau\|_{F_*} = O(1)$, $0 < \tau < \infty$, gives us that

$$\begin{aligned}\int_0^1 g(s)s^{1-\alpha}\left(\int_0^s Q_f(t)t^{\alpha-1}\,dt\right)ds &= \int_0^1 g(s)s^{-1}\left(\int_0^1 Q_f(s\tau)\tau^{\alpha-1}\,d\tau\right)ds \\ &= \int_0^1 \tau^{\alpha-1}\left(\int_0^1 Q_f(s\tau)g(s)\,\frac{ds}{s}\right)d\tau \\ &\leqslant \int_0^1 \|\sigma_\tau\|_{F_*}\tau^{\alpha-1}\,d\tau < \infty.\end{aligned}$$

The right-hand side of (9) is equivalent to

$$\sum_{s=0}^{\infty} 2^{s\alpha}\left(\int_0^{2^{-s}} Q_f(t)t^{\alpha-1}\,dt\right)b_s,$$

so, since the function $\int_0^s Q_t(T)t^{\alpha-1}\,dt$ is increasing, we get that $A_1 < \infty$. Consider now the quantity A_2, with $\varepsilon = k - \beta$ first set equal to the number in condition (S_k):

$$A_2 = \sum_{s=0}^{\infty} b_s(\varphi(2^{-s}))^{-1}2^{-s\varepsilon}2^{s(k+\varepsilon)}\sum_{N=0}^{s-1} 2^{N(a+k)}\|q_N\|_E$$

$$= \sum_{s=0}^{\infty} 2^{-s\varepsilon}\left(\sum_{N+0}^{s-1}\frac{2^{aN}\|q_N\|_E}{\varphi(2^{-N})}2^{N\varepsilon}\right) \asymp \sum_{s=0}^{\infty} b_s 2^{s\varepsilon}\int_{2^{-s+1}}^{1}\frac{Q_f(t)}{t^{1+\varepsilon}}\,dt.$$

As above, we can show that

$$h^{\varepsilon}\int_h^1 \frac{Q_f(t)}{t^{1+\varepsilon}}\,dt \in F_*.$$

Therefore, since the function $\int_h^1 Q_f(t)t^{-1-\varepsilon}\,dt$ is decreasing, we get that $A_2 < \infty$.

The proof of the theorem is completed.

REMARK. In what follows we use expansions in EFET 2^{s+3}; this clearly affects only the equivalence constants in Theorem 2.

The expression for the norm ${}^{(0)}\|\cdot\|$ gives us that the step function

$$\sum_{s=0}^{\infty}\frac{2^{as}\|q_s\|_E}{\varphi(2^{-s})}\kappa_{(2^{-s},2^{-s+1}]}(t)$$

must belong to F_*. Using the same arguments as in the proof of Theorem 1, we get the following assertion as a consequence of Theorem 2.

COROLLARY 1. *The norm* ${}^{(1)}\|\cdot\|_B$ *can be given in the form*

$$ {}^{(1)}\|f\|_B = \left\|\left\{\frac{2^{as}\|q_s\|_E}{\varphi(2^{-s})}\right\}_0^{\infty}\right\|_{F^d}. \tag{10}$$

It is the form (10) we shall use for ${}^{(1)}\|\cdot\|_B$ in what follows.

COROLLARY 2. *The known theorems on constructive descriptions of the spaces* $B_{p\theta}^{a,\varphi}$ *and* $B_{p\theta}^{r}$ *(see* [1]–[3]*) are obtained when* $E = L_p$ *and* $F_* = L_{\theta *}$.

We remark that the condition $\varphi k \in (S)\cap(S_k)$ is necessary for the norms $\|\cdot\|_B$ and ${}^{(1)}\|\cdot\|_B$ to be equivalent even when $E = L_p$ and $F_* = L_{\theta *}$ (see [3]).

§3. Main theorems

Consider the functions $q_s(x) = r_s(x)\exp[i2^{s+2}\sum_1^n x_j]$, where $r_s(x)$ is an EFET 2^s. Clearly, $q_s(x)$ is an EFET 2^{s+3} and, moreover, $|q_s| = |r_s|$.

LEMMA 1. *Suppose that the numbers* c_s $(s = 0, 1, \ldots)$ *are such that the series*

$$f(x) = \sum_{s=0}^{\infty} c_s q_s(x)$$

converges weakly in $S'(R^n)$, *while the series*

$$f_1(y) = f(y, 0) = \sum_{s=0}^{\infty} c_s q_s(y, 0)$$

converges in $E(R^m)$. *Then*

$$\|c_s q_s(y, 0)\|_{E(R^m)} \ll \mathscr{E}_{2^s}(f_1)_{E(R^m)} \qquad (m \leqslant n).$$

PROOF. The support of the Fourier transform $\tilde{q}_s$ of q_s is contained in the cube with faces from $2^{s+2} - 2^s$ to $2^{s+2} + 2^s$, i.e., from $3 \cdot 2^s$ to $5 \cdot 2^s$. Consider the infinitely differentiable function $\mu_0(\lambda)$ of a single variable given by

$$\mu_0(\lambda) = \begin{cases} 1, & 3 \leqslant \lambda \leqslant 5, \\ 0, & \lambda \leqslant 5/2, \lambda \geqslant 11/2, \end{cases}$$

along with the functions $\mu_s(\lambda) = \mu_0(\lambda/2^s)$. Then $V_s(t) = \overline{\mu_s(\lambda)} = 2^s V_0(2^s t)$, where $V_0(t) = \mu_0(\lambda)$, so that $\|V_s\|_{L_1} = O(1)$ uniformly in s. Further, let $W_s(y) = \prod_{j=1}^m V_s(x_j)$ and $\mu_s(\xi) = \prod_{j=1}^m \mu_s(\xi_j)$. Since $f_1 \in E(R^m) \hookrightarrow (L_1 + L_\infty)(R^m) \hookrightarrow S'(R^m)$, it follows that

$$\tilde{f}_1(\xi) = \sum_{s=0}^{\infty} c_s \tilde{q}_s(\xi, 0).$$

Using the facts that $\mu_s \equiv 1$ on supp $\tilde{q}_s$ and $\mu_s \equiv 0$ on supp $q_{s'}$ for $s' \neq s$, we get that $(\xi \in R^m)$

$$c_s \tilde{q}_s = \left(\sum_{k=0}^{\infty} c_k \tilde{q}_k \right) \mu_s = \tilde{f}_1(\xi) \mu_s(\xi). \tag{11}$$

If now q_{2^s} is an arbitrary function in $\mathfrak{M}_{2^s, E(R^m)}$, then $\tilde{q}_{2^s} \cdot \mu_s \equiv 0$ and $c_s q_s = (f_1 - q_{2^s}) * W_s$ by (11). Therefore, using the fact that E is an interpolation space between L_1 and L_∞ and properties of the function W_s, we get that

$$\|c_s q_s(y, 0)\|_E \ll \|f_1 - q_{2^s}\|_E,$$

and this proves the lemma.

THEOREM 3 (imbeddings with different metrics and different dimensions). *If* $\varphi_1 \in (S) \cap (S_k)$ *and the space* E_1 *is maximal, then the inclusion*

$$\{\alpha_s\}_1^\infty = \left\{ 2^{s(b-a)} \frac{\varphi_1(2^{-s})}{\varphi_2(2^{-s})} T_{2^s} \right\}_1^\infty \in F_2^d \ominus F_1^d \tag{12}$$

suffices for the imbedding([4])

$$B_{E_1, F_{1*}}^{a, \varphi_1}(R^n) \hookrightarrow B_{E_2, F_{2*}}^{b, \varphi_2}(R^m) \qquad (m \leqslant n). \tag{13}$$

But if $\varphi_i \in (S) \cap (S_k)$ $(i = 1, 2)$, *and the space* E_2 *is maximal, then for any* E_1 *the inclusion* (12) *is necessary for the inclusion* (13).

PROOF. *Sufficiency.* For brevity denote the spaces in (13) by B_1 and B_2, respectively. Let $f \in B_1$. Then for this function the norm ${}^{(1)}\|f\|_{B_1}$ is finite. We fix z in the

([4]) The trace of a function on $R^m(z)$ is understood in the sense of the space $E_2(R^m)$ as in [1], §6.4.

expansion $f = \Sigma_0^\infty q_s$. Proposition 5 gives us that, for all $\{b_s\}_1^\infty \in (F_2^d)'$, $b_s \geqslant 0$,

$$\sum_{s=0}^{\infty} \|q_s(\cdot, z)\|_{E_2} 2^{bs}(\varphi_2(2^{-s}))^{-1} b_s \ll \sum_{s=0}^{\infty} \|q_s\|_{E_1} T_{2^s} 2^{bs}(\varphi_2(2^{-s}))^{-1} b_s$$

$$= \sum_{s=0}^{\infty} \frac{2^{as}\|q_s\|_{E_1}}{\varphi_1(2^{-s})} \alpha_s b_s < \infty.$$

Arguing as in the proof of Theorem 2, we get that $\Sigma_0^\infty q_s(y, z)$ converges in $E_2(R^m)$ for any fixed $z \in R^{n-m}$ to some function $f_1(y, z) \in E_2(R^m)$. Moreover, this series converges in $E_1(R^n)$ to $f(y, z)$. We show that $f = f_1$ almost everywhere in the sense of n-dimensional Lebesgue measure. To do this it suffices to show that if $f_k(y, z) \to f(y, z)$ in $(L_1 + L_\infty)(R^n)$ and $f_k(y, z) \to f_1(y, z)$ for any $z \in R^{n-m}$ in $(L_1 + L_\infty)(R^m)$, then $f = f_1$ a.e. in the sense of n-dimensional Lebesgue measure. Indeed, the convergence $\|f_k - f\|_{(L_1 + L_\infty)(R^n)} \to 0$ implies the existence of a representation $f_k - f = (f_k^{(0)} - f^{(0)}) + (f_k^{(1)} - f^{(1)})$ such that

$$\|f_k^{(0)} - f^{(0)}\|_{L_1(R^n)} + \|f_k^{(1)} - f^{(1)}\|_{L_\infty(R^n)} \to 0$$

and of a subsequence $\{k_j\} \uparrow \infty$ such that

$$\left\|\left(f_{k_j}^{(0)} - f^{(0)}\right)(\cdot, z)\right\|_{L_1(R^m)} + \left\|\left(f_{k_j}^{(1)} - f^{(1)}\right)(\cdot, z)\right\|_{L_\infty(R^m)} \to 0$$

for almost all $z \in R^{n-m}$, Q.E.D.

Thus, the function $f_1(y, z)$, which is equivalent to f, belongs for any $z \in R^{n-m}$ to $B_2(R^m)$, and

$$\|f_1(\cdot, z)\|_{B_2(R^m)} \ll \|\alpha\|_{F_2^d \ominus F_1^d}$$

with a constant not depending on z. It remains to show that, for any particular $z \in R^{n-m}$, f_1 is the trace of f on $R^m(z)$, i.e., to establish that

$$\|\Delta_{x_j h} f_1\|_{E_2(R^m)} \to 0 \quad \text{as } h \to 0 \qquad (m + 1 \leqslant j \leqslant n).$$

To do this we use the obvious estimates

$$\|\Delta_{x_j h} q_s\|_{E_2(R^m)} \ll |h| 2^s \|q_s\|_{E_2(R^m)} \ll |h| 2^s T_{2^s} \|q_s\|_{E_1(R^n)},$$

$$\|\Delta_{x_j h} q_s\|_{E_2(R^m)} \leqslant 2\|q_s\|_{E_2(R^m)} \ll T_{2^s}\|q_s\|_{E_1(R^n)}.$$

Then

$$\|\Delta_{x_j h} f_1\|_{E_2} \ll \sum_{s=0}^{k-1} |h| 2^s T_{2^s} \|q_s\|_{E_1} + \sum_{s=0}^{\infty} T_{2^s}\|q_s\|_{E_1} = A.$$

As in Theorem 2, it can be shown that $\|q_s\|_{E_1} \ll \varphi_1(2^{-s})/2^{sa}$; therefore, since $\alpha_s = O(1)$, we get that

$$A \ll \sum_{s=0}^{k-1} |h| 2^s \frac{\varphi_2(2^{-s})}{\varphi_1(2^{-s}) s} 2^{s(a-b)} \frac{\varphi_1(2^{-s})}{2^{sa}} + \sum_{s=k}^{\infty} \frac{\varphi_2(2^{-s})}{\varphi_1(2^{-s})} 2^{s(a-b)} \frac{\varphi_1(2^{-s})}{2^{sa}}$$

$$\asymp |h| \int_{2^{-k}}^{1} \frac{\varphi_2(t) t^b}{t^2} dt + \int_{2}^{2^{-k}} \frac{\varphi_2(t) t^b}{t} dt.$$

From this it follows easily that

$$\|\Delta_{x_j,h} f_1\|_{E_2(R^m)} = O\begin{pmatrix} \varphi_2(|h|) & \text{for } b = 0, \beta < 1, \\ \varphi_2(|h|)\ln(|h|^{-1}) & \text{for } b = 0, \beta = 1, \\ \varphi(|h|^{1/b+\beta})|h|^{b/b+\beta} & \text{for } b + \beta > 1, \end{pmatrix}, \qquad (14)$$

where β is the number in the condition $\varphi_2 \in (S_k)$.

We mention that the choice of the "lower" indices of the Besov spaces affects at most the constants in the estimate (14) which is important for applications.

Necessity. Assume that (12) does not hold. This means that there is a sequence $\{a_s\}_1^\infty \in F_1^d$, $a_s \geqslant 0$, such that $\{a_s\alpha_s\}_1^\infty \notin F_2^d$. We take two cases.

I. The sequence $\{\alpha_s\}_1^\infty$ is unbounded. We show that in this case the "smallest" space $B_{E_1,L_{1*}}^{a,\varphi_1}(R^n)$ cannot be imbedded in the "largest" space $B_{E_2,L_{\infty*}}^{b,\varphi_2}(R^m)$ (see the corollary to Theorem 2), i.e., we construct a function $f_0 \in B_{E_1,L_{1*}}^{a,\varphi_1}(R^n)$ whose trace on $R^m(0)$ exists in the sense of E_2 but does not belong to $B_{E_2,L_{\infty*}}^{b,\varphi_2}(R^m)$.

Consider a sequence $\{\beta_s\}_1^\infty \in l_1$ such that $\{\gamma_s\}_1^\infty = \{\alpha_s\beta_s\}_1^\infty \notin l_\infty$. Choose a subsequence $\{s_i\} \uparrow \infty$ such that $\gamma_{s_i} \geqslant 2^{2i}$ and $2^{\alpha s_i/2} \geqslant 2^{2i}$, where α is the number in the condition $\varphi_2 \in (S)$. If now $\varepsilon_i = \min\{1, 2^{\alpha s_i/2}/\gamma_{s_i}\}$, then $\{\varepsilon_i\}_1^\infty \in l_\infty$, and

$$2^{2i} \leqslant \gamma_{s_i}\varepsilon_i \leqslant 2^{\alpha s_i/2}. \qquad (15)$$

We take the function $r_s(x)$ to be an EFET 2^s such that

$$\|r_s(\cdot, 0)\|_{E_2} \asymp T_{2^s}\|r_s\|_{E_1},$$

and as in Lemma 1 we write

$$q_s(x) = r_s(x)\exp\left[i2^{s+2}\sum_{j=1}^n x_j\right] \qquad \left(\|q_s\|_{E_i} = \|r_s\|_{E_i}, i = 1, 2\right).$$

Moreover, we introduce the functions

$$\bar{q}_s(x) = \begin{cases} 2^{-as_i}\varphi_1(2^{-s_i})\beta_{s_i}\varepsilon_i\dfrac{q_{s_i}(x)}{\|q_{s_i}\|_{E_1(R^n)}} & \text{for } s = s_i, \\ 0 \quad \text{for } s \neq s_i, \end{cases}$$

and the function $f_0(x) = \sum_0^\infty \bar{q}_s(x)$. Then

$$\sum_{s=0}^\infty \frac{2^{a(s+3)}\|\bar{q}_s\|_{E_1}}{\varphi_1(2^{-s-3})} \ll \sum_{s=0}^\infty \frac{2^{sa}\|\bar{q}_s\|_{E_1}}{\varphi_1(2^{-s})} = \sum_{i=0}^\infty \beta_{s_i}\varepsilon_i < \infty,$$

and so $f_0 \in B_1$.

Let us show that $f_0(y, 0)$ is the trace on $R^m(0)$ of a function not belonging to $B_2(R^m)$. Using (15), we get the following estimate in a way analogous to the preceding $(m + 1 \leqslant j \leqslant n)$:

$$\begin{aligned}\left\|\Delta_{x_j,h}\bar{q}_{s_i}(\cdot, z)\right\|_{E_2} &\ll |h|2^{s_i}2^{-as_i}\varphi_1(2^{-s_i})T_{2^{s_i}}\beta_{s_i}\varepsilon_i \\ &\leqslant |h|2^{s_i}2^{-s_ib}\varphi_2(2^{-s_i})\gamma_{s_i}\varepsilon_i \leqslant |h|2^{s_i}2^{-s_ib}\varphi_2(2^{-s_i})2^{\alpha s_i/2},\end{aligned}$$

$$\left\|\Delta_{x_j,h}\bar{q}_{s_i}(\cdot, z)s\right\|_{E_2} \leqslant 2\left\|q_{s_i}(\cdot, z)\right\|_{E_2} \ll 2^{-bs_i}\varphi_2(2^{-s_i})2^{\alpha s_i/2}.$$

This means that $\Sigma_0^\infty q_s(y,0)$ converges absolutely in $E_2(R^m)$. Moreover,

$$\|\Delta_{x_j h} f_0(\cdot,0)\|_{E_2(R^m)} \ll |h| \sum_{j=0}^{N-1} 2^{s_i} 2^{-s_i b} \varphi_2(2^{-s_i}) 2^{\alpha s_i/2} + \sum_{i=N}^{\infty} 2^{-bs_i} \varphi_2(2^{-s_i}) 2^{\alpha s_i/2}$$

$$\ll |h| \sum_{i=0}^{N-1} 2^{-s_i(b+\alpha/2-1)} + \sum_{i=N}^{\infty} 2^{-s_i(b+\alpha/2)},$$

which implies that $\|\Delta_{x_j h} f_0(\cdot,0)\|_{E_2} \to 0$ as $|h| \to 0$.

On the other hand, by (15) and the choice of the functions $r_s(x)$,

$$\sup_s \frac{2^{bs}\|\bar{q}_s(\cdot,0)\|_{E_2}}{\varphi_2(2^{-s})} = \sup\{\gamma_{s_i}\varepsilon_i\} = \infty.$$

Using Lemma 1 and the remark after Theorem 2, we get the implication (13) ⇒ (12).

II. $\{\alpha_s\}_1^\infty \in l_\infty$, but $\{\alpha_s a_s\}_1^\infty \notin F_2^d$. Consider the function

$$f_0(x) = \sum_{s=0}^{\infty} c_s q_s(x) = \sum_{s=0}^{\infty} \bar{q}_s(x),$$

where the functions $q_s(x)$ are defined as in the proof of the preceding case, and

$$c_s = 2^{-as}\varphi_1(2^{-s})\left(\|q_s\|_{E_1(R^n)}\right)^{-1} a_s, \qquad s = 0,1,\ldots.$$

It is not hard to compute that ${}^{(1)}\|f_0\|_{B_1} < \infty$. We have the estimates ($m+1 \leqslant j \leqslant n$)

$$\|\Delta_{x_j h}\bar{q}_s(\cdot,0)\|_{E_2} \ll |h| 2^s 2^{-as}\varphi_1(2^{-s}) T_{2^s} a_s,$$

$$\|\Delta_{x_j h}\bar{q}_s(\cdot,0)\|_{E_2} \ll 2\|\bar{q}_s(\cdot,0)\|_{E_2} \ll 2^{-as}\varphi_1(2^{-s}) T_{2^s} a_s,$$

from which, using the boundedness of the sequence $\{a_s b_s\}_1^\infty$ and properties of $\varphi_2(t)$, we get (in a way analogous to the preceding) that $f_0(y,0) \in E_2(R^m)$ and that it is the trace of the function $f_0(x)$ on $R^m(0)$. But, on the other hand,

$$\|\bar{q}_s(\cdot,0)\|_{E_2} \asymp \varphi_1(2^{-s}) T_{2^s} 2^{-as} a_s,$$

and since $\{a_s\alpha_s\}_1^\infty \notin F_2^d$, the arguments in the preceding case give us that $\|f_0(\cdot,0)\|_{B_2} = \infty$. The theorem is proved.

THEOREM 4 (imbeddings with different metrics). *Suppose that* $\varphi_2 \in (S) \cap (S_k)$, *the space* E_1 *is maximal, and* E_2 *is arbitrary. Then the relation*

$$\{\beta_s\}_1^\infty = \left\{2^{s(b-a)}\frac{\varphi_1(2^{-s})}{\varphi_2(2^{-s})}\theta(2^{sn})\right\}_1^\infty \in F_2^d \ominus F_1^d \tag{16}$$

suffices for the imbedding (13) *when* $m = n$. *But if* $\varphi_i \in (S) \cap (S_k)$, *the space* E_2 *is maximal, and* E_1 *is arbitrary, then* (13) ⇒ (16) *for* $m = n$.

REMARK. If one of the conditions in Proposition 4 holds, then the requirement that the space be maximal is superfluous for the implication (16) ⇒ (13); this follows from the fact that the quantity $\theta(2^{sn})$ is the same for the imbeddings $\mathfrak{M}_{\mu,E_1} \hookrightarrow \mathfrak{M}_{\mu,E_2}$ and $\mathfrak{M}_{\mu,E_1} \hookrightarrow \mathfrak{M}_{\mu,E_2''}$ (the spaces E_2 and E_2'' have the same fundamental function).

Thus, we have determined that the constructive description of the imbedding (13) requires knowledge of the space $F_2^d \ominus F_1^d$, or at least the answer to the question of when a given sequence belongs to $F_2^d \ominus F_1^d$. Below we present some results in this direction.

THEOREM 5. 1. *If* $F_1^d \hookrightarrow F_2^d$, *then* (13) *holds if and only if* $\{\alpha_s\}_1^\infty \in l_\infty$.

2. *If* $F_i^d = l_{\theta_i}$, $\theta_1 \geqslant \theta_2$, *then* (13) *holds if and only if* $\{\alpha_s\}_1^\infty \in l_{\theta_1\theta_2/(\theta_1-\theta_2)}$.

3. *If* $F_i^d = l_{r_i\theta_i}$, $\theta_1 \geqslant \theta_2$, *and* $\theta_1/r_1 \geqslant \theta_2/r_2$, *then*

$$\sum_{s=0}^{\infty} \left(\alpha_s^* s^{1/r_2 - 1/r_1 + 1/\theta_1 - 1/\theta_2}\right)^{\theta_1\theta_2/(\theta_1-\theta_2)} < \infty \Rightarrow (13).$$

If $\{\alpha_s\}_1^\infty \downarrow$, *then the reverse implication also holds.*

4. *If* F_i^d *is the sequence space* m_{ψ_i} *of Marcinkiewicz, and* $\psi_1(s) \in (S)$, *then* $\alpha_s^* = O(\psi_2(s)/\psi_1(s)) \Rightarrow (13)$. *But if* $\{\alpha_s\}_1^\infty \downarrow$, *then the reverse implication is also valid.*

5. *The imbedding* (13) *holds independently of the indices* F_{i*} *if and only if* $\{\alpha_s\}_1^\infty \in l_1$.

The proof of this theorem is entirely based on the specific nature of the concrete sequence spaces, and sometimes (for example, assertion 3) requires fairly long computations. The author hopes to devote a separate note to this proof. We mention only that assertions 2 and 1 with $E_i = L_{p_i}$ and $F_{i*} = L_{\theta_i *}$ give us the imbedding theorem for the spaces $B_{p\theta}^{a;\varphi}$ in [3], while assertion 5 gives us a generalization (and a converse) of the classical imbedding $B_{p\theta_1}^{r+\varepsilon} \hookrightarrow B_{p\theta_2}^{r}$ $(1 \leqslant \theta_1, \theta_2 \leqslant \infty)$ when the spaces $L_{\theta_i *}$ are replaced by arbitrary maximal symmetric spaces F_{i*}.

Combining Theorems 3 and 4 with Theorem 5, we can get concrete constructive conditions for imbeddings of generalized Besov spaces. We do not present these here for lack of space.

REMARK. The proofs and inequalities given for EFET with a vector-valued $\mu = (\mu_1, \dots, \mu_n)$ (see [13]) enable us to carry over all the statements here to the case of anisotropic Besov spaces $B_{a,F_*}^{a;\bar\varphi}$. But the case of complete anisotropy of $B_{\bar E,F_*}^{a;\bar\varphi}$, where $\bar E$ is an ideal space on R^n with mixed norm, still requires further investigation.

The author sincerely thanks M. L. Gol'dman for useful discussions.

BIBLIOGRAPHY

1. S. M. Nikol'skiĭ, *Approximation of functions of several variables and embedding theorems*, 2nd rev. aug. ed, "Nauka", Moscow, 1977; English transl. of 1st ed., Springer-Verlag, 1974.

2. A. S. Dzhafarov, *Imbedding theorems for classes of functions with differential properties in norms of special spaces*, Akad. Nauk Azerbaĭdzhan. SSR Dokl. **21** (1965), no. 2, 10–14. (Russian)

3. M. L. Gold'dman, *On imbedding of generalized Hölder classes*, Zametki **12** (1972), 325–336; English transl. in Math. Notes **12** (1972).

4. ______, *A description of the traces for certain function classes*, Trudy Mat. Inst. Steklov. **150** (1979), 99–126; English transl. in Proc. Steklov Inst. Math. **1981**, no. 4 (150).

5. G. A. Kalyabin, *A description of traces for anisotropic spaces of Triebel-Lizorkin type*, Trudy Mat. Inst. Steklov. **150** (1979), 160–173; English transl. in Proc. Steklov Inst. Math. **1981**, no. 4 (150).

6. A. P. Calderón, *Intermediate spaces and interpolation, the complex method*, Studia Math. **24** (1964), 113–190.

7. Yu. A. Brudnyĭ and V. K. Shalashov, *Lipschitz spaces of functions*, Dokl. Akad. Nauk SSSR **197** (1971), 18–20; English transl. in Soviet Mat. Dokl. **12** (1971).

8. K. K. Golovkin, *On equivalent norms for fraction spaces*, Trudy Mat. Inst. Steklov. **66** (1962), 364–383; English transl. in Amer. Math. Soc. Transl. (2) **81** (1969).

9. Yu. A. Brudnyĭ, *An extension theorem for a family of function spaces*, Zap. Nauchn. Sem. Leningrad. Otdel. Mat. Inst. Steklov. (LOMI) **56** (1976), 170–1173; English transl. in J. Soviet Math. **14** (1980), no. 2.

10. V. S. Klimov, *Imbedding theorems and geometric inequalities*, Izv. Akad. Nauk SSSR Ser. Mat. **40** (1976), 645–671; English transl. in Mat. USSR Izv. **10** (1976).

11. S. G. Kreĭn, Yu. I. Petunin and E. M. Semenov, *Interpolation of linear operators*, "Nauka", Moscow, 1978; English transl., Amer. Math. Soc., Providence, R.I., 1982.

12. M. Z. Berkolaĭko, *Imbedding theorems for generalized Besov and Hölder spaces*, Dokl. Akad. Nauk SSSR **251** (1980), 521–525; English transl. in Soviet Math. Dokl. **21** (1980).

13. M. Z. Berkolaĭko and V. I. Ovchinnikov, *Inequalities for entire functions of exponential type in symmetric spaces*, Trudy Mat. Inst. Steklov. **161** (1983), 3–17; English transl. in this issue.

14. E. A. Gorin, *G. E. Shilov's investigations in the theory of commutative Banach algebras and their subsequent development*, Uspekhi Mat. Nauk **33** (1978), no. 4 (202), 169–188; English transl. in Russian Math. Surveys **33** (1978).

Translated by H. H. McFADEN

Труды Матем. Инст. Стеклов.
161(1983)

Proceedings of the Steklov
Institute of Mathematics
1984, Issue 3

ON THE DENSENESS OF THE COMPACTLY SUPPORTED FUNCTIONS IN A WEIGHTED SOBOLEV SPACE

UDC 517.581.235

O. V. BESOV

ABSTRACT. It is established that $C_0^\infty(G)$ is dense in the space $W_{p,\sigma}^{(l)}(G)$ with the norm $\sum_{|\alpha|\leqslant l}\|\sigma D^\alpha u\|_{L_p(G)}$, $1<p<\infty$, where G is a bounded domain in R^n with Lipschitz boundary, and the weight function is equivalent to $\sigma(r(x))$ near ∂G, where $r(x)$ is a regularized distance from the corresponding part of the boundary. The function $\sigma(t)$ on $(0,1)$ is subject to certain integral constraints and can degenerate at zero.

Bibliography: 18 titles.

1. For a positive integer l and $1\leqslant p<\infty$ we consider the space $W_{p,\sigma}^{(l)}(G)$ of functions u having Sobolev generalized derivatives $D^\alpha u$, $|\alpha|\leqslant l$, on a bounded domain G in the n-dimensional Euclidean space E^n, with the norm

$$\|u\|_W=\sum_{|\alpha|\leqslant l}\|\sigma D^\alpha u\|_{p,G}=\sum_{|\alpha|\leqslant l}\left\{\int_G|\sigma(x)D^\alpha u(x)|^p\,dx\right\}^{1/p}. \tag{1}$$

The boundary ∂G of G is assumed to be an $(n-1)$-dimensional Lipschitz manifold. More precisely, *it is assumed that G satisfies the following conditions*:

a) *For each* k, $1\leqslant k\leqslant K<\infty$, *an orthogonal transformation* $T_k(E^n\to E_y^n)$ *is given.*

b)

$$\Gamma_k=T_k^{-1}\{y=(y',y_n):y'\in D_k\subset E_{y'}^{n-1},y_n=\gamma_k(y')\},$$
$$G_k=T_k^{-1}\{y=(y',y_n):y'\in D_k,\gamma_k(y')<y_n<\gamma_k(y')+\eta\},$$
$$\Gamma_k^\delta=T_k^{-1}\{y=(y',y_n):y'\in D_k^\delta\subset E_{y'}^{n-1},y_n=\gamma_k(y')\},$$
$$G_k^\delta=T_k^{-1}\{y=(y',y_n):y'\in D_k^\delta,\gamma_k(y')<y_n<\gamma_k(y')+\eta-\delta\},$$

where $0<\delta<\eta$, D_k *is a bounded open subset of* $E_{y'}^{n-1}$, $D_k^\delta=\{y':\ y'\in D_k,$ $\rho(y',\partial D_k)>\delta\}$, *and the function* γ_k *satisfies the Lipschitz condition*

$$|\gamma_k(y')-\gamma_k(z')|\leqslant M|y'-z'|,\qquad y',z'\in D_i.$$

c) $G=\bigcup_0^K G_k=\bigcup_0^K G_k^\delta$, $\partial G=\bigcup_1^K\Gamma_k$, *where* G_0 *is an open bounded set and* $G_0^\delta=\{x:$ $x\in G_0,\rho(x,\partial G_0)>\delta\}$.

1980 *Mathematics Subject Classification*. Primary 41A30; Secondary 46E35.

0081-5438/84 $1.00 + $.25 per page

Let $r_k(y)$, $y \in T_kG_k$ $(1 \leqslant k \leqslant K)$, denote an l times continuously differentiable function such that the following estimates are satisfied on T_kG_k $(c_i > 0)$:

a) $c_1\rho(y, T_k\Gamma_k) \leqslant r_k(y) \leqslant c_2\rho(y, T_k\Gamma_k)$;

b) $|D^\alpha r_k(y)| \leqslant c_3(\rho(y, T_k\Gamma_k))^{1-|\alpha|}$, $\forall\, \alpha$, $|\alpha| \leqslant l$;

c) $\partial r_k(y)/\partial y_n \geqslant c_4$.

Such a function $r_k(y)$ is called a *regularized distance* from $T_k\Gamma_k$.

REMARK. A regularized distance $r_k(y)$ is not unique. Along with every such $r_k(y)$ the same requirements are satisfied, for example, by $\xi(y)r_k(y)$ if $|\xi^{(\beta)}| \leqslant c_\beta r_k(y)^{-|\beta|}$ for $|\beta| \leqslant l$ and if $\partial\xi(y)/\partial y \geqslant (b_0 - c_4c_0)r_k(y)^{-1}$ for some c_0, $b_0 > 0$.

Everywhere below it is assumed that $1/p + 1/q = 1$.

The weight function $\sigma(x)$ is assumed to satisfy the following conditions:

I. $\sigma(x) > 0$ *on* G, *and* $\sigma \in L_p^{\mathrm{loc}}(G)$.

II. $\sigma(T_ky) \sim \sigma_k(r_k(y))$ *for* $y \in T_kG_k$, $1 \leqslant k \leqslant K$, *where* $1/\sigma_k \in L_q^{\mathrm{loc}}((0,1])$.

III. $m_1 = \cdots = m_k$, *where* m_k *denotes the largest integer in the interval* $[0, l-1]$ *such that*

$$\left\{\int_0^1 t^{(l-m_k-1)q}\sigma_k(t)^{-q}\,dt\right\}^{1/q} < \infty,$$

or $m_k = -1$ *if no such integer exists. The common value of all the* m_k*'s is denoted by* $m(\sigma)$.

We are interested in the problem of existence of boundary values (traces) of derivatives of functions in $W_{p,\sigma}^{(l)}(G)$ on ∂G and the possibility of approximating $W_{p,\sigma}^{(l)}(G)$-functions arbitrarily well by compactly supported functions in G or by infinitely differentiable compactly supported functions in G. Moreover, we study the possibility of a constructive determination of the approximating functions by means of cutoff functions, i.e., of using an approximation method not depending on the individual function.

The following two theorems are established below.

THEOREM 1. *Suppose that conditions* I, II, *and* III *are satisfied for* $p \in [1, \infty)$. *Then for all* $u \in W_{p,\sigma}^{(l)}(G)$ *the trace of each of the derivatives* $D^\alpha u$, $|\alpha| \leqslant m(\sigma)$, *on* ∂G *exists in the sense of* L_p*-convergence.*

To formulate the next theorem we define for $m = m(\sigma)$ and $1 \leqslant k \leqslant K$ the following functions on $(0, 1)$:

$$\Phi_k(t) = 0 \quad \text{if } m - 1.$$

$$\Phi_k(t) = \left\{\int_0^1 s^{(l-m-1)q}\sigma_k(s)^{-q}\,ds\right\}^{1/q} \quad \text{if } 0 \leqslant m \leqslant l-1, \tag{2}$$

$$\Psi_k(t) = 0 \quad \text{if } m = l-1,$$

$$\Psi_k(t) = \left\{\int_t^1 s^{(l-m-2)q}\sigma_k(s)^{-q}\,ds\right\}^{1/q} \quad \text{if } -1 \leqslant m \leqslant l-2, \tag{3}$$

$$\Omega_k^*(t)^q = \frac{t^{(l-m-2)q}\sigma_k(t)^{-q}}{t^{-q}\Phi_k^q(t) + \Psi_k^q(t)}. \tag{4}$$

Further, let $\Omega_k(t)$ be some function satisfying the conditions

$$0 < \Omega_k(t) \leqslant \Omega_k^*(t), \qquad \int_0^1 \Omega_k^q(t)\,dt = \infty, \tag{5}$$

$$\varlimsup_{\varepsilon \downarrow 0} \frac{\min_E \int_{[\varepsilon,1]\cap E} \Omega_k^q(t)\,dt}{\int_\varepsilon^1 \Omega_k^q(t)\,dt} > 0, \tag{6}$$

where $E = \bigcup_1^{3l}[2^{-s_j}, 2^{-s_j+1}]$, $s_j \in N$, or the condition (stronger than (6))

$$\varliminf_{s\to\infty} \int_{2^{-s}}^{2^{-s+1}} \Omega_k^q(t)\,dt < \infty. \tag{6'}$$

On a function σ satisfying condition I, II, III *we impose the additional condition*

IV. *for* $1 \leqslant k \leqslant K$, $m = m(\sigma)$, *some* $\delta \in (0,1)$, *and all* ε, h', h'', h $(0 < \varepsilon < h'' < h < h' < \delta)$

$$\begin{aligned}\Phi_k(h)&\left[\int_h^{h'} t^{(1-l+m)p}\sigma_k(t)^p\,dt\right]^{1/p}\left[\int_{h'}^{\delta} t^{-1}\Omega_k^1(t)\,dt\right]^{1/q}\\ &+\Psi_k(h)\left[\int_\varepsilon^{h''} t^{(2-l+m)p}\sigma_k^p(t)\,dt\right]^{1/p}\left[\int_{h''}^{h} t^{-1}\Omega_k^q(t)\,dt\right]^{1/q}\\ &\leqslant M < +\infty,\end{aligned}$$

or the stronger condition

IV′. *for* $1 \leqslant k \leqslant K$, $m = m(\sigma)$, *and some* $\delta(0,1)$

$$\sup_{h\in(0,\delta)}\left[\int_0^h t^{(1-l+m)p}\sigma_k(t)^p\,dt\right]^{1/p}\left[\int_h^\delta t^{(l-m-2)q}\sigma_k(t)^{-q}\omega_k(t)^q\,dt\right]^{1/q} < \infty,$$

where

$$0 < \omega_k(t) = \Omega_k(t)/\Omega_k^*(t) \leqslant 1.$$

THEOREM 2. *Assume conditions* I, II, III, *and* IV (*or* IV′) *for* $p \in (1,\infty)$. *Then the closure of* $C_0^\infty(G)$ *in* $W_{p,\sigma}^{(l)}(G)$ *coincides with the set of all functions* $u \in W_{p,\sigma}^{(l)}(G)$, *such that*

$$D^\alpha u|_{\partial G} = 0 \quad \text{for } |\alpha| \leqslant m(\sigma).$$

These theorems are closely related to results of Poulsen [1], [2], who obtained the corresponding assertions for the spaces of functions defined on the unit cube $Q = Q' \times (0,1] = (0,1)^{n-1} \times (0,1]$, with the norm

$$\left\|\sigma\frac{\partial^l u}{\partial x_n^l}\right\|_{p,Q} + \sum_{i=0}^{l-1}\left\|\frac{\partial^i u(\cdot,1)}{\partial x_n^i}\right\|_{p,Q'} < \infty, \tag{7}$$

where $\sigma = \sigma(x_n)$ satisfies only fairly general conditions of type I, II, and III. In [1] the case with $l = 1$, $p = 2$, and a continuous weight is considered, and the approximation by infinitely differentiable compactly supported functions is realized by means of multiplication by a cutoff function depending only on x_n. In [2] the more general case of functions from an interval to a Banach space is considered, and the possibility of arbitrarily good approximation by infinitely differentiable compactly supported functions is established by methods of functional analysis without concrete construction of approximating functions.

Here we study the space of functions defined on a domain with nonsmooth (Lipschitz) boundary and having derivatives with respect to all variables up to order l and a finite norm (1). The question of denseness of the set of infinitely differentiable compactly supported functions. The weight function σ, which satisfies conditions I, II, and III, and certain new conditions IV in addition to those in [2].

We mention some other results involving the existence of traces of functions (and traces of their derivatives) on the manifolds where the weight is degenerate and involving the denseness of the set of compactly supported functions in the subspace of a weighted space consisting of the functions with zero traces for the derivatives that make sense. In most previous articles the weight is a function of the distance to the manifold where it degenerates, or a function equivalent to such a function.

Similar results for $m = l - 1 = 0$ were obtained in [3] when $l = 1$ and the manifold where the weight degenerates is twice continuously differentiable. The case of arbitrary dimensions $s \leqslant n - 1$ of the manifold where the weight degenerates is also considered there.

The denseness of $C_0^\infty(G)$ in $W_{p,\sigma}^{(l)}(G)$ for a domain $G \subset E^n$ with locally Lipschitz boundary was established in [4] and [5] for $\sigma(\rho) = \rho^\alpha$ with $\alpha \leqslant -1/p$ and with $\alpha \leqslant -1/p$ and with $\alpha > l - 1/p$. The same was obtained in [6] for a nonincreasing function σ with the properties that $\inf \sigma(t) > 0$, $\int_0^1 \sigma(t)^p\, dt = \infty$, and $\sigma(t) \leqslant c_1\sigma(ct)$ for some $c > 1$ and $c_1 > 0$ and any $t \in (0, 1)$. The case where $m = l - 1$ and the weight is a continuous nondecreasing function of the distance to the subspace $E^s \subset E^n (s \leqslant n - 1)$ was considered in [7]. Cases of the weight in [2] with arbitrary m and degenerating on the hyperplane E^{n-1} were considered in [8] for a weighted function class whose norm contains only nonmixed derivatives. For $\sigma(\rho) = \rho^\alpha$, $\alpha > -1/p$, Theorems 1 and 2 were obtained for the case of an infinitely differentiable boundary ∂G in [9] (§3.3.3) (by interpolation theory methods), and for the case with $-1/p < \alpha < l - 1/p$ and an $[l - \alpha - 1/p + 1]$ times continuously differentiable boundary ∂G in [10]. With the help of multiplication by a cutoff function and averaging it is proved in [11] that the smooth compactly supported functions are dense in the space of functions with finite seminorm $\sum_{|\alpha|=l} \|\sigma D^\alpha u\|_{p, R_n^+}$ on the half-space R_n^+. Here $\sigma = \sigma(x_n)$ is monotonically increasing, $m(\sigma) = l - 1$, and σ has certain regularity properties. In the same paper the case of degeneration of the weight on a plane of arbitrary dimension is also considered. The denseness of the smooth compactly supported functions in a weighted space with seminorm coinciding with the first term in (7) was established in [12] under certain general constraints on the weight, which depends on x. We mention further the results in [4], [5], and [13], where the closure of the set of infinitely differentiable compactly supported functions is described with the help of weighted L_p-norms of lower derivatives of functions with a more strongly degenerate weight than for the higher derivatives. This is done in [4] and [5] for power weights, and in [13] for a more general sufficiently smooth weight when $p = 2$.

The trace of a function on ∂G (in other words, its limit or boundary values on ∂G) is understood here in the sense of convergence in L_p. We make this concept precise (see also [14] and [15]).

DEFINITION. Suppose that the functions F and f are defined on G and ∂G, respectively. We say that f is the *trace* of F on ∂G ($F|_{\partial G} = f$) if for each k,

$1 \leqslant k \leqslant K$, there exists on $G_k \cup \Gamma_k$ a function F_k such that

a) $F_k(x) = F(x)$ almost everywhere on G_k,

b) $F_k(x) = f(x)$ on Γ_k,

c) $\int_{T_k\Gamma_k} |f(T_k^{-1}y)^p\, dy' < \infty$, and

d) $\int_{T_k\Gamma_k} |\Gamma_h F_k(T_k^{-1}y)|^p\, dy' \to 0$ as $h \downarrow 0$, where, by definition,

$$\int_{T_k\Gamma_k} \varphi(y)\, dy' = \int_{D_k} \varphi(y', \gamma_k(y'))\, dy',$$
$$\Delta_h\varphi(y) = \varphi(y', y_n + h) - \varphi(y', y_n).$$

The concept of the trace of a function in this definition is thus connected with the choice of the representation of ∂G in the form $\partial G = \bigcup_1^k \Gamma_k$ and with the orthogonal transformation T_k defined for each k. This representation and these transformations are assumed to be fixed.

If the function F is continuous on $\overline{G}$, then its restriction to ∂G is obviously the trace of F on ∂G.

The trace of a function is defined to within equivalence with repsect to the $(n-1)$-dimensional measure on the sets D_k. We give some examples of weight functions satisfying conditions I–IV. Here it will be assumed that $\sigma_1 = \cdots = \sigma_K$ and $m = m(\sigma)$, and the subscript k will be omitted.

EXAMPLE 1. $\sigma(t) = t^{l-\mu-1/p}$, $\mu \in (-\infty, \infty)$. Condition III is satisfied with $m = -1$ for $\mu < 0$, $m = [\mu - 0]$ for $0 \leqslant \mu \leqslant l$, and $m = l - 1$ for $\mu > l$.

In conditions IV and IV′ we take $\Omega = \Omega^*$ and $\omega = 1$. For $\mu < 0$ and for fractional $\mu \in (0, l)$ Condition IV′ is satisfied. Condition IV holds for integers $\mu \in [0, l-1]$. In the verification, the second term in the inequality in Condition IV can be estimated directly. The first term is equal to zero for $\mu = 0$, while $\Phi(h) = c_1 h$ and $\Psi(h) \sim c_2(\ln(1/n))^{1/q}$, for integers $\mu \in [1, l-1]$, and the desired estimate of the first term is easily obtained after multiplying the integrands in its second and third factors by $t^{-\nu}$ and t^{ν} $(\nu > 0)$, respectively. If $\mu \geqslant l$, then $\Psi = 0$ and $\Phi(h) = ch^{\mu-l+1}$, and Condition IV holds. Moreover, if $\mu > l$, then the estimate is made directly, while if $\mu = l$, then it is made after multiplying the integrands in the second and third factors by $t^{-\nu}$ and t^{ν} $(\nu > 0)$, respectively.

EXAMPLE 2. $\sigma(t) = t^{l-\mu-1/p}\zeta(t)$, $\zeta(t_2) \geqslant c\zeta(t_1)$ for $t_2 > t_1 > 0$, $ct^{\nu} \leqslant \zeta(t)$, $\nu > 0$, $c > 0$, and $\mu < 0$ or $[\mu] < \mu - \nu < \mu < [\mu] + 1$. Condition III is satisfied with $m = -1$ for $\mu < 0$ and $m = [\mu]$ for other μ. Condition IV′ holds; to check this it suffices to replace $\zeta(t)$ by $c^{-1}\zeta(h) \geqslant \zeta(t)$.

EXAMPLE 3.

$$\sigma(t) = t^{l-\mu-1/p}\zeta_j(t), \quad \zeta_j(t) = \left(\ln\frac{H}{t}\right)^{s_1}\left(\ln\ln\frac{H}{t}\right)^{s_2}\cdots\left(\ln\cdots\ln\frac{H}{t}\right)^{s_j},$$

and $\zeta_j(1) > 0$, where $\mu < 0$ or $0 \leqslant [\mu] < \mu < l$. This example is the special case of Example 2 with $\zeta(t) = t^{\nu}\zeta_j(t)$, $0 < \nu < \mu - [\mu]$.

2. We first establish some auxiliary assertions for the one-dimensional case which contain the main idea in the arguments. Here x, y, t, etc., will denote numerical variables.

LEMMA 1. *Suppose that* $1 < q < \infty$, $f(x) > 0$ *on* $(0,1]$, $\int_x^1 f(t)\,dt < \infty$ *on* $(0,1)$, $\int_0^1 t^q f(t)\,dt < \infty$, *and* $\int_0^1 f(t)\,dt = \infty$. *Then the integral*

$$A = \int_0^1 \frac{f(x)\,dx}{x^{-q}\int_0^x t^q f(t)\,dt + \int_x^1 f(t)\,dt} = \int_0^1 \frac{f(x)\,dx}{F(x)}$$

diverges.

PROOF. Let

$$G(x) = \int_0^x t^q f(t)\,dt, \quad H(x) = \int_x^1 f(t)\,dt, \quad F(x) = x^{-q}G(x) + H(x).$$

Suppose first that $H(x) \geqslant x^{-q}G(x)$ on $(0, a)$ for some $a \in (0,1)$. Then

$$\int_0^1 \frac{f(x)\,dx}{F(x)} \geqslant \frac{1}{2}\int_0^a \frac{f(x)\,dx}{H(x)} = -\frac{1}{2}\int_0^a d\ln H(x) = +\infty.$$

Suppose that $x^{-q}(G(x)) \geqslant H(x)$ on $(0, a)$ for some $a \in (0,1)$. Then

$$\int_0^1 \frac{f(x)\,dx}{F(x)} \geqslant \frac{1}{2}\int_0^a \frac{x^q f(x)\,dx}{G(x)} = \frac{1}{2}\int_0^a d\ln G(x) = +\infty.$$

It remains to consider the case where there is a sequence $\{x_k\}$ with the properties $0 < x_k < 1$, $x_{k+1} < x_k$ and $x_k \to 0$ $(k \to \infty)$, such that

$$F(x_k) = 2H(x_k), \qquad k \in \mathbf{N}.$$

Since $H(x_k)\uparrow +\infty$ as $k \to \infty$, it can be assumed without loss of generality (by passing to a subsequence if necessary) that

$$H(x_{k+1}) \geqslant 2H(x_k), \qquad k \in \mathbf{N}.$$

Then, since $F(x)$ is monotone,

$$\int_0^1 \frac{f(x)\,dx}{F(x)} \geqslant \sum_{k=1}^{\infty}\int_{x_{k+1}}^{x_k}\frac{f(x)\,dz}{F(x)} \geqslant \sum_{k=1}^{\infty}\frac{1}{2H(x_{k+1})}\int_{x_{k+1}}^{x_k} f(x)\,dx$$
$$= \sum_{k=1}^{\infty}\frac{H(x_{k+1}) - H(x_k)}{2H(x_{k+1})} \geqslant \sum_{k=1}^{\infty}\frac{1}{4} = +\infty,$$

which concludes the proof of the lemma.

For a function $\Omega \in L_q(\varepsilon, \delta)$ with $\Omega(x) > 0$ for $x \in [\varepsilon, \delta] \subset (0,1)$, we introduce

$$\Omega^q_{\varepsilon,\delta} = \inf_{P_{l-1}} \frac{\int_\varepsilon^\delta |P_{l-1}(x)|^q x^{(1-l)q}\Omega^q(x)\,dx}{\int_0^\delta \Omega^q(x)\,dx},$$

where the infimum is taken over all polynomials P_{l-1} of degree $l-1$ with leading coefficient 1.

LEMMA 2. *Suppose that* $1 < q < \infty$, $\Omega(t) > 0$ *on* $(0,1]$, $\Omega^q \subset L_{\mathrm{loc}}((0,1])$ *and* $\int_0^1 \Omega^q(t)\,dt = \infty$. *Suppose, further, that for* $\delta = 1$

$$a_\delta = \varlimsup_{\varepsilon\downarrow 0} \frac{\min_E \int_{[\varepsilon,\delta]\setminus E}\Omega^q(t)\,dt}{\int_\varepsilon^\delta \Omega^q(t)\,dt} > 0,$$

where $E = \bigcup_1^{3l}[2^{-s_j}, 2^{-s_j+1}]$, $s_j \in \mathbf{N}$, *or, what is stronger,*

$$\int_{2^{-s}}^{2^{-s+1}} \Omega^q(t)\,dt \leqslant M < \infty, \quad \forall s \in \mathbf{N}. \tag{8}$$

Then there exists a $c_0 > 0$ *such that for any* $\delta \in (0,1)$ *the inequality* $\Omega^q_{\varepsilon,\delta} \geqslant c_0$ *holds for all* ε *in some subset of* $(0, \delta)$ *with zero as a limit point, and for all sufficiently small* $\varepsilon \in (0, \delta)$ *if* (8) *holds.*

PROOF. Condition (8) is obviously stronger than the preceding condition. It is also obvious that $a_\delta > 0$ for any $\delta \in (0,1]$ under the conditions of the lemma. Let $0 < \kappa \leqslant 1/4$ and let x_i be the roots of the polynomial P_{l-1} such that $\operatorname{Re} x_i > 0$. Then the inequality $|1 - (\operatorname{Re} x_i)/x| < \kappa$ implies that $1/2 < 1/(1+\kappa) < x/\operatorname{Re} x_i < 1/(1-\kappa) < 3/2$. Therefore,

$$x^{1-l}|P_{l-1}(x)| \geqslant \prod_i \left|1 - \frac{x_i}{x}\right| \geqslant \kappa^{l-1} \quad \text{on } (0, +\infty) \setminus \tilde{E},$$

where $\tilde{E} = \bigcup [\frac{1}{2} \operatorname{Re} x_i, \frac{3}{2} \operatorname{Re} x_i]$. Taking into account that each of the closed intervals making up $\tilde{E}$ is covered by three adjacent closed intervals of the form $[2^{-s}, 2^{-s+1}]$, we get from the conditions of the lemma that for E

$$\begin{aligned}\int_\varepsilon^\delta |P_{l-1}(x)|^q x^{(1-l)q} \Omega^q(x)\, dx &\geqslant \min_E \kappa^{(l-1)q} \int_{[\varepsilon,\delta]\setminus E} \Omega^q(x)\, dx \\ &\geqslant \kappa^{(l-1)q} \frac{a_\delta}{2} \int_\varepsilon^\delta \Omega^q(t)\, dt\end{aligned}$$

at least for all ε in some sequence in $(0, \delta)$ converging to zero, and this gives us the first assertion of the lemma. But if (8) holds, then the assertion of the lemma follows from the estimates

$$\begin{aligned}\int_\varepsilon^\delta |P_{l-1}(x)|^q x^{(1-l)q} \Omega^q(x)\, dx &\geqslant \min_E \kappa^{(l-1)q} \int_{[\varepsilon,\delta]\setminus E} \Omega^q(x)\, dx \\ &\geqslant \kappa^{(l-1)q} \left[\int_\varepsilon^\delta \Omega^q(x)\, dx - 3M(l-1)\right].\end{aligned}$$

Lemma 1 shows that there exists a function Ω_k satisfying the conditions in (5). The right-hand side of (4) can be taken as Ω_k. A function Ω_k chosen from the conditions (5) will also satisfy Condition (6′) if it is replaced by the truncation $\min[\Omega(t), a_s]$ on the intervals $(2^{-s}, 2^{-s+1})$ where $\int_{2^{-s}}^{2^{-s+1}} \Omega^q(t)\, dt > 1$, with a_s such that

$$\int_{2^{-s}}^{2^{-s+1}} \min[\Omega(t), a_s]^q\, dt = 1.$$

We shall use two inequalities of Hardy below ($1 \leqslant p \leqslant \infty$, $1/p + 1/q = 1$):

$$\left[\int_0^\infty \left|\varphi(t) \int_0^t f(s)\, ds\right|^p dt\right]^{1/p} \leqslant B_1 \left[\int_0^\infty |\psi(t) f(t)|^p\, dt\right]^{1/p},$$

$$\left[\int_0^\infty \left|\varphi(t) \int_t^\infty f(s)\, ds\right|^p dt\right]^{1/p} \leqslant B_2 \left[\int_0^\infty |\psi(t) f(t)|^p\, dt\right]^{1/p},$$

where

$$B_1 = p^{1/p} q^{1/q} \sup_{h>0} \left[\int_h^\infty |\varphi(t)|^p\, dt\right]^{1/p} \left[\int_0^h |\psi(t)|^{-q}\, dt\right]^{1/q},$$

$$B_2 = p^{1/p} q^{1/q} \sup_{h>0} \left[\int_0^h |\varphi(t)|^p\, dt\right]^{1/p} \left[\int_h^\infty |\psi(t)|^{-q}\, dt\right]^{1/q}.$$

The finiteness of B_1 (B_2) is a necessary condition for the validity of the first (second) Hardy inequality with some constant instead of B_1 (B_2). The first of these inequalities is contained in definitive form in articles by several authors (Tomaselli, Talenti, Artola), and a simpler proof is given by Muchkenhoupt [16]. The second inequality can be derived from the first by duality arguments [16].

For a function $\sigma = \sigma(t)$, $t > 0$, we introduce the following conditions:

I_1. *$\sigma(t) > 0$ for $t > 0$, $\sigma(t) = 1$ for $t > 1$, and $\sigma \in L_p^{\mathrm{loc}}((0,1])$.*

II_1. *$1/\sigma \in L_q^{\mathrm{loc}}((0,1])$.*

We say also that $\sigma(t)$ satisfies Condition III_1, IV_1, or IV_1' if as $\sigma_1(t)$ it satisfies the respective Condition III, IV, *or* IV′ *with* $K = 1$. Here the index $k = 1$ is omitted in the statements of these conditions and the notation for the corresponding functions. In what follows we omit the index k also for the functions Φ, Ψ, Ω^*, and Ω constructed from σ according to (2)–(6).

Let $V = V_{0,\sigma,0}^{(l)}((0,1))$ ($l \in N$, $1 \leqslant p < \infty$) denote the Banach space of functions u having on $(0,\infty)$ a generalized derivative u^l (i.e., having a locally absolutely continuous derivative $u^{(l-1)}$) and equal to zero on $[1, \infty)$, with the norm

$$\|u\|_V = \sum_{k=0}^{l} \|\sigma u^{(k)}\|_{p(0,1)}, \tag{9}$$

where σ satisfies conditions I_1 and II_1.

According to [2], the limit values $u^{(k)}(0)$ of the functions $u \in V$ exist for all k with $0 \leqslant k \leqslant m$. By $\overset{0}{V} = \overset{0}{V}{}_{p,\sigma,0}^{(l)}((0,1))$ we denote the subspace V of functions u such that $u^{(k)}(0) = 0$ ($k = 0,\dots,m$).

LEMMA 3. *Suppose that the function σ satisfies conditions I_1, II_1, and IV_1 (IV_1') for $1 < p < \infty$. Then for each $0 < \delta < 1$ there exists a function $\eta = \eta(t)$ concentrated on $[0, \delta]$ such that $\eta(t) = 1$ for $0 \leqslant t < \varepsilon = \varepsilon(\delta) < \delta$ and for any $u \in \overset{0}{V}$*

$$\|\sigma\eta^{(i)}t^{-j}u^{(s)}\|_p \to 0 \quad \text{as } \delta \to 0,$$

where $t(x) = x$, $i, j, s \geqslant 0$, $i + j + s \leqslant l$, and $j = 0$ for $i = 0$.

PROOF. Suppose that $0 < \varepsilon < \delta < 1$, the function $\eta = \eta(t)$ has on $(0,\infty)$ an absolutely continuous derivative $\eta^{(l-1)}$ concentrated on $[0,\delta]$, and $\eta(t) = 1$ for $0 \leqslant t \leqslant \varepsilon$. We represent $u^{(s)}$ for $0 \leqslant s \leqslant m$ in the form

$$u^{(s)}(x) = \int_0^x \frac{(x-z)^{-s+m}}{(m-s)!} \int_{+\infty}^z \frac{(z-y)^{l-m-2}}{(l-m-2)!} \frac{d^{l-s}}{dx^{l-s}} u^{(s)}(y)\,dy\,dz \tag{10}$$

and for $m < s \leqslant l-1$ in the form

$$u^{(s)}(x) = \int_{+\infty}^x \frac{(x-y)^{l-s-1}}{(l-s-1)!} \frac{d^{l-s}}{dx^{l-s}} u^{(s)}(y)\,dy. \tag{11}$$

From (10) we get that for $0 \leqslant s \leqslant m$

$$\begin{aligned} |u^{(s)}(x)| &\leqslant C_1 \int_0^x z^{(m-s)} \int_z^\infty y^{l-m-2} |u^{(l)}(y)|\,dy\,dz \\ &\leqslant C_2 x^{m-s} \int_0^x y^{l-m-1} |u^{(l)}(y)|\,dy \\ &\quad + C_2 x^{m-s+1} \int_x^\infty y^{l-m-2} |u^{(l)}(y)|\,dy. \end{aligned} \tag{12}$$

From (11) we get that for $m < s \leqslant l-1$

$$|u^{(s)}(x)| \leqslant C_3 \int_x^\infty y^{l-s-1} |u^{(l)}(y)|\, dy \leqslant C_3 x^{m-s+1} \int_x^\infty y^{l-m-2} |u^{(l)}(y)|\, dy. \tag{13}$$

For $i \geqslant 1$ let us use (12) and (13) along with Hardy's inequalities. Here we use the first Hardy inequality with

$$\varphi(t) = \sigma(t)|\eta^{(i)}(t)|t^{i-l+m}, \qquad \psi(t) = \sigma(t)t^{-l+m+1}$$

and the second Hardy inequality with

$$\varphi(t) = \sigma(t)|\eta^{(i)}(t)|t^{i-l+m+1}, \qquad \psi(t) = \sigma(t)t^{-l+m+2}.$$

The result is that for $i \geqslant 1$

$$\|\sigma\eta^{(i)}t^{-j}u^{(s)}\|_p \leqslant \|\sigma\eta^{(i)}t^{-l+s+i}u^{(s)}\|_p \leqslant C \sup_{h>0} A_i(h)\|\sigma u^{(l)}\|_p, \tag{14}$$

where

$$A_i(h) = \left[\int_h^\infty |\sigma(t)\eta^{(i)}(t)t^{i-l+m}|^p\, dt\right]^{1/p} \Phi(h) + \left[\int_0^h |\sigma(t)\eta^{(i)}(t)t^{i-l+m+1}|^p\, dt\right]^{1/p} \Psi(h). \tag{15}$$

An immediate consequence for $i = j = 0$ is

$$\|\sigma\eta u^{(s)}\|_p \leqslant \max_{[0,\delta]} |\eta|\, \|\chi_\delta \sigma u^{(s)}\|_p, \tag{16}$$

where χ_δ is the indicator function of $[0, \delta]$.

Thus, the proof of Lemma 3 reduces to the construction of a function η such that

$$\sum_{i=1}^{l} \sup_{h>0} A_i(h) = \sum_{i=1}^{l} \sup_{\varepsilon \leqslant h \leqslant \delta} A_i(h) \to 0 \quad (\delta \to 0), \qquad \overline{\lim_{\delta\to 0}}\, \max|\eta| < \infty. \tag{17}$$

Obviously, for $1 \leqslant i \leqslant l$

$$A_i^p(h) \leqslant 2^{p-1} \int_\varepsilon^\delta |\eta^{(i)}(t)t^{i-1}|^p\, \Omega(t)^{-p}\, dt, \tag{18}$$

where Ω satisfies the conditions of Lemma 2 and

$$0 < \Omega^q(t) \leqslant \Omega^{*q}(t) = \frac{t^{(l-m-2)q}\sigma(t)^{-q}}{t^{-q}\Phi^q(t) + \Psi^q(t)} \quad \text{on } (0, \delta). \tag{19}$$

We look for tl e function $\eta^{(l)}(x)$ as the solution of the isoperimetric variational problem of minimizing the integral

$$J_l = \int_\varepsilon^\delta |\eta_l(x)|^p x^{(l-1)p} \Omega(x)^{-p}\, dx$$

under the additional conditions

$$\int_\varepsilon^\delta (\delta - x)^{l-1} \eta_l(x)\, dx = -(l-1)!, \tag{20}$$

$$\int_\varepsilon^\delta (\delta - x)^{k-1} \eta_l(x)\, dx = 0 \qquad (k = 1, \ldots, l-1). \tag{21}$$

These conditions are necessary and sufficient for the existence of a function η with a given derivative η_l of order l and such that $\eta(x) = 1$ for $x < \varepsilon$ and $\eta(x) = 0$ for $x > \delta$. Moreover, extending η_l by zero outside (ε, δ), we get that

$$\eta(x) = 1 + \int_\varepsilon^x \frac{(x-t)^{l-1}}{(l-1)!}\eta_l(t)\,dt, \qquad \eta^{(l)}(x) = \eta_l(x),$$
$$\eta(x) = 1 \quad \text{for } x < \varepsilon, \qquad \eta(x) = 0 \quad \text{for } x > \delta. \tag{22}$$

Solving the variational problem, we find that on (ε, δ)

$$\eta_l(x) = \left(\sum_0^{l-1} \lambda_j x^j\right)\left|\sum_0^{l-1} \lambda_j x^j\right|^{q-2} x^{(1-l)q}\Omega^q(x), \qquad x \in (\varepsilon, \delta), \tag{23}$$

where the λ_j are found from (20) and (21). Then

$$J_l = \int_\varepsilon^\delta |\eta_l(x)|^p x^{(l-1)p}\Omega^{-p}(x)\,dx = \int_\varepsilon^\delta \left|\sum_0^{l-1}\lambda_j x^j\right|^q x^{(1-l)q}\Omega^q(x)\,dx \tag{24}$$

and the relations

$$\int_\varepsilon^\delta (\delta - x)^{l-1}\left|\sum_0^{l-1}\lambda_j x^j\right|^{q-1} \operatorname{sgn}\left(\sum_0^{l-1}\lambda_j x^j\right) x^{(1-l)q}\Omega^q(x)\,dx = -(l-1)!,$$

$$\int_\varepsilon^\delta (\delta - x)^{k-1}\left|\sum_0^{l-1}\lambda_j x^j\right|^{q-1} \operatorname{sgn}\left(\sum_0^{l-1}\lambda_j x^j\right) x^{(1-l)q}\Omega^q(x)\,dx = 0$$
$$(k + 0, \dots, l-1)$$

or their equivalents

$$\int_\varepsilon^\delta x^{l-1}\left(\sum_0^{l-1}\lambda_j x^j\right)\left|\sum_0^{l-1}\lambda_j x^j\right|^{q-2} x^{(1-l)q}\Omega^q(x)\,dx = (-1)^l(l-1)!, \tag{25}$$

$$\int_\varepsilon^\delta x^{k-1}\left(\sum_0^{l-1}\lambda_j x^j\right)\left|\sum_0^{l-1}\lambda_j x^j\right|^{q-2} x^{(1-l)q}\Omega^q(x)\,dx = 0$$
$$(k = 0, \dots, l-1) \tag{26}$$

are valid. These last relations give us that

$$J_l = \int_\varepsilon^\delta \lambda_{l-1} x^{l-1}\left(\sum_0^{l-1}\lambda_j x^j\right)\left|\sum_0^{l-1}\lambda_j x^j\right|^{q-2} x^{(1-l)q}\Omega^q(x)\,dx$$
$$= (-1)^l(l-1)!\lambda_{l-1}. \tag{27}$$

Hence,

$$J_l = |\lambda_{l-1}|^q \int_\varepsilon^\delta |P_{l-1}(x)|^q x^{(1-l)q}\Omega^q(x)\,dx = (-1)^l(l-1)!\lambda_{l-1}, \tag{28}$$

where $P_{l-1}(x)$ is some polynomial of degree $(l-1)$ with leading coefficient 1 and with coefficients depending on ε and δ.

It follows from (28) that

$$J_l = [(l-1)!]^{p/q}\left[\int_\varepsilon^\delta |P_{l-1}(x)|^q x^{(1-l)q}\Omega^q(x)\,dx\right]^{-p/q}, \tag{29}$$

and, by (8) and Lemma 2,

$$J_l \leqslant [(l-1)!]^{p/q}\Omega_{\varepsilon,\delta}^{-p}\left(\int_\varepsilon^\delta \Omega^q(t)\,dt\right)^{-p/q} \to 0 \tag{30}$$

for $0 < \varepsilon = \varepsilon(\delta) < \delta$ as $\delta \to 0$, where $\varepsilon(\delta)$ is a function such that $\int_{\varepsilon(\delta)}^\delta \Omega^q(t)\,dt \to \infty$ as $\delta \to 0$.

Thus, for the function η in (22) constructed form the η_l in (23),

$$A_l^p(h) \leqslant 2^{p-1}J_l \to 0 \quad \text{for } \delta \to 0. \tag{31}$$

From (22), (24), and (30) we get with the help of Hölder's inequality that for $x \in (0\,\delta)$

$$|\eta(x)| = \left|\int_x \frac{(x-t)^{l-1}}{(l-1)!}\eta_l(t)\,dt\right| \leqslant \frac{1}{(l-1)!}\int_\varepsilon^\delta t^{l-1}|\eta_l(t)|dt$$

$$\leqslant \frac{1}{(l-1)!}J_l^{1/p}\left[\int_\varepsilon^\delta \Omega^q(t)\,dt\right]^{1/q} \leqslant [(l-1)!]^{-1/p}\Omega_{\varepsilon,\delta}^{-1}. \tag{32}$$

Thus,

$$\max_{[0,\delta]}|\eta| \leqslant [(l-1)!]^{-1/p}c_0^{-1/q} \tag{33}$$

for any sufficiently small $\varepsilon > 0$, where c_0 is taken from Lemma 2.

We estimate the $A_i(h) = A_i'(h) + A_i''(h)$ in (15) for $1 \leqslant i \leqslant l-1$, also in terms of the J_l in (24). Taking into account that

$$(l-i-1)!|\eta^{(i)}(x)| = \left|\int_x^\delta (x-y)^{l-i-1}\eta_l(y)\,dy\right| \leqslant \int_x^\delta y^{l+i-1}|\eta_l(y)|\,dy,$$

and using the second inequality of Hardy, we get that

$$\begin{aligned}
&(l-i-1)!A_i'(h)\\
&\quad\leqslant \Phi(h)p^{1/p}q^{1/q}\\
&\quad\times \sup_{n'\in(h,\delta)}\left[\int_h^{h'} s^{(i-1+m)p}\sigma(x)^p\,dx\right]^{1/p}\\
&\quad\times\left[\int_{h'}^\delta x^{-iq}\Omega^q(x)\,dx\right]^{1/q}\left[\int_\varepsilon^\delta |\eta_l(x)|^p x^{(l-1)p}\Omega^{-p}(x)\,dx\right]^{1/p}\\
&\quad= \Phi(h)p^{1/p}q^{1/q}\\
&\quad\times \sup_{h'\in(h,\delta)}\left[\int_h^{h'} x^{(1-l+m)p}\sigma(x)^p\,dx\right]^{1/p}\left[\int_{h'}^\delta x^{-1}\Omega^q(x)\,dx\right]^{1/q}J_l^{1/p}.
\end{aligned} \tag{34}$$

Similarly, estimation of $A_i''(h)$ gives us that

$$\begin{aligned}
&A_i(h) \leqslant p^{1/p}q^{1/q}\Phi(h)\\
&\quad\times \sup_{h'\in(h,\delta)}\left[\int_h^{h'} x^{(1-l+m)p}\sigma^p(x)\,dx\right]^{1/p}\left[\int_{h'}^\delta x^{-q}\Omega^q(x)\,dx\right]^{1/q}J_l^{1/p}\\
&\quad+p^{1/p}q^{1/q}\Psi(h)\\
&\quad\times \sup_{h''\in(\varepsilon,h)}\left[\int_\varepsilon^{h''} x^{(2-l+m)p}\sigma^p(x)\,dx\right]^{1/p}\left[\int_{h''}^h x^{-q}\Omega^q(x)\,dx\right]^{1/q}J_l^{1/p}.
\end{aligned} \tag{35}$$

Let us make the estimate (33) coarser in order to simplify its form. We use the monotonicity of $\Phi(h)$ and $\Psi(h)$, replacing them in (35) by $\Phi(x)$ and $\Psi(x)$ under the integral signs in the second factors. This gives us

$$A_i(h) \leqslant p^{1/p}q^{1/q} \sup_{h' \in (\varepsilon,\delta)} \left[\int_\varepsilon^{h'} x^{(1-l+m)p}\sigma^p(x)\,dx\right]^{1/p} \times \left[\int_{h'}^{\delta} x^{(l-m-2)q}\sigma^{-q}(x)\omega^q(x)\,dx\right]^{1/q} J_l^{1/p}$$
$$+p^{1/p}q^{1/q} \sup_{h'' \in (\varepsilon,\delta)} \left[\int_\varepsilon^{h''} x^{(2-l+m)p}\sigma^p(x)\,dx\right]^{1/p} \times \left[\int_{h''}^{\delta} x^{(l-m-3)q}\sigma(x)^{-q}\omega^q(x)\,dx\right]^{1/q} J_l^{1/p}, \quad (36)$$

where $0 < \omega(x) = \Omega(x)/\Omega^*(x) \leqslant 1$.

The derivation of (34) shows that condition IV_1' is stronger than IV_1. Lemma 3 follows from (31) and (35) or (36), which lead to (17) in view of condition IV_1 or IV_1'.

LEMMA 4. *Under the conditions of Lemma* 3, *for any function* $u \in \overset{0}{V}$ *and any function* η *in* (22)

$$\|u-(1-\eta)u\|_V \to 0 \quad \text{as } \delta \to 0,$$

and $C_0^\infty((0,1])$ *is dense in* V.

The proof of the first assertion reduces to use of the Leibniz formula for differentiation of the product $(1-\eta)u$ and Lemma 3 with $j=0$. To prove the second assertion it suffices to show that every function $v \in \overset{0}{V}$ with $v(t)=1$ outside $[\delta, 1]$ $(0<\delta<1)$ can be approximated in the norm (9) with any accuracy by functions in $C_0^\infty((0,1])$, i.e., by infinitely differentiable functions concentrated on $(0,1]$. Let σ be such a function,$^{1/p}$ and let $\zeta \in C^\infty(0,\infty)$ be such that $\zeta = 0$ on $(0,1/2)$; and $\zeta=1$ on $(1,\infty)$. Consider the function

$$v(x,N) = \zeta\left(\frac{x}{\delta}\right)\int_{+\infty}^x \frac{(x-t)^{l-1}}{(l-1)!} v_l(t,N)\,dt,$$

where $v_l(t,N) = v^{(l)}(t)$ for $|v^{(l)}(t)| \leqslant N$ and $v_l(t,N)=0$ for $|v^{(l)}(t)| > N$. Obviously, $v(x,N)=0$ on $(0,\delta/2)$ and on $(1,+\infty)$, and $v_l(x,N) \to v^{(l)}(x)$ almost everywhere on $(0,1]$, while

$$v^{(s)}(x,N) = \frac{d^s}{dx^s}v(x,N) \rightrightarrows v^{(s)}(x)$$

on $(0,1]$ as $N \to +\infty$, $0 \leqslant s < l$. From this, by I_1 and Lebesgue's theorem on passing to the limit under the integral sign,

$$\|v(\cdot)-v(\cdot,N)\|_V \to 0 \quad \text{as } N \to +\infty.$$

The question thus reduces to the arbitrarily precise approximation of $v(t,N)$ in the norm (9) by functions in $C_0^\infty((0,1])$. This can be done by averaging with a sufficiently small parameter $h>0$:

$$v_h(x,N) = \frac{1}{h}\int \omega\left(\frac{t}{h}\right)v(x+t,N)\,dt,$$

where $\omega \in C_0^\infty((\frac{1}{2}, 1))$ and $\int \omega(t)\,dt = 1$. For a proof it suffices to take into account the boundedness of the functions $v^{(s)}(x, N)$ and $v_h^{(s)}(x, N)$, the almost everywhere convergence $v_h^{(s)}(x, N) \to v^{(s)}(x, N)$ as $h \to 0$ for $s = 0, 1, \ldots, l$, conditions I_1, and Lebesgue's theorem on passing to the limit under the integral sign.

3. We now establish some properties of a geometric nature for a Lipschitz manifold.

LEMMA 5. *Let*

$$\Gamma = \{x = (x', x_n): x = \gamma(x'),\, x' \in D\},$$

where D is an open set in E^{n-1}, and suppose that

$$|\gamma(x') - \gamma(z')| \leqslant M|x' - z'|, \qquad x', z' \in D,$$
$$G = \{x = (x', x_n): \gamma(x') < x_n < 1 + \gamma(x'),\, x' \in D\}.$$

Then the following estimates hold for $\rho(x) = \rho(x, \Gamma)$, $x \in G$ and $\gamma(x') < x_n + \Delta x_n < 1 + \gamma(x')$:

$$\frac{x_n - \gamma(x')}{\sqrt{1 + M^2}} \leqslant \rho(x) \leqslant x_n - \gamma(x'),$$
$$\frac{1}{2\sqrt{1 + M^2}} \leqslant \frac{\Delta\rho}{\Delta x_n} \leqslant 1.$$

PROOF. All these estimates follows from simple geometric relations. We explain only the derivation of the left-hand one of the last inequalities. Let

$$\rho_2 = \rho(x', \gamma(x') + x_n), \qquad \rho_1 = \rho(x', \gamma(x') + x_n - \Delta x_n), \quad \Delta_n > 0,$$

and let $\Delta x_n/\rho_2$ be sufficiently small. Obviously, if $\rho_2 = |x - x^0|$ and $x^0 \in \Gamma$, then

$$\cos(e_n, x - x^0) \geqslant 1/\sqrt{1 + M^2},$$

and the question reduces to the consideration of simple relations between the elements of a triangle.

LEMMA 6. *Let*

$$\Gamma = \{x = (x', x_n): x_n = \gamma(x'),\, x' \in D \subset E^{n-1}\},$$
$$G = \{x = (x', x_n): \gamma(x') < x_n < 1 + \gamma(x'),\, x' \in D\},$$

where D is an open subset of E^{n-1}, and the function γ satisfies the Lipschitz condition

$$|\gamma(x') - \gamma(z')| \leqslant M|x' - z'|, \qquad x', z' \in D.$$

Then there exists an infinitely differentiable function $r(x)$ on G such that the following inequalities hold on G for some positive constants c_1, c_2, b_α, and b:

a) $c_1\rho(x, \Gamma) \leqslant r(x) \leqslant c_2\rho(x, \Gamma)$;

b) $|D^\alpha r(x)| \leqslant b_\alpha(\rho(x, \Gamma))^{1-|\alpha|} \quad \forall\, |\alpha| > 0$;

c) $\partial r(x)/\partial x_n \geqslant b > 0$.

PROOF. A function $r(x)$ satisfying a) and b) and called the *regularized distance* was constructed in [17] for a bounded domain with locally Lipschitz $(n-1)$-dimensional boundary, and in [18] (Chapter VI, §2.1) for an arbitrary closed $\Gamma \subset E^n$ and $G = E^n \setminus \Gamma$. Another construction of a regularized distance was given in [15]

(§20.10). We consider a modification of the latter. It will be assumed that $D = E^{n-1}$. This can always be achieved by extending the function γ to E^{n-1} with preservation of the Lipschitz condition (however, for what follows it would suffice to use the local variant of the lemma with the same construction of $r(x)$). Suppose that

$$\mu \in C_0^\infty(\{x': |x'| < 1\}), \qquad \mu(x') = 1 \quad \text{for } 2|x'| < \varepsilon,$$
$$\mu_\varepsilon(x') = \varepsilon^{1-n}\mu(x'/\varepsilon),$$

where $\varepsilon = \varepsilon(M, \max|\operatorname{grad}\mu|) > 0$ is chosen to be sufficiently small, depending on the Lipschitz constant M. Let

$$r(x) = \int_\Gamma \mu_\varepsilon\left(\frac{x'-y'}{|x-y|}\right)\frac{dy'}{|x-y|^{n-2}}, \tag{37}$$

where $\int_\Gamma \varphi(y)\,dy' = \int_D \varphi(y'), \gamma(y')\,dy'$ by definition. Note that

$$\mu_\varepsilon(x'/|x|) = 0 \quad \text{for } |x'| \geqslant \varepsilon|x|. \tag{38}$$

The equivalence $r(x) \sim \rho(x, \Gamma)$ follows from the obvious equality

$$\int_{y_n=0} \mu_\varepsilon\left(\frac{x'-y'}{|x-y|}\right)\frac{dy'}{|x-y|^{n-2}} = a|x_n|, \qquad a > 0,$$

which is obtained by the change of variables $y' = |x_n|z'$. Condition b) can be verified by differentiating under the integral sign. Let us establish c). Without loss of generality it will be assumed that $x' = 0$ and $\gamma(0) = 0$. Then $x_n > 0$ for $x = (0, x_n) \in G$, and after the substitution $y' = x_n z'$ we have

$$r(x) = \int_{E_{n-1}} \mu_\varepsilon\left(\frac{-z'}{|(z', 1-\gamma(x_n z')/x_n)|}\right)\frac{x_n\,dz'}{|(z', 1-\gamma(x_n z')/x_n)|^{n-2}}. \tag{39}$$

By (38), for $0 < \varepsilon M < 1/4$ the integrand is nonzero at most for those z' satisfying

$$|z'| < \varepsilon\sqrt{|z'|^2 + (1 + M|z'|)^2} \leqslant \varepsilon + \varepsilon M|z'|,$$

hence satisfying $|z'| < 2\varepsilon$.

We note that for almost all z'

$$\left|\frac{\partial}{\partial x_n}\frac{\gamma(x_n z')}{x_n}\right| \leqslant \frac{|\gamma(x_n z')|}{x_n^2} + \frac{1}{x_n}|z'|\sup|\operatorname{grad}\gamma| \leqslant 2M\frac{|z'|}{x_n} \leqslant \frac{4M\varepsilon}{x_n}.$$

Hence,

$$\begin{aligned}
&\varepsilon^{n-1}\left|\frac{\partial}{\partial x_n}\mu_\varepsilon\left(\frac{-z'}{|(z', 1-\gamma(x_n z')/x_n)|}\right)\right| \\
&\quad\leqslant \frac{3\max|\operatorname{grad}\mu|}{\varepsilon}\,\frac{|z'|(1-\gamma(x_n z')/x_n)}{|(z', 1-\gamma(x_n z')/x_n)|^3}\,\frac{4M\varepsilon}{x_n} \\
&\quad\leqslant 48\max|\operatorname{grad}\mu|\frac{M\varepsilon}{x_n} \leqslant \frac{A\varepsilon}{x_n},
\end{aligned} \tag{40}$$

$$\left|\frac{\partial}{\partial x_n}\left|\left(z', 1 - \frac{\gamma(x_n z')}{x_n}\right)\right|^{2-n}\right| \leqslant (n-2)2^{n-1}\frac{4M\varepsilon}{x_n} \leqslant \frac{A\varepsilon}{x_n}. \tag{41}$$

Differentiating $r(x)$ in the form (39) under the integral sign by the Leibniz formula, we get a sum of three terms $\partial r(x)/\partial x_n = \Sigma_1^3 B_i$, where

$$B_3 = \int_{|z'|<2\varepsilon} \mu_\varepsilon\left(\frac{z'}{|(z',1-\gamma(x_n z')/x_n)|}\right)\frac{dz'}{|(z',1-\gamma(x_n z')/x_n)|^{n-2}}$$

$$\geqslant \int_{4|z'|<\varepsilon} \varepsilon^{1-n}2^{n-2}\,dz' \geqslant 3b > 0$$

with b not depending on ε. For sufficiently small $\varepsilon > 0$ the terms B_1 and B_2 do not exceed b in modulus because of the respective estimates (40) and (41). Thus, $\partial r(x)/\partial x_n \geqslant b > 0$.

Let $\{e_k\}_0^K$ be a partition of unity on G corresponding to the covering $\{G_k\}_0^K$, i.e., a collection of functions on G with the following properties:

a) $0 \leqslant e_k(x) \leqslant 1$ on G;

b) $e_k(x) = 0$ on $G \setminus G_k^{(\delta/2)}$;

c) $\Sigma_0^K e_k(x) = 1$ on G;

d) $|D^\alpha e_k(x)| \leqslant C_\alpha < \infty$ ($|\alpha| \geqslant 0$).

There is such a partition of unity (see, for example, [9]). Replacing $u(x)$ by $u(x)\Sigma_0^K e_k(x)$, we get

$$\|u\|_W \leqslant \sum_0^K \|e_k u\|_W \leqslant C\|u\|_W, \tag{42}$$

where C does not depend on u.

4. The proof of Theorem 1 is based on an extension of the corresponding one-dimensional arguments in [2]. Suppose that $u \in W_{p,\sigma}^{(l)}(G)$ and $v = D^\alpha u$, $|\alpha| = m \leqslant m(\sigma)$. We show that v has a trace on each Γ_k, $1 \leqslant k \leqslant K$. It is assumed that $u = 0$ outside the ε-neighborhood of ∂G for sufficiently small $\varepsilon > 0$; this can be achieved with the help of a partition of unity. Let k be fixed and let

$$\tilde{v}(y) = v(T_k^{-1}y), \qquad \sigma_k(y) = \sigma_k(r_k(y)) = \sigma(T_k^{-1}y)$$

for $y \in T_k G_k$.

Since $\|u\|_W < \infty$, for $|\beta| = l - m$ we have that

$$\tilde{v}^{(\beta)}\sigma_k \in L_p(T_k G_k).$$

From this, using Hölder's inequality, Fubini's theorem, and Lemmas 5 and 6, we obtain

$$\int_{D_k}\left[\int_0^\varepsilon t^{l-m-1}\left|\tilde{v}^{(\beta)}(y',\gamma_k(y')+t)\right|dt\right]^p dy'$$

$$\leqslant \int_{D_k}\int_0^\varepsilon \left|\tilde{v}^{(\beta)}(y',\gamma_k(y')+t)\sigma_k(y',\gamma_k(y')+t)\right|^p dt\,dy'$$

$$\times \sup_{y'\in D_k}\left\{\int_0^\varepsilon t^{(l-m-1)q}\sigma_k^{-q}(y',\gamma_k(y')+t)\,dt\right\}^{p/q}$$

$$\leqslant C\int_{T_k G_k}\left|\tilde{v}^{(\beta)}(y)\sigma_k(y)\right|^p dy < \infty. \tag{43}$$

It follows from this (see [14], §4.1) that $\tilde{v}$ (after a possible modification of it on a set of n-dimensional measure zero) has for almost every y' a locally absolutely continuous derivative $\tilde{v}_{y_n}^{(l-m-1)}$ on $(\gamma_k(y'), \gamma_k(y') + \varepsilon]$ whose derivative with respect to y_n (which exists almost everywhere and is locally integrable) satisfies

$$\int_0^\varepsilon t^{l-m-1}\left|\tilde{v}_{y_n}^{(l-m)}(y', \gamma_k(y') + t)\right| dt < \infty.$$

We use Taylor's formula for a subset of D_k obtained as the difference of D_k and an appropriate null set:

$$\tilde{v}(y', \gamma_k(y') + h) = (-1)^{l-m} \int_h^\varepsilon \frac{(t-h)^{l-m-1}}{(l-m-1)!} \tilde{v}_{y_n}^{(l-m)}(y', \gamma_k(y') + t)\, dt; \tag{44}$$

with its help we get from the finiteness of the left-hand side that for $0 < \mu < \nu \leqslant \varepsilon$

$$\begin{aligned} &\|\tilde{v}(\cdot, \gamma_k(\cdot) + \mu) - \tilde{v}(\cdot, \gamma_k(\cdot) + \nu)\|_{p, D_k}^p \\ &\quad \leqslant \int_{D_k} \left[\int_0^\varepsilon \lambda(t, \mu, \nu) t^{l-m-1}\left|\tilde{v}_{y_n}^{(l-m)}(y', \gamma_k(y') + t)\right| dt\right]^p dy', \end{aligned} \tag{45}$$

where

$$0 \leqslant (l-m-1)!\lambda(t, \mu, \nu) \leqslant 1 - \chi(t-\nu)\left(\frac{t-\nu}{t}\right)^{l-m-1} \leqslant 1,$$

χ being the Heaviside function. From this we get that for $0 < \mu < \nu = \varepsilon$

$$\tilde{v}(\cdot, \gamma_k(\cdot) + \mu) \in L_p(D_k).$$

If $0 < \mu < \nu \to 0$, then $\lambda(t, \mu, \nu) \to 0$ for each $t \in (0, \varepsilon)$, so the right-hand side of (45) tends to zero by Lebesgue's theorem on passing to the limit under the integral sign. From (45) and the completeness of $L_p(D_k)$ it now folflows that $\tilde{v}(\cdot, \gamma_k(\cdot) + t)$ has a limit $\tilde{v}(\cdot, \gamma_k(\cdot))$ in $L_p(D_k)$ as $t \to 0$, i.e., the trace $v|_{\Gamma_k} = D^\alpha u|_{\Gamma_k}$ exists. By (45) and (43),

$$\|\tilde{v}(\cdot, \gamma_k(\cdot))\|_{p, D_k} \leqslant C\left\{\int_{T_k G_k} \left|\tilde{v}_{y_n}^{(l-m)}(y)\sigma_k(y)\right|^p dy\right\}^{1/p}. \tag{46}$$

It remains to show that the traces of $v = D^\alpha u$ on Γ_k and on Γ_s coincide on $\Gamma_k \cap \Gamma_s$ $(k \neq s)$. Note that $v \in W_1^{(1)}(G)$. Indeed, in view of the representation

$$\begin{aligned} &\frac{\partial \tilde{v}(y', \gamma_k(y') + h)}{\partial y_i} \\ &\quad = (-1)^{l-m-1} \int_h^\varepsilon \frac{(t-h)^{l-m-2}}{(l-m-2)!} \frac{\partial}{\partial y_i} \frac{\partial^{l-m-1}}{\partial y_n^{l-m-1}} \tilde{v}(y', \gamma_k(y') + t)\, dt \end{aligned}$$

we have the estimate

$$\int_{D_k}\int_0^{\varepsilon}\left|\frac{\partial\tilde{v}}{\partial y_i}(y',\gamma_k(y')+h)\right|dh\,dy'$$
$$\leqslant \sum_{|\beta|=l-m}\int_{D_k}\int_0^{\varepsilon}\int_h^{\varepsilon} t^{l-m-2}|D^{\beta}\tilde{v}(y',\gamma_k(y')+t)|dt\,dh\,dy'$$
$$= \sum_{|\beta|=l-m}\int_{D_k}\int_0^{\varepsilon} t^{l-m-1}|D^{\beta}\tilde{v}(y',\gamma_k(y')+t)|dt\,dy'$$
$$\leqslant \sum_{|\beta|=l-m} \operatorname{meas}^{1/q} D_k\left\{\int_{D_k}\left[\int_0^{\varepsilon} t^{l-m-1}|D^{\beta}\tilde{v}(y',\gamma_k(y')+t)|dt\right]^p dy'\right\}^{1/p},$$

and the representation (44) leads to the estimate

$$\int_{D_k}\int_0^{\varepsilon}|\tilde{v}(y',\gamma_k(y')+h)|dh\,dy'$$
$$\leqslant \operatorname{meas}^{1/q} D_k\left\{\int_{D_k}\left[\int_0^{\varepsilon} t^{l-m-2}|\tilde{v}_n^{(l-m)}(y',\gamma_k(y')+t)|dt\right]^p dy'\right\}^{1/p}.$$

But it is known that a function $v \in W_1^{(1)}(G)$ has a trace on ∂G in the L_1-sense (see, for example, [15]). For each k this trace must coincide with the trace of v on Γ_k in the L_p-sense, and the agreement on $\Gamma_k \cap \Gamma_s$ of the traces on Γ_k and on Γ_s in the L_p-sense follows from the agreement of these traces in the L_1-sense. The theorem is proved.

REMARK. It has been shown even in the one-dimensional case $G = (0,1)$ [2] that if

$$\textstyle\int_0^1 t^{(l-m-1)q}\sigma(t)^{-q}\,dt = \infty,$$

then the trace $d^m u/dx^m|_{x=0}$ does not exist for some functions $u \in W_{p,\sigma}^{(l)}(G)$.

Suppose now that D_1 is an open bounded set in E^{n-1}, and

$$G_1 = \{x = (x', x_n)\colon \gamma(x') < x_n < \gamma(x') + 1,\ x' \in D_1\} \in E^n,$$
$$\Gamma = \{x\colon x_n - \gamma(x'),\ x' \in D_1\},$$

where the function γ satisfies a Lipschitz condition on D_1, and let $\varepsilon_0 > 0$. Denote by $W_0 = W_{p,\sigma,0}^{(l)}(G_1)$ the space of functions u defined on G_1 and equal to zero outside

$$G_1^{(\varepsilon_0)} = \{x\colon x \in G_1,\ \rho(x, \partial G_1 \setminus \Gamma) > \varepsilon_0\},$$

with the norm

$$\|u\|_{W_0} = \sum_{|\alpha|\leqslant l}\left\{\int_{G_1}|\sigma(x)D^{\alpha}u(x)|^p\,dx\right\}^{1/p}. \tag{47}$$

LEMMA 7. *Suppose that* $1 < p < \infty$, *and* $\sigma(x) = \sigma(r(x))$ *satisfies on* G_1 *conditions* I, II, III, *and* IV (*or* IV′) *with* $k = 1$, $T_1 = E$, $r(x) = r_1(x)$ *and* $m_1 = m(\sigma)$. *Then a function* u *in* W_0 *can be approximated with arbitrary accuracy in the norm* (47) *by compactly supported functions in* G_1 *if and only if*

$$u^{(\alpha)}|_{\Gamma} = 0 \quad \text{for } |\alpha| \leqslant m_1 = m(\sigma). \tag{48}$$

PROOF. Suppose first that for $u \in W_0$ there exists a sequence $\{u_j\}_1^{\infty}$ of compactly supported functions in G_1 such that $\|u - u_j\|_{W_0} \to 0$ $(j \to \infty)$. Then, by (46) and the fact that $D^\alpha u_j|_\Gamma = 0$, we have for $|\alpha| \leqslant m(\sigma)$ that

$$\|D^\alpha u(\cdot, \gamma(\cdot))\|_{p,D_1} = \|D^\alpha(u - u_j)(\cdot, \gamma(\cdot))\|_{p,D_1} \to 0 \qquad (j \to \infty),$$

which implies (48).

Suppose now that (48) holds for a $u \in W_0$. We approximate this function by the compactly supported function $u_\delta(x) = [1 - \eta(r(x))]u(x)$, in G_1, where η is taken from (22) and, in particular, $\eta(t) = 1$ for $t < \varepsilon$, $\eta(t) = 0$ for $t > \delta$, $0 < \varepsilon = \varepsilon(\delta) < \delta < 1$, and $r(x)$ is a regularized distance with properties a), b) and c) in Lemma 6.

We show that the norm of the difference, namely,

$$\|u - u_\delta\|_{W_0} = \sum_{|\alpha| \leqslant l} \|\sigma(r(\cdot)) D^\alpha[\eta(r(\cdot))u(\cdot)]\|_{p,G_1} \tag{49}$$

can be made arbitrarily small by suitable choices of the parameters δ and ε, and the function η. Differentiating according to the Leibniz formula, we get that (49) can be estimated in terms of a sum of the integrals $J_{i,j,\alpha}$,

$$J^p_{i,j,\alpha} = \int_{D_1} \int_{\gamma(x')}^{1+\gamma(x')} \left|\sigma(r(x))\eta^{(i)}(r(x))r(x)^{-j}u^{(\alpha)}(x)\right|^p dx_n\, dx', \tag{50}$$

where $i + j + |\alpha| \leqslant l$, $j \geqslant 0$, $i \geqslant 0$, $|\alpha| \geqslant 0$, and $j = 0$ for $i = 0$. Let $u(x) = 0$ for $x' \in D_1$ and $x_n \geqslant 1 + \gamma(x')$. Making the substitution $x_n = \gamma(x') + t$ in the inner integral in (50), we get (taking into account the equivalence $t \sim r$ by virtue of Lemmas 4 and 5) the estimate

$$J^p_{i,j,\alpha} \leqslant C \int_{D_1} \int_0^\infty \times \left|\sigma(r(x', \gamma(x') + t))\eta^{(i)}(r(x', \gamma(x') + t))t^{-j}u^{(\alpha)}(x', \gamma(x') + t)\right|^p dt\, dx'.$$

We estimate the inner integral (with respect to t) by using inequality (14), and we replace the variable of integration t by r in the factors $A_i(h)$ on the right-hand side of (14) (the dependence of t on r and the reciprocal dependence satisfy Lipschitz conditions by Lemmas 5 and 6). As a result, for $i \geqslant 1$

$$J^p_{i,j,\alpha} \leqslant C_1 \sup_{h>0} A^p_i(h) \sum_{|\beta|=l} \|\sigma u^{(\beta)}\|^p_{p,G_1}, \tag{51}$$

where

$$A_i(h) = \left[\int_h^\infty \left|\sigma(r)\eta^{(i)}(r)r^{i-l+m}\right|^p dr\right]^{1/p} \Phi(h) + \left[\int_0^h \left|\sigma(r)\eta^{(i)}(r)r^{i-l+m+1}\right|^p dr\right]^{1/p} \Psi(h). \tag{52}$$

For $i = j = 0$ we get immediately that

$$J_{0,0,\alpha} = \|\sigma(r)\eta(r)u^{(\alpha)}\|_{p,G_1} \leqslant \max_{[0,\delta]} |\eta|\, \|\chi_\delta(r)\sigma(r)u^{(\alpha)}\|_{p,G_1}, \tag{53}$$

where χ_δ is the indicator function of $[0, \delta]$. By (17), (51) and (53), a function $u \in W_0$ satisfying (48) can be approximated in the norm (47) arbitrarily well by the function $u_\delta \in W_0$ for sufficiently small $\delta > 0$. This establishes the lemma.

In fact, under the conditions of Lemma 7 any function in W_0 such that (48) holds can be approximated arbitrarily well in the norm (47) by functions in $C_0^\infty(G_1)$, i.e., by infinitely differentiable compactly supported functions in G_1. This follows from the next lemma.

LEMMA 8. *Suppose that* $1 \leqslant p < \infty$, $\sigma \in L_0((0,1))$, *and the function* $u(x) \in W_0$ *has compact support in* G_1, *where* G_1 *and* W_0 *are the same as in Lemma* 7. *Then* $u(x)$ *can be approximated arbitrarily well in the norm* (47) *by functions in* $C_0^\infty(G_1)$.

PROOF. Let $r = r(x)$ be a regularized distance satisfying the assertion of Lemma 6 and appearing in the definition of the weight in (47) ($\sigma(x) = \sigma(r(x))$). Consider the mapping G_1: $(x', x)_n \leftrightarrow (x', r)$, which is one-to-one and Lipschitz in both directions (i.e., it and its inverse satisfy Lipschitz conditions), by Lemma 6. Let $u(x', x_n) = \tilde{u}(x', r)$. The estimates in Lemma 6 show that on the set of functions $u \in W_0$ concentrated on a fixed compact set $K \subset G_1$ the norm (47) is equivalent to the norm

$$\sum_{|\alpha| \leqslant l} \left\{ \int_{\alpha_n > 0} |\sigma(x_n) D^\alpha \tilde{u}(x)|^p |\tilde{u}(x)|^p \, dx \right\}^{1/p}. \tag{54}$$

In particular, the change of variables in the integral is justified by citing results of Rademacher and Kudryavtsev mentioned in [15] (§20.8) (or, by passing to iterated integrals, the question of the validity of the change of variables can be reduced to the one-dimensional case for a mapping that is Lipschitz in both directions). Due to this, it will be assumed from the very beginning of the proof and without loss of generality that $\Gamma \subset \partial G_1$ is part of the hyperplane $x_n = 0$ and that the norm in W_0 has the form and u will be approximated by functions concentrated on a fixed compact subset of G_1.

Suppose that $\omega \in C_0^\infty(\{x' \in E^{n-1}, |x'| < 1\})$, $\omega_h(x') = h^{1-n}\omega(x'/h)$, and $u_h(x)$ is the averaging of u with respect to the variable x' by means of the kernel ω_h for sufficiently small $h > 0$ (see, for example, [15]). Since $u_h \to u$ in the norm (54′) as $h \to 0$, the proof of the lemma reduces to the possibility of approximating u_h arbitrarily well in the norm (54′) by functions in $C_0^\infty(K)$, where $K = \overline{K} \subset G_1$. We represent $u_h(x)$ ($u(x) = 0$ for $x_n \geqslant 1$) in the form

$$u_h(x) = u_h(x') = \int_{+\infty}^{x_n} \frac{(x_n - t)^{l-1}}{(l-1)!} \left(\frac{\partial^l}{\partial x_n^l} u \right)_h (x', t) \, dt$$

and consider

$$v(x, N) = \zeta(x_n) \int_{+\infty}^{x_n} \frac{(x_n - t)^{l-1}}{(l-1)!} (v_{(l,N)})_h (x', t) \, dt, \tag{55}$$

where $v_{(l,N)}(x) = \partial^l u(x)/\partial x_n^l$ for $|\partial^l u(x)/\partial x_n^l| \leqslant N$, $v_{(l,N)}(x) = 0$ for $|\partial^l u(x)/\partial x_n^l| > N$, $\zeta \in C_0^\infty(0, \infty)$, and $\zeta(x_n) = 1$ if $u_h(x) \neq 0$.

It is not hard to see that $v(\cdot, N) \to u_h$ in the norm (54′) as $N \to +\infty$. to achieve this, in computing the derivatives with respect to x' we must transfer them to the averaging kernel and then take the limit under the Lebesgue integral sign as in the proof of Lemma 4. The question is thereby reduced to approximation of a function $v(x, N)$ of the form (55) by functions in $C_0^\infty(K)$, where $K = \overline{K} \subset G_1$, and this can be done as in Lemma 4 by averaging $v(x, N)$ with respect to x_n by means of the kernel $\frac{1}{\varepsilon}\omega(x_n/\varepsilon)$, where $\omega \in C_0^\infty((-1,1))$ and $\varepsilon > 0$ is sufficiently small. The lemma is proved.

The proof of Theorem 2 with the help of a partition of unity reduces to the application of Lemmas 7 and 8.

BIBLIOGRAPHY

1. Effe Thue Poulsen, *Boundary value properties connected with some improper Dirichlet integrals*, Math. Scand. **8** (1960), 5–14.

2. ______, *Boundary values in function spaces*, Math. Scand. **10** (1962), 45–52.

3. G. N. Yakovlev, *On the denseness of the compactly supported functions in weighted spaces*, Dokl. Akad. Nauk SSSR **170** (1966), 797–798; English transl. in Soviet Math. Dokl. **7** (1966).

4. O. V. Besov, Ya. Kadlets [J. Kadlec] and A. Kufner, *On certain properties of weighted classes*, Dokl. Akad. Nauk SSSR **171** (1966), 514–516; English transl. in Soviet Math. Dokl. **7** (1966).

5. Jan Kadlec and Alois Kufner, *Characterization of functions with zero traces by integrals with weight functions*, I; II, Časopis Pěst. Mat. **91** (1966) 463–471; **92** (1967), 16–28.

6. O. V. Besov and A. Kufner, *On the denseness of the smooth functions in weighted spaces*, Czechoslovak Math. J. **18** (1968), 178–188. (Russian)

7. L. D. Kudryavtsev, *On the denseness of the compactly supported functions in weighted spaces*, Dokl. Akad. Nauk SSSR **239** (1978), 46–49; English transl. Soviet Math. Dokl. **19** (1978).

8. P. I. Lizorkin, *On the closure of the set of compactly supported functions in the weighted space* $W^l_{p,\Phi}$, Dokl. Akad. Nauk SSSR **239** (1978), 789–792; English transl. in Soviet Math. Dokl. **19** (1978).

9. Hans Triebel, *Interpolation theory, function spaces, differential operators*, VEB Deutscher Verlag Wiss., Berlin, 1977, and North-Holland, Amsterdam 1978.

10. P. I. Lizorkin and S. M. Nikol'skiĭ, *Coercive properties of elliptic equations with degeneration and a generalized right-hand side*. Trudy Mat. Inst. Steklov. **161** (1983), 157–183; English transl. in this issue.

11. L. D. Kudryavtsev, *On the construction of sequences of compactly supported functions approximating functions in weighted classes*, Trudy Mat. Inst. Steklov. **156** (1980), 121–129; English transl. in Proc. Steklov Inst. Math. **1983**, no. 2 (156).

12. L. N. Domysheva, *On the denseness of compactly supported functions in weighted spaces*, Trudy Mat. Inst. Steklov. **161** (1983), 106–111; English transl. in this issue.

13. K. Kh. Boĭmatov, *Separation theorems, weighted spaces, and their applications to boundary value problems*, Dokl. Akad. Nauk SSSR **247** (1979), 532–535; English transl. in Soviet Math. Dokl. **20** (1979).

14. S. M. Nikol'skiĭ, *Approximation of functions of several variables and embedding theorems*, "Nauka", Moscow, 1969; English transl., Springer-Verlag, 1974.

15. O. V. Besov, V. P. Il'in and S. M. Nikol'skiĭ, *Integral representations of functions and imbedding theorems*, "Nauka", Moscow, 1975; English transl., Vols. 1, 2, Wiley, 1979.

16. Benjamin Muckenhoupt, *Hardy's inequality with weights*, Studia Math. **44** (1972), 31–38.

17. Jindřich Nečas, *Sur une méthode pour résoudre les équations aux dérivées partielles du type elliptique, voisine de la variationelle*, Ann. Scuola Norm. Sup. Pisa (3) **16** (1962), 305–326.

18. Elias M. Stein, *Singular integrals and differentiability properties of functions*, Princeton Univ. Press, Princeton, N. J., 1970.

Translated by H. H. McFADEN

Труды Матем. Инст. Стеклов.
161(1983)

Proceedings of the Steklov
Institute of Mathematics
1984, Issue 3

ON THE INTERCONNECTION OF NORMS OF OPERATORS IN PERIODIC AND NONPERIODIC FUNCTION SPACES

UDC 517.51

V. I. BURENKOV AND M. L. GOL′DMAN

ABSTRACT. Under minimal assumptions about the normed function spaces considered, definitions are given for periodic analogues of them, it is determined when these analogues are normed spaces, and equivalent ways of norming them are studied. In a very general situation, estimates of the form $\|S_a^*\| \leqslant c\|S\|$ and converse estimates of the form $\|S\| \leqslant \overline{\lim}_{a\to\infty}\|S_a^*\|$ are derived for the norms of the "periodic" operators S_a^* (a is the period) in terms of the norm of the "nonperiodic" operators S. For a large class of operators having the property of being local it is proved that $\|S\| = \lim_{a\to\infty}\|S_a^*\| = \sup_{a>0}\|S_a^*\|$. Separate consideration is given to the question of estimates for the norms of imbedding operators, trace and extension operators, and operators acting in spaces of entire functions and spaces of trigonometric polynomials. Applications of the results to general periodic and nonperiodic anisotropic Nikol′skiĭ-Besov spaces are given.

Bibliography: 30 titles.

INTRODUCTION

In analysis situations often arise where various results known for function spaces on R^n must be carried over to the periodic case, or conversely. It suffices to think of imbedding theorems and of trace and extension theorems and their periodic analogues, of multiplier theorems for Fourier series and integrals (of the Marcinkiewicz or Mikhlin-Hörmander type), of estimates for entire function and trigonometric polynomials of the type of Nikol′skiĭ, Bernstein, Bohr and Favard, etc., of questions of equiconvergence of Fourier series and integral expansions, and, more generally, of estimates of the norms of operators acting in nonperiodic spaces, and their periodic analogues (see, for example, the books of Nikol′skiĭ [1] and of Stein and Weiss [2]).

The usual way of solving the problem of carrying a known result over to the periodic case (or, conversely, from the periodic to the nonperiodic case) involves the parallel creation in this case of an analogous technical apparatus enabling us to obtain the expected analogue of the known result. If the original result was fairly complicated, then the way of proving the analogous result is just as complicated. Moreover, difficulties of a theoretical character sometimes arise when properties we have used, for example, in the periodic case do not have nonperiodic analogues. The authors of this article encountered such a situation in studying the question of the existence (more precisely, the nonexistence) of a bounded linear extension operator

1980 *Mathematics Subject Classification*. Primary 46E35; Secondary 30D15, 42B15, 46E30, 46F05

from $L_p(R^m)$ to a Besov space, namely,

$$T: L_p(R^m) \to B^r_{p1}(R^n), \qquad 1 \leqslant p < \infty, \tag{1}$$

in the limiting case of the trace theorem: $r = (r_1, \dots, r_n)$, $r_i > 0$, and $p = \sum_{m+1}^n 1/r_i$ (see [3]). A nonlinear operator (1) was constructed, and, moreover, we have succeeded in proving in the periodic case that a bounded linear operator analogous to (1) does not exist. The proof was based on properties of periodic spaces (the possibility of taking any linear operator and constructing a linear operator that is invariant under translation by an averaging of translations over a period; the finite-dimensionality of the space of trigonometric polynomials of a definite degree, and so on) that do not have nonperiodic analogues. The analogous result for the nonperiodic case was obtained with the help of the assertion that if a bounded linear extension operator (1) exists, then so does its bounded linear periodic analogue. Since such an analogue does not exist, it is also impossible to construct the linear operator (1).

This result served as a starting point of our subsequent investigations here on the interconnection of periodic and nonperiodic spaces. It suggests another way of getting an analogue of a known result in the periodic or nonperiodic case: direct derivation of one result from the other. In many cases this way turns out to be considerably simpler because, even though the results themselves are fairly complicated, there is a close interconnection between them that goes beyond the bounds of simple analogy. There are certainly results in the literature that were obtained in just this way. It suffices to think of the subtle theorems in the book [2] of Stein and Weiss on the interconnection of periodic and nonperiodic Fourier multipliers. We trace precisely this natural path in a very general situation in the present article. The results given here enable us in many specific cases to obtain needed estimates at once by referring to assertions in this paper. In more complicated situations our results show what direction to take in carrying over estimates from the periodic to the nonperiodic case or conversely.

Under minimal assumptions about the nonperiodic function space $Z(R^n)$, a periodic analogue $Z^*(R^n, a)$ of it is defined in Chapter I; $Z^*(R^n, a)$ consists of functions of period $a > 0$ in all variables. We consider the question of whether $Z^*(R^n, a)$ can be normed and study a number of equivalent norms on this space which are defined with the help of extensions to functions in $Z(R^n)$ (see §1.1) and with the help of multiplications by compactly supported or rapidly decreasing functions (see §1.2). The definition of the space $Z^*(R^n, a)$ and the equivalent norms in it are suitable for studying the interconnection between this space and the operators acting in it, and the space $Z(R^n)$ and the operators acting in it.

Chapter II deals with the question of estimating the norms of operators acting in the periodic and nonperiodic spaces. General conditions are given here which ensure reciprocal estimates of the norms of the operators S: $Z(R^n) \to W(R^m)$ and S^*_a: $Z^*(R^n, a) \to W^*(R^m, a)$ (see §§2.1 and 2.2). Moreover, for a large class of local operators we not only obtain reciprocal estimates of the norms but also study the interconnection between the operators S^*_a and S; in particular, we obtain the relation

$$\|S\| = \lim_{a \to \infty} \|S^*_a\| = \sup_{a > 0} \|S^*_a\|$$

(see §2.3). The problem of constructing local operators S_a^* from a known local operator S is considered along with the inverse problem (see §2.4).

In Chapter III the trace operators and extension operators are considered separately in the periodic and nonperiodic cases. The concept of a trace operator is formulated in §3.1 in such a way that it includes various concrete definitions of trace used in the study of the trace problem in various spaces. We establish the interconnection between traces in the periodic and nonperiodic cases, as well as theorems on estimating norms of trace operators developing the theorems in §§2.3 and 2.4. Then in §3.2 we give the concept of an extension operator and consider the interconnection between periodic and nonperiodic extension operators.

In Chapter IV various generalizations of results in Chapters I–III are indicated, comments are made on the individual assertions, and concrete examples of periodic and nonperiodic function spaces are considered. For example, in §4.3 we consider the interconnection between estimates for entire functions and for trigonometric polynomials. Using results from Chapters I–III, we give periodic analogues of certain new estimates obtained by Magaril-Il'yaev [4] and by Berkolaĭko and Ovchinnikov [5] for entire functions. Periodic and nonperiodic Sobolev, Nikol'skiĭ, and Besov spaces are considered in §4.4. Using results in Chapters I–III along with known properties of these spaces (in particular, additivity properties of the norms) studied by Nikol'skiĭ [6], Yakovlev [7], Burenkov [8]–[10], and Kuznetsov [11]–[14], we get a number of new equivalent norms for the periodic spaces, and we formulate for them some imbedding theorems which follow from known theorems and results in Chapters I–III.

Chapter V deals with the application of our results to general anisotropic Nikol'skiĭ-Besov spaces. These spaces include the known anisotropic Nikol'skiĭ-Besov spaces as well as generalizations of them involving moduli of continuity of a more general form. They were introduced by Gol'dman [15] and independently (in a somewhat different form) by Kalyabin [16]. Using the main properties studied by Gol'dman (see [17] and §5.1) of general anisotropic Besov spaces (completeness, translation invariance of the norms, boundedness of the operator of multiplication by a function in $C^M(R^n)$, and denseness of the compactly supported functions), we can apply the results from Chapters I–III to them. Here it is also necessary to show that the periodic space introduced in Chapter I coincides with the natural periodic analogue of a general anisotropic Besov space. This is done in §5.2. Some imbedding theorems for periodic spaces following from known theorems and results in Chapters I–III are given as an illustration. We should also mention Triebel's paper [18], in which there is another approach to the treatment of periodic Besov spaces.

The numbering of formulas is consecutive within one confines of a single chapter. In referring to formulas in another chapter we preface the formula numbers with the chapter number.

The results in Chapters I–IV were obtained jointly by V. I. Burenkov and M. L. Gol'dman; those in Chapter V were obtained by the latter author.

Chapter I

Equivalent Norms in Periodic Spaces

§1.1. Definition of a periodic space by means of extensions

1°. Notation. Preliminaries. Let $Z(R^n)$ denote a normed linear space of functions (or generalized functions) on the n-dimensional Euclidean space R^n, with norm

$\|\cdot\|_{Z(R^n)}$. For definiteness it can be assumed that $Z(R^n) \subset S'(R^n)$, the Schwartz space of generalized functions (the tempered distributions). In many cases (but not always) the space $Z(R^n)$ will consist of compactly supported functions.

Everywhere below Π denotes a finite open rectangular parallelepiped in R^n with faces parallel to the coordinate planes, and $\rho(\Pi) = (\rho_1(\Pi), \dots, \rho_n(\Pi))$ is the vector with coordinates equal to the lengths of the corresponding edges of Π. Further, let

$$Z(\Pi) = \left\{ f \in S'(R^n) : \exists g \in Z(R^n),\, g = f \text{ on } \Pi, \right.$$
$$\left. \cdot \|f\|_{Z(\Pi)} = \inf_{\substack{g \in Z(R^n) \\ g = f|_\Pi}} \|g\|_{Z(R^n)} < \infty \right\}, \tag{1}$$

and let

$$\check{Z}(R^n) = \{ f \in S'(R^n) : f \in Z(\Pi)\ \forall \Pi \subset R^n \}. \tag{2}$$

Thus, $\check{Z}(R^n)$ consists of all the generalized functions whose restrictions to any parallelepiped Π can be extended to $Z(R^n)$-functions on R^n. The topology in $\check{Z}(R^n)$ is given by the system of seminorms $\|\cdot\|_{Z(\Pi)}$, $\Pi \subset R^n$.

Let $\overline{Z}(R^n)$ denote the completion of $Z(R^n)$ with respect to $\check{Z}(R^n)$ (see [19], Chapter 1), with the norm

$$\|f\|_{\overline{Z}(R^n)} = \inf\left\{ c : \exists f_\nu \in Z(R^n),\ \lim_{\nu \to \infty} f_\nu = f \text{ in } \check{Z}(R^n),\ \|f_\nu\|_{Z(R^n)} \leqslant c \right\}$$
$$= \inf\left\{ c_1 : \exists f_\nu \in Z(R^n),\ \lim_{\nu \to \infty} f_\nu = f \text{ in } \check{Z}(R^n),\ \overline{\lim_{\nu \to \infty}} \|f_\nu\|_{Z(R^n)} = c_1 \right\}.$$

It is easy to see that

$$\sup_{\Pi \subset R^n} \|f\|_{Z(\Pi)} \leqslant \|f\|_{\overline{Z}(R^n)} \leqslant \|f\|_{Z(R^n)}. \tag{3}$$

Indeed, the right-hand inequality is obvious; we prove the left-hand one. Let $f \in \overline{Z}(R^n)$. For any sequence $f_\nu \in Z(R^n)$ such that $f_\nu \to f$ $(\nu \to \infty)$ in $\check{Z}(R^n)$ and any $\Pi \subset R^n$ we have that $f_\nu \to f$ $(\nu \to \infty)$ in $Z(\Pi)$, consequently,

$$\|f\|_{Z(\Pi)} = \lim_{\nu \to \infty} \|f_\nu\|_{Z(\Pi)} \leqslant \overline{\lim_{\nu \to \infty}} \|f_\nu\|_{Z(R^n)},$$

and so

$$\sup_{\Pi \subset R^n} \|f\|_{Z(\Pi)} \leqslant \overline{\lim_{\nu \to \infty}} \|f_\nu\|_{Z(R^n)},$$

which yields the required inequality, because the sequence $f_\nu \to f$ was arbitrary.

Of course, $Z(R^n) \subset \overline{Z}(R^n) \subset \check{Z}(R^n)$, but in many cases $\check{Z}(R^n)$ is considerably larger and, in particular, includes periodic functions. For $a > 0$ let $Z^*(R^n, a)$ denote the set of functions in $\check{Z}(R^n)$ having period a in all arguments. The linear space $Z^*(R^n, a)$ plays the role of a periodic analogue of $Z(R^n)$. For example, let $Z(R^n)$ be the set of mesaurable compactly supported functions with norm $\|f\|_{L_p(R^n)} < \infty$. Then $\check{Z}(R^n) = L_p^{\mathrm{loc}}(R^n)$, $\overline{Z}(R^n) = L_p(R^n)$, and $Z^*(R^n, a)$ is the set of a-periodic functions in $L_p^{\mathrm{loc}}(R^n)$, which coincides with the periodic space $L_p^*(T_a)$. Here and below, T_a is the period cube:

$$T_a \equiv T_a^n = \{ x = (x_1, \dots, x_n) \in R^n : |x_i| < a/2,\ i = 1, \dots, n \}. \tag{4}$$

For $y \in R^n$ let E_y denote the translation operator(¹)

$$(F_y f)(x) = f(x+y), \qquad f \in S'(R^n), \tag{5}$$

and let norms of this operator be defined by

$$\|E_y\| = \sup_{\substack{f \in Z(R^n) \\ f \neq 0}} \left[\|E_y f\|_{Z(R^n)} / \|f\|_{Z(R^n)}\right], \tag{6}$$

$$\|E_y\|^*_\Pi = \sup_{\substack{f \in Z^*(R^n, a) \\ \|f\|_{Z(\Pi)} \neq 0}} \left[\|E_y f\|_{Z(\Pi - y)} / \|f\|_{Z(\Pi)}\right]. \tag{7}$$

Here $\Pi - y = \{z \in R^n \colon z = x - y, x \in \Pi\}$. Note that

$$\|E_y\|^*_\Pi \leqslant \|E_y\|. \tag{8}$$

We prove this inequality, assuming that $\|E_y\| < \infty$. If $g \in Z(R^n)$ and $g = f$ on Π, then $E_y g \in Z(R^n)$ and $E_y g = E_y f$ on $\Pi - y$ and, hence,

$$\|E_y f\|_{Z(\Pi - y)} \leqslant \|E_y g\|_{Z(R^n)} \leqslant \|E_y\| \cdot \|g\|_{Z(R^n)}.$$

Taking the infimum over such g, we get

$$\|E_y f\|_{Z(\Pi - y)} \leqslant \|E_y\| \cdot \|f\|_{Z(\Pi)}, \tag{9}$$

which is what was required.

2°. On norming of the space $Z^*(R^n, a)$. For $f \in Z^*(R^n, a)$ the seminorms $\|f\|_{Z(\Pi)}$ are monotone with respect to Π: $\|f\|_{Z(\Pi_1)} \leqslant \|f\|_{Z(\Pi_2)}$ if $\Pi_1 \subset \Pi_2$. The space $Z^*(R^n, a)$ is locally convex; we can take $\{\| \cdot \|_{Z(T_{ka})}\}_{k=1,2,\ldots}$ as a defining system of seminorms. In the general case these seminorms are not mutually equivalent. We give an example. Let $n = 1$, $a = 2$, $T_a = (-1, 1)$ and $\|f\|_{Z(R^1)} = \int_{-\infty}^{\infty} |f(x)| \rho(x)\, dx$, where the weight $\rho(x) = (|x| - 2l)^{-l-1}$ for $2l < |x| < 2l + 2$, and $\rho(\pm 2l) = 1$, $l = 0, 1, \ldots$. Then, obviously,

$$\|f\|_{Z(T_{(k+1)a})} = \int_{-2k-1}^{2k+1} |f(x)| \rho(x)\, dx.$$

Consider the function $f_m(x)$ equal to 0 for $|x| < 1/m$ and to 1 for $1/m \leqslant |x| \leqslant 1$, and periodic with period 2 ($m = 1, 2, \ldots$). For it

$$\|f'_m\|_{Z(T_{(2k+1)a})} \sim 4m^k/k \qquad (m \to \infty),$$

which shows that the seminorms are not equivalent for different k.

Of greatest interest is the case where for $Z^*(R^n, a)$ the norms $\| \cdot \|_{Z(\Pi)}$ are equivalent for different $\Pi \supset \overline{T}_a$, and $Z^*(R^n, a)$ is a normable space. Conditions on $Z(R^n)$ ensuring that $Z^*(R^n, a)$ is normable are given below.

3°. Conditions on the space $Z(R^n)$. We first formulate some variants of a requirement that the norm in $Z(R^n)$ be translation invariant.

1(a) *For any* $\Pi \subset R^n$ *with* $\rho(\Pi) > a$(²) *and any* $l \in \mathbf{Z}^n$

$$\|E_{la}\|^*_\Pi < \infty. \tag{10}$$

(¹)In the case of generalized functions the definitions of a translate, of the property of periodicity, and of the coincidence of functions on an open set should be understood in the sense of the theory of generalized functions.

(²)That is, $\rho_j(\Pi) > a$ for $j = 1, \ldots, n$.

1(b) *For any* $\Pi \subset R^n$ *with* $\rho(\Pi) \geqslant a$ *and any* $l \in \mathbf{Z}^n$, *condition* (10) *holds.*

Here (and everywhere below) $\mathbf{Z}^n$ denotes the set of vectors in R^n with integer coordinates.

1(c) $\sup_{|y| \leqslant h} \|E_y\| < \infty$ *for any* $h > 0$.

1(d) $A = \sup_{y \in R^n} \|E_y\| < \infty$.

The following implications are obvious (see also (7)):

$$1(\mathrm{d}) \Rightarrow 1(\mathrm{c}) \Rightarrow 1(\mathrm{b}) \Rightarrow 1(\mathrm{a}). \tag{11}$$

In condition 1(d) we have $A \geqslant \|E_0\| = 1$; if $A = 1$, then the requirement 1(d) is equivalent to the condition

$$\|E_y f\|_{Z(R^n)} = \|f\|_{Z(R^n)} \quad \forall y \in R^n \quad \forall f \in Z(R^n)$$

(translation invariance of the norm). Under condition 1(c) if $f \in \check{Z}(R^n)$, then $E_y f \in \check{Z}(R^n)$ for any $y \in R^n$, and for any $\Pi \subset R^n$ we have the estimate (9), which implies that the operator E_y: $\check{Z}(R^n) \to \check{Z}(R^n)$ is continuous (its derivation goes through also for $f \in \check{Z}(R^n)$).

Below we present some variants of an additivity requirement for the norms $\|\cdot\|_{Z(\Pi)}$. Parallelepipeds Π_1 and Π_2 will be said to be extensions of each other if their projections on some coordinate plane coincide and $\Pi_1 \cap \Pi_2 \neq \varnothing$.

2(a) *If* Π_1 *and* Π_2 *are extensions of each other and* $\rho(\Pi_i) > a$ $(i = 1, 2)$, *then for any* $f \in Z^*(R^n, a)$

$$\|f\|_{Z(\Pi_1 \cup \Pi_2)} \leqslant c(\Pi_1, \Pi_2)\left(\|f\|_{Z(\Pi_1)} + \|f\|_{Z(\Pi_2)}\right). \tag{12}$$

2(b) *Under condition* 2(a) *it is required in addition that for any* $\alpha > 0$ *and* $0 < \beta < 1$

$$c(\Pi_1, \Pi_2) \leqslant c_1(\alpha, \beta) \tag{13}$$

for all parallelepipeds Π_1 *and* Π_2 *such that* $(j = 1, \dots, n)$

$$\rho_j(\Pi_1 \cap \Pi_2) \geqslant \alpha, \qquad \frac{\rho_j(\Pi_1 \cap \Pi_2)}{\min\{\rho_j(\Pi_1), \rho_j(\Pi_2)\}} \geqslant \beta.$$

2(c) *If* $\overline{\Pi}_1$ *and* $\overline{\Pi}_2$ *are extensions of each other and* $\rho(\Pi_i) \geqslant a$ $(i = 1, 2)$, *then for any* $f \in Z^*(R^n, a)$

$$\|f\|_{Z((\overline{\Pi}_1 \cup \overline{\Pi}_2)^0)} \leqslant c(\Pi_1, \Pi_2)\left(\|f\|_{Z(\Pi_1)} + \|f\|_{Z(\Pi_2)}\right). \tag{14}$$

Here A^0 *is the set of interior points of* A.

2(d) *Under the conditions of* 2(c) *it is required in addition that for any* $\gamma > 1$

$$c(\Pi_1, \Pi_2) \leqslant c_2(\gamma) \tag{15}$$

for all parallelepipeds Π_1 *and* Π_2 *such that* $(j = 1, \dots, n)$

$$\gamma^{-1}\rho_j(\Pi_2) \leqslant \rho_j(\Pi_1) \leqslant \gamma\rho_j(\Pi_2).$$

Simple conditions sufficient for 2(a) and 2(b) are given in §1.2. The question of the validity of 2(a)–2(d) for the spaces $B^l_{p,\theta}(\Pi)$ has been studied by Nikol'skiĭ [6], Yakovlev [7], Burenkov [8]–[10], and Kuznetsov [11]–[14].

4°. Theorems on equivalent norms in $Z^*(R^n, a)$.

THEOREM 1. 1. *The norms* $\|f\|_{Z(\Pi)}$, $f \in Z^*(R^n, a)$, *are equivalent for different* Π *with* $\rho(\Pi) > a$ *if and only if* 1(a) *and* 2(a) *hold.*

2. *The norms* $\|f\|_{Z(\Pi)}$, $f \in Z^*(R^n, a)$, *are equivalent for different* Π *with* $\rho(\Pi) \geqslant a$ (*in particular, they are equivalent to* $\|f\|_{Z(T_a)}$) *if and only if* 1(b) *and* 2(c) *hold.*

PROOF. 1. We prove the necessity part of the first assertion. Let $f \in Z^*(R^n, a)$. For $l \in \mathbf{Z}^n$

$$\|E_{la} f\|_{Z(\Pi - la)} = \|f\|_{Z(\Pi - la)} \leqslant c(\Pi - la, \Pi)\|f\|_{Z(\Pi)},$$

because $\rho(\Pi - la) = \rho(\Pi) > a$, and, consequently, the norms $\|\cdot\|_{Z(\Pi - la)}$ and $\|\cdot\|_{Z(\Pi)}$ are equivalent. This implies 1(a) (see (10) and (7)).

Next, if Π_1 and Π_2 are extensions of each other, then $\rho(\Pi_1 \cup \Pi_2) \geqslant \rho(\Pi_1) > a$, and, by the equivalence of the corresponding norms,

$$\|f\|_{Z(\Pi_1 \cup \Pi_2)} \leqslant c(\Pi_1 \cup \Pi_2, \Pi_1)\|f\|_{Z(\Pi_1)} \leqslant c_1(\Pi_1, \Pi_2)\left(\|f\|_{Z(\Pi_2)} + \|f\|_{Z(\Pi_2)}\right),$$

which implies 2(a). The necessity part in the second assertion of the theorem can be proved similarly.

2. To prove the sufficiency part in the first assertion it can be assumed without loss of generality (because the norms are monotone) that $\overline{\Pi}_1 \subset \overline{\Pi}_2$ and $\rho(\Pi_1) > a$, and the resulting estimate is

$$\|f\|_{Z(\Pi_2)} \leqslant c(\Pi_1, \Pi_2)\|f\|_{Z(\Pi_1)}. \tag{16}$$

Let e_s be the unit vector directed along the sth coordinate axis. Let $\Pi^0 = \Pi_1$ and $\Pi^s = \bigcup_{|l| \leqslant L} \Pi^{s-1}(l)$, where $\Pi^{s-1}(l) = \Pi^{s-1} + lae_s$ (arithmetic sum), for $s = 1, \ldots, n$, and L is a fixed positive integer.

The parallelepipeds $\Pi^{s-1}(l)$ and $\Pi^{s-1}(l+1)$ are extensions of each other. Therefore, Π^1 is a parallelepiped obtained from Π^0 by taking a union of translates of it by la along e_1, Π^2 is obtained from Π^1 by taking a union of translates of it by la along e_2, and so on. As a result we get a parallelepiped Π^n which contains Π_2 for sufficiently large L.

The periodicity of f implies that for $s = 1, \ldots, n$

$$\|f\|_{Z(\Pi^{s-1}(l))} = \|E_{-lae_s} f\|_{Z(\Pi^{s-1} + lae_s)} \leqslant \sup_{|l| \leqslant L} \|E_{-lae_s}\|^*_{\Pi^{s-1}} \|f\|_{Z(\Pi^{s-1})}.$$

Using inequality (12) for the parallelepipeds $K^{s-1}(l) = \bigcup_{m=-L}^{l} \Pi^{s-1}(m)$ and $\Pi^{s-1}(l+1)$, $l = -L, \ldots, L-1$, we get

$$\|f\|_{Z(\Pi^s)} \leqslant B \sum_{|l| \leqslant L} \|f\|_{Z(\Pi^{s-1}(l))} \leqslant B(2L+1) \sup_{|l| \leqslant L} \|E_{-lae_s}\|^*_{\Pi^{s-1}} \|f\|_{Z(\Pi^{s-1})}. \tag{17}$$

And using (17) for $s = 1, \ldots, n$, we get (16).

3. To prove the sufficiency part in the second assertion of the theorem we assume that $\Pi_1 = T_a$ and set $\Pi^s = (\overline{\bigcup_{|l| \leqslant L} \Pi^{s-1}(l)})^0$ without changing the rest of the notation. The parallelepipeds $\Pi^{s-1}(l)$ and $\Pi^{s-1}(l+1)$ are not extensions of each other, but their closures have this property. Therefore, we again get (16) by using (14).

Conditions for the normability of the space $Z^*(R^n, a)$ are obtained in assertion 1 of Theorem 1. We choose one of the (equivalent) norms $\|\cdot\|_{Z(\Pi)}$, $\rho(\Pi) > a$, as the norm in $Z^*(R^n, a)$:

$$\|f\|Z^{(1)}_{Z^*(R^n, a)} = \|f\|_{Z(T_{3a/2})} \tag{18}$$

Under the conditions of assertion 2 this norm is equivalent to

$$\|f\|^{(2)}_{Z^*(R^n, a)} = \|f\|_{Z(T_a)} \tag{19}$$

for $f \in Z^*(R^n, a)$.

THEOREM 2. 1. *Suppose that* 1(d) *and* 2(b) *hold, and* $a_0 > 0$, $1 < \lambda_1 < \lambda_2 < \infty$. *Then for* $f \in Z^*(R^n, a)$ *the norms* $\|f\|_{Z(\Pi)}$ *are mutually equivalent for different* Π *such that* $\lambda_1 a \leqslant \rho(\Pi) \leqslant \lambda_2 a$ (*in particular, they are equivalent to the norm* $\|f\|^{(1)}_{Z^*(R^n,a)}$) *with constants not depending on* a *for all* $a \geqslant a_0$.

2. *If* 1(d) *and* 2(d) *hold, then assertion* 1 *is valid also for* $\lambda_1 = 1$ (*in particular, the norms* $\|f\|_{Z(\Pi)}$ *are equivalent to* $\|f\|^{(2)}_{Z^*(R^n,a)}$).

PROOF. Without loss of generality it can be assumed that $\lambda_1 a \leqslant \rho(\Pi_1) \leqslant \rho(\Pi_2) \leqslant \lambda_2 a$ and $\Pi_1 \cap \Pi_2 \neq \varnothing$, and we get (16) with a constant not depending on $a \geqslant a_0$. Using the proof of Theorem 1, we observe that now

$$\rho\big(K^{s-1}(l) \cap \Pi^{s-1}(l+1)\big) \geqslant (\lambda_1 - 1)a_0$$

and

$$\frac{\rho_i\big(K^{s-1}(l) \cap \Pi^{s-1}(l+1)\big)}{\min\{\rho_i(K^{s-1}(l)), \rho_i(\Pi^{s-1}(l+1))\}} \geqslant \frac{(\lambda_1 - 1)}{\lambda_2} \qquad (i = 1, \dots, n).$$

Therefore, by (17), the desired inequality follows by taking into account that $L \leqslant \lambda_1/\lambda_1$ and

$$\sup_{|l| \leqslant L} \|E_{-lae_s}\|^*_{\Pi^{s-1}} \leqslant \sup_{y \in R^n} \|E_y\| < \infty.$$

To prove the second assertion it suffices to argue as above and note that

$$\rho\big(\Pi^{s-1}(l+1)\big) \leqslant \rho\big(K^{s-1}(l)\big) \leqslant \frac{(2L+1)\lambda}{\lambda_1}\rho\big(\Pi^{s-1}(l+1)\big).$$

§1.2. Equivalent norms connected with multiplication by compactly supported functions

1°. Conditions on the space $Z(R^n)$. Here we shall give more stringent conditions on $Z(R^n)$ than the additivity requirements in §1.1; they ensure the existence of other, simpler equivalent norms on $Z^*(R^n, a)$.

Recall that the condition $\varphi \in C^M(R^n)$, $M \geqslant 0$ an integer, means that $\varphi(x)$ is continuous along with its derivatives through order M and that the norm is

$$\|\varphi\|_{C^M(R^n)} = \sum_{|\alpha| \leqslant M} \sup_{x \in R^n} |D^\alpha \varphi(x)| < \infty, \tag{20}$$

where

$$\alpha = (\alpha_1, \dots, \alpha_n) \in \mathbf{Z}^n, \qquad \alpha \geqslant 0, |\alpha| = \alpha_1 + \cdots + \alpha_n$$

and

$$D^\alpha \varphi(x) = \partial^{|\alpha|}\varphi/\partial x_1^{\alpha_1} \cdots \partial x_1^{\alpha_n}.$$

The set of compactly supported functions in $C^M(R^n)$ is denoted by $C_0^M(R^n)$, and $C_0^\infty(R^n)$ denote the set of infinitely differentiable compactly supported functions. We give two variants of the requirements on $Z(R^n)$.

3(a) *Any function* $\varphi(x) \in C_0^\infty(R^n)$ *is a multiplier in* $Z(R^n)$, *i.e., the operator* $\mathscr{P}_\varphi$ *of multiplication by* $\varphi(x)$ *is bounded on* $Z(R^n)$.

3(b) *There exists an integer* $M \geqslant 0$ *such that any* $\varphi \in C_0^M(R^n)$ *is a multiplier in* $Z(R^n)$, *and the norm of* $\mathscr{P}_\varphi$ *has the estimate*

$$\|\mathscr{P}_\varphi\| \leqslant B\|\varphi\|_{C^M(R^n)}, \tag{21}$$

where B *does not depend on* φ.

REMARK 1. Suppose that condition 3(a) holds. Then any function $\varphi(x) \in C_0^\infty(R^n)$ is a multiplier from $\check{Z}(R^n)$ to $Z(R^n)$, and for any $\Pi \subset R^n$ with $\operatorname{supp}\varphi \subset \Pi$

$$\|f\varphi\|_{Z(R^n)} \leqslant \|\mathscr{P}_\varphi\| \cdot \|f\|_{Z(\Pi)}, \qquad f \in \check{Z}(R^n), \tag{22}$$

and

$$\|f\varphi\|_{Z(R^n)} \leqslant \|\mathscr{P}_\varphi\| \cdot \|f\|_{\bar{Z}(R^n)}, \qquad f \in \bar{Z}(R^n), \tag{23}$$

where $\|\mathscr{P}_\varphi\|$ is the norm of the operator $\mathscr{P}_\varphi$: $Z(R^n) \to Z(R^n)$. Moreover, any compactly supported function $f \in \check{Z}(R^n)$ belongs to $Z(R^n)$.

Indeed, suppose that $f \in \check{Z}(R^n)$, $\varphi \in C_0(R^n)$ and $\operatorname{supp}\varphi \subset \Pi_0$. For any $\Pi \supset \Pi_0$ and any $g \in Z(R^n)$ with $g = f|_\Pi$ we have, by property 3(a), that $f\varphi \equiv g\varphi \in Z(R^n)$ and

$$\|f\varphi\|_{Z(R^n)} \leqslant \|\mathscr{P}_\varphi\| \cdot \|g\|_{Z(R^n)}.$$

Taking the infimum over such functions g gives us (22), and (23) follows from (22) and (3). If $f \in \check{Z}(R^n)$ is compactly supported, then by choosing a $\varphi \in C_0^\infty(R^n)$ such that $\varphi = 1$ on $\operatorname{supp} f$, we have that $f \equiv f\varphi \in Z(R^n)$ according to what was proved above.

REMARK 2. Suppose that condition 3(b) holds. Then:

1) Any function $\varphi(x) \in C_0^M(R^n)$ is a multiplier from $\check{Z}(R^n)$ to $Z(R^n)$, and for any $\Pi \subset R^n$ with $\operatorname{supp}\varphi \subset \Pi$

$$\|f\varphi\|_{Z(R^n)} \leqslant B\|\varphi\|_{C^M(R^n)}\|f\|_{Z(\Pi)}, \qquad f \in \check{Z}(R^n).$$

In particular, for $f \in \bar{Z}(R^n)$ and $\varphi \in C_0^M(R^n)$

$$\|f\varphi\|_{Z(R^n)} \leqslant B\|\varphi\|_{C^M(R^n)}\|f\|_{\bar{Z}(R^n)}. \tag{24}$$

2) Any function $\varphi(x) \in C^M(R^n)$ is a multiplier in $\check{Z}(R^n)$, and for any $\Pi \subset R^n$

$$\|f\varphi\|_{Z(\Pi)} \leqslant B\|\varphi\|_{C^M(R^n)}\|f\|_{Z(\Pi)}, \qquad f \in \check{Z}(R^n).$$

The proof of the first part is the same as in Remark 1 (the estimate (12) must also be taken into account). We prove that $f\varphi \in \check{Z}(R^n)$ for $f \in \check{Z}(R^n)$ and $\varphi \in C^M(R^n)$. For any $\Pi \subset R^n$ we choose $g \in Z(R^n)$ with $g = f|_\Pi$ and $\psi \in C_0^M(R^n)$ with $\psi = \varphi|_\Pi$; here $\psi = \psi_\varepsilon$ can be chosen for any $\varepsilon > 0$ so that

$$\|\psi\|_{C^M(R^n)} \leqslant (1+\varepsilon)\|\varphi\|_{C^M(R^n)}.$$

Then $g\psi \in Z(R^n)$, $g\psi = f\varphi|_\Pi$, and, consequently, by 3(b),

$$\|f\varphi\|_{Z(\Pi)} \leqslant \|g\psi\|_{Z(R^n)} \leqslant B\|\varphi\|_{C^M(R^n)}(1+\varepsilon)\|g\|_{Z(R^n)}.$$

Taking the infimum over such g and using the arbitrariness of ε, we get the required estimate from this.

Let us now show that conditions 3(a) and 3(b) are stronger than the corresponding additivity requirements of §1.1.

LEMMA 1. 1) 3(a) $\Rightarrow$ 2(a); 2) 3(b) $\Rightarrow$ 2(b).

PROOF. 1. We prove the first part of the lemma. Let $\varphi_1, \varphi_2 \in C_0^\infty(R^n)$ be such that $\varphi_1(x) + \varphi_2(x) = 1$ for $x \in \Pi \equiv \Pi_1 \cup \Pi_2$, and $\operatorname{supp}\varphi_i \cap (\Pi \setminus \Pi_i) = \varnothing$ for $i = 1, 2$. Then for $f \in \check{Z}(R^n)$

$$\|f\|_{Z(\Pi)} = \|f\varphi_1 + f\varphi_2\|_{Z(\Pi)} \leqslant \|f\varphi_1\|_{Z(\Pi)} + \|f\varphi_2\|_{Z(\Pi)}.$$

For any $g \in Z(R^n)$ with $g = f|_{\Pi_i}$ we have that $f\varphi_i = g\varphi_i|_{\Pi}$, and hence

$$\|f\varphi_i\|_{Z(\Pi)} \leqslant \|g\varphi_i\|_{Z(R^n)} \leqslant \|\mathscr{P}_{\varphi_i}\| \cdot \|g\|_{Z(R^n)}.$$

Taking the infimum over such g, we get the estimate

$$\|f\varphi_i\|_{Z(\Pi)} \leqslant \|\mathscr{P}_{\varphi_i}\| \cdot \|f\|_{Z(\Pi_i)} \qquad (i = 1, 2).$$

Consequently,

$$\|f\|_{Z(\Pi)} \leqslant \max_{i=1,2} \|\mathscr{P}_{\varphi_i}\| \left(\|f\|_{Z(\Pi_1)} + \|f\|_{Z(\Pi_2)} \right) \tag{25}$$

for $f \in \check{Z}(R^n)$, which is what was required.

2. Let us prove the second part of the lemma. If $\rho_j(\Pi_1 \cap \Pi_2) \geqslant \alpha$ $(j = 1, \ldots, n)$, then φ_1 and φ_2 can be taken to be functions in $C_0^M(R^n)$ such that

$$\|\varphi_i\|_{C^M(R^n)} \leqslant c(M, n)\alpha^{-M}, \qquad i = 1, 2.$$

Consequently, $\|\mathscr{P}_{\varphi_i}\| \leqslant Bc(M, n)\alpha^{-M}$ in (25), and this gives us 2(b).

REMARK 3. If 3(b) holds with $M = 0$ and with $(L_\infty)_0(R^n)$ instead of $C_0(R^n)$, then 2(c) and 2(d) hold. In this case it suffices to take φ_1 and φ_2 to be the characteristic functions of the respective parallelepipeds Π_1 and $\Pi \setminus \Pi_1$.

2°. A periodic partition of unity and estimates of compactly supported multipliers. Everywhere below for a function $\varphi(x)$ we write

$$(Q_a\varphi)(x) \equiv (Q_a^{(n)}\varphi)(x) = \sum_{l \in \mathbf{Z}^n} \varphi(x + al) \tag{26}$$

for the corresponding periodic extension of $\varphi(x)$.

LEMMA 2 (a periodic partition of unity). *Suppose that* $\psi(x) \geqslant 0$ *for* $x \in R^n$, $\psi(x) \geqslant h > 0$ *for* $x \in T_a$, *and* $\psi \in C_0^M(T_{\gamma a})$ $(\gamma > 1)$. *Then the functions* $\psi_1(x + la)$, $l \in \mathbf{Z}^n$, *where*

$$\psi_1(x) = \psi(x)/\chi(x), \qquad \chi(x) = (Q_a\psi)(x), \tag{27}$$

form a periodic partition of unity, i.e., $(Q_a\psi_1)(x) \equiv 1$. *Moreover,* $\psi_1 \in C_0^M(T_{\gamma a})$, supp ψ_1 = supp ψ, *and*

$$\|\psi_1\|_{C^M(R^n)} \leqslant c(M, n)h^{-M-1}\gamma^{nM}\|\psi\|_{C^M(R^n)}^{M+1},$$

where $c(M, n)$ *depends only on* M *and* n.

PROOF. For each x the sum in $\chi(x)$ contains a finite number of terms, corresponding to $l \in L(\psi) = \{l \in \mathbf{Z}^n\colon\ T_{\gamma^1} \cap \operatorname{supp} E_{la}\psi \neq \varnothing\}$. This number is $|L(\psi)| \leqslant (2[\gamma] + 1)^n$; therefore,

$$\|\chi\|_{C^M(R^n)} \leqslant (2[\gamma] + 1)^n \|\psi\|_{C^M(R^n)}.$$

Moreover, $\chi(x) \geqslant h$ for $x \in R^n$. According to the differentiation rules,

$$\|\psi_1\|_{C^M(R^n)} \leqslant n^M \|\psi\|_{C^M(R^n)} \|1/\chi\|_{C^M(R^n)},$$

$$\|1/\chi\|_{C^M(R^n)} \leqslant c_1(M, n)h^{-M-1}\|\chi\|_{C^M(R^n)}^M.$$

Thus,

$$\|1/\chi\|_{C^M(R^n)} \leqslant c_2(M, n)h^{-M-1}\gamma^{nM}\|\psi\|_{C^M(R^n)}^M. \tag{28}$$

The lemma follows from these estimates.

LEMMA 3. *Suppose that* $Z(R^n)$ *satisfies* 1(c) *and* 3(a), $\varphi, \psi \in C_0(R^n)$, *and* $Q_a\psi \equiv 1$ *or* $\psi(x) \geqslant h > 0$ *on* T_a. *Then for any* $f \in Z^*(R^n, a)$

$$\|f\varphi\|_{Z(R^n)} \leqslant c(\varphi, \psi)\|f\psi\|_{Z(R^n)}. \tag{29}$$

PROOF. 1. Suppose first that $\psi \in C_0^\infty(R^n)$ and $Q_a\psi \equiv 1$. Note that the right-hand side of (29) is finite (see Remark 1). Using the periodicity of $f(x) \in Z^*(R^n, a)$, i.e., the condition that $E_{al}f = f$, $l \in \mathbf{Z}^n$, we get that

$$\|f\varphi\|_{Z(R^n)} = \Big\| \sum_{l\in\mathbf{Z}^n} f\varphi E_{al}\psi\Big\|_{Z(R^n)} \leqslant \sum_{l\in L(\varphi,\psi)} \|\varphi E_{al}(f\psi)\|_{Z(R^n)}$$

$$\leqslant \|\mathcal{P}_\varphi\| \sum_{l\in L(\varphi,\psi)} \|E_{al}\|\cdot\|f\psi\|_{Z(R^n)}.$$

Here $L(\varphi, \psi) = \{l \in \mathbf{Z}^n\colon \operatorname{supp}\varphi \cap \operatorname{supp} E_{al}\psi \neq \varnothing\}$ is a finite set. We have obtained (29).

2. Suppose now that $0 \leqslant \psi(x) \in C_0(R^n)$ and $\psi(x) \geqslant h > 0$ for $x \in T_a$. Construct $\psi_1(x)$ according to Lemma 2 and apply to it the first step of the proof. Then

$$\|f\varphi\|_{Z(R^n)} \leqslant c(\varphi, \psi_1)\|f\psi_1\|_{Z(R^n)}.$$

We choose $\xi(x) \in C_0^\infty(R^n)$ with $\xi(x) = 1$ on $\operatorname{supp}\psi$ and get that $\xi/\chi \in C_0^\infty(R^n)$; so $\psi_1(x) \equiv (\xi/\chi)(x)\psi(x)$. Hence,

$$\|f\psi_1\|_{Z(R^n)} \leqslant \|\mathcal{P}_{\xi/\chi}\|\cdot\|f\psi\|_{Z(R^n)}.$$

These estimates give us (29). The condition $\psi(x) \geqslant 0$ does not restrict the generality: otherwise we can multiply ψ by a function $u \in C_0^\infty(R^n)$ with $u = 1$ on T_a such that $u\psi \geqslant 0$, apply (29) with $u\psi$ instead of ψ, and then use property 3(a).

3°. Multipliers in the space $C^M(R^n, a)$. Let $M \geqslant 0$ be an integer, and let $a > 0$. We consider the space $C^M(R^n, a)$ of all functions $\varphi(x) \in C^M(R^n)$ such that

$$\|\varphi\|_{C^M(R^n,a)} = \sum_{l\in\mathbf{Z}^n} \|\varphi\|_{C^M(T_{a+1}+al)} < \infty. \tag{30}$$

It is not hard to see that the spaces $C^M(R^n, a)$ have the following properties.

1. For any function $\varphi \in C_0^M(R^n)$

$$\|\varphi\|_{C^M(R^n,a)} \leqslant |L_a(\varphi)|\cdot\|\varphi\|_{C^M(R^n)}. \tag{31}$$

Here

$$L_a(\varphi) = \{l \in \mathbf{Z}^n\colon \operatorname{supp}\varphi \cap (T_{a+1} + al) \neq \varnothing\} \tag{32}$$

is a finite set, and $|L_a(\varphi)|$ is the number of elements in it.

2. If $\varphi(x) = \Pi_{j=1}^n \xi_j(x_j)$ and $\xi_j \in C^M(R^1, a)$, then $\varphi \in C^M(R^n, a)$, and

$$\|\varphi\|_{C^M(R^n,a)} \leqslant \prod_{j=1}^n \|\xi\|_{C^M(R^1,a)}. \tag{33}$$

3. If $\xi_j(x_j) = \eta_j(|x_j|)$ in the assumptions of 2 and the functions $|\eta_j^{(k)}(u)|$, $k = 0, 1, \dots, M$, are monotonically decreasing for $u \geqslant u_0$ ($j = 1, \dots, n$), then the condition $\varphi \in C^M(R^n, a)$ is equivalent to the condition that $\varphi \in W_1^M(R^n)$, and for such φ

$$c_1\|\varphi\|_{C^M(R^n,a)} \leqslant \|\varphi\|_{C^M(R^n)} + \|\varphi\|_{W_1^M(R^n)} \leqslant_2 \|\varphi\|_{C^M(R^n,a)},$$

where c_1 and c_2 depend only on a, u_0, and n. For $n = 1$ in particular, $(1 + x^2)^{-\alpha/2} \in C^M(R^1, a) \Leftrightarrow \alpha > 1$ for any integer $M \geqslant 0$ and any $a > 0$.

4. Suppose that: $a \geqslant a_0 > 0$, $0 \leqslant \psi \in C_0^M(T_{a+1})$, $\psi = 1$ on T_a, $T\|\psi\|_{C^M(R^n)} \leqslant c(M, n)$, which does not depend on a, and $\{\psi_0(x + al)\}_{l \in \mathbf{Z}^n}$ is the corresponding partition of unity (see Lemma 2). For $\varphi \in C^M(R^n, a)$ let $\varphi_l(x) = \varphi(x)\psi_0(x + al)$, $l \in \mathbf{Z}^n$. Then

$$c_3\|\varphi\|_{C^M(R^n,a)} \leqslant \sum_{l \in \mathbf{Z}^n} \|\varphi_l\|_{C^M(R^n)} \leqslant c_4\|\varphi\|_{C^M(R^n,a)}, \tag{34}$$

where the constants c_3 and c_4 depend only on M, n, and a_0.

LEMMA 4. *Suppose that $a_0 > 0$, $Z(R^n)$ satisfies* 1(d) *and* 3(b), *$\varphi, \psi \in C^M(R^n, a)$, and $\psi(x) > 0$ on T_a. Then* (29) *holds for any $f \in Z^*(R^n, a)$, and if $\psi(x) \geqslant 0$ $x \in T_{a+1}$ and $\psi(x) \geqslant h > 0$ for $x \in T_a$, then for all $a \geqslant a_0$*

$$c(\varphi, \psi) \leqslant c_1(M, n, a_0)B^2Ah^{-M-1}\|\psi\|^M_{C^M(T_{a+1})}\|\varphi\|_{C^M(R^n,a)}. \tag{35}$$

Here A, M, and B are the constants in 1(d) *and* 3(b).

PROOF. We confine ourselves to deriving (35). The general case ($\psi(x) > 0$ on T_a) can be handled similarly, except that the constants in the estimates will depend on ψ and a, and we get (29) without (35).

1. Let $0 \leqslant u \in C_0^M(T_{a+1})$ be such that $u = 1$ on T_a and $\|u\|_{C^M(R^n)} \leqslant d(M, n)$ (which does not depend on a). Define $\psi_0(x) = \psi(x)u(x) \geqslant 0$; then $\psi_0 \in C_0^M(T_{a+1})$, $\psi_0 \geqslant h$ on T_a, and

$$\|\varphi_0\|_{C^M(R^n)} \leqslant d_1(M, n)\|\psi\|_{C^M(R^n)}. \tag{36}$$

From $\psi_0(x)$ we construct an a-periodic partition of unity, defining (as in Lemma 2) $\chi(x) = (Q_a\psi_0)(x) \geqslant h$ and $\psi_1(x) = \psi_0(x)/\chi(x)$. Then $Q_a\psi_1 \equiv 1$ and, moreover, $\psi_1(x) = \psi_2(x)\psi(x)$, where $\psi_2(x) = u(x)/\chi(x) \in C_0^M(R^n)$. If $a \geqslant a_0 > 0$, then $\gamma = (a + 1)/a \leqslant 1 + 1/a_0$ in Lemma 2. Therefore, using (28) (with ψ_0 instead of ψ) along with (36), we finally get

$$\|\psi_2\|_{C^M(R^n)} \leqslant d_1(M, n, a_0)h^{-M-1}\|\psi\|^M_{C^M(T_{a+1})}. \tag{37}$$

Thus, if $f \in Z^*(R^n, a)$ and $\|f\psi\|_{Z(R^n)} < \infty$ (i.e., $f\psi \in \overline{Z}(R^n)$; see (3) and the second part of Remark 2), then $f\psi_1 = \psi_2(f\psi) \in Z(R^n)$, and, by (23) and (37),

$$\|f\psi_1\|_{Z(R^n)} \leqslant d_1(M, n, a_0)Bh^{-M-1}\|\psi\|^M_{C^M(T_{a+1})}\|f\psi\|_{Z(R^n)}. \tag{38}$$

2. We get (29) first for the case of a compactly supported function $\varphi \in C_0^M(R^n)$. Since

$$f(x)\varphi(x) \equiv \sum_{l \in L_a(\varphi)} \varphi(x)E_{al}(f\psi_1)(x),$$

where $L_a(\varphi)$ is defined by (32), it follows from 3(b) and 1(d) that

$$\|f\varphi\|_{Z(R^n)} \leqslant \sum_{l \in L_a(\varphi)} \|\mathscr{P}_\varphi\| \cdot \|E_{al}(f\psi_1)\|_{Z(R^n)} \leqslant B\|\varphi\|_{C^M(R^n)}A|L_a(\varphi)|\|f\psi_1\|_{Z(R^n)}.$$

This and (38) give us inequality (29) with

$$c(\varphi, \psi) \leqslant d_1(M, n, a_0)B^2Ah^{-M-1}|L_a(\varphi)|\|\psi\|^M_{C^M(T_{a+1})}\|\varphi\|_{C^M(R^n)}. \tag{39}$$

3. Now let $\varphi \in C^M(R^n, a)$. Consider the functions $\varphi_l(x) \in C_0^M(R^n)$ with $\operatorname{supp}\varphi_l \subset T_{a+1} + al$, $l \in \mathbf{Z}^n$, introduced in the formulation of property 4 of the spaces

$C^M(R^n, a)$. For them, $|L_a(\varphi)| \leqslant d_2(n, a_0)$ when $a \geqslant a_0$. Applying (29) and (39) to them, we get, using (34), that

$$\|f\varphi\|_{Z(R^n)} = \Big\|f \sum_{l\in\mathbf{Z}^n} \varphi_l\Big\|_{Z(R^n)} \leqslant \sum_{l\in\mathbf{Z}^n} \|f\varphi_l\|_{Z(R^n)}$$

$$\leqslant d_3(M, n, a_0) B^2 A h^{-M-1} \|\psi\|^M_{C^M(T_{a+1})} \|\varphi\|_{C^M(R^n,a)} \|f\psi\|_{Z(R^n)}. \tag{40}$$

Note that $f\varphi_l \in Z(R^n)$ (see Remark 2), and, by (40), the series $\sum_{l\in\mathbf{Z}^n} f\varphi_l$ converges to $f\varphi$ in the norm $\|\cdot\|_{Z(R^n)}$; hence $f\varphi$ belongs to the closure of $Z(R^n)$ in this norm. Inequality (40) makes sense if $\|f\psi\|_{Z(R^n)} < \infty$. For $\psi \in C_0^M(R^n)$ this follows from Remark 2. But if ψ is any function in $C^M(R^n, a)$, then this follows from (40) with ψ instead of φ and with some function $\tilde{\psi} \in C_0^M(R^n)$ instead of ψ. The lemma is proved.

4°. Theorem on equivalent norms. For $f \in Z^*(R^n, a)$ let

$$\|f\|_{Z_\varphi(R^n)} = \|f\varphi\|_{Z(R^n)}. \tag{41}$$

THEOREM 3. *Suppose that* 1(c) *and* 3(a) *hold. Then, for* $f \in Z^*(R^n, a)$,

(a) *the norms* $\|f\|_{Z_\varphi(R^n)}$ *are equivalent for different* $\varphi \in C_0^\infty(R^n)$ *such that* $\varphi(x) > 0$ *on* T_a (*or such that* $Q_a\varphi \equiv 1$); *and*

(b) *for the indicated* φ *they are equivalent to the norms* $\|f\|_{Z(\Pi)}$ *when* $\rho(\Pi) > a$ (*in particular, to the norm* $\|f\|^{(1)}_{Z^*(R^n,a)}$).

PROOF. Assertion (a) follows at once from Lemma 3. We prove (b) first for $\Pi \supset \overline{T}_a$. Choose functions $\varphi, \psi \in C_0^\infty(R^n)$ with $\varphi = \psi = 1$ on T_a such that $\operatorname{supp}\varphi \subset \Pi$ and $\psi = 1$ on Π.

Since $f\psi \in Z(R^n)$ and $f\psi = f$ on Π, it follows that $\|f\|_{Z(\Pi)} \leqslant \|f\psi\|_{Z(R^n)}$. Moreover, (22) is valid. It remains to use (a). For an arbitrary parallelepiped Π with $\rho(\Pi) > a$ we must also take into account Theorem 1. Theorem 3 is proved.

Under the requirements 1(d) and 3(b) it is possible to essentially supplement the assertion about equivalence of norms. Let $\Phi_a = \Phi_a(n, h, a_0, M)$ be the set of functions $\varphi(x)$ of the form

$$\varphi(x) = \prod_{j=1}^n \mu_a(x_j), \qquad a \geqslant a_0, \tag{42}$$

where

$$\begin{aligned} &1) \quad \mu_a(t) \geqslant h > 0 \quad \text{for } |t| \leqslant a/2, \\ & \mu_a(t) \geqslant 0 \quad \text{for } |t| \leqslant (a+1)/2; \\ &2) \quad \|\mu_a\|_{C^M(R^1,a)} \leqslant c(a_0, M, h) \end{aligned} \tag{43}$$

(here $c(a_0, M, h)$ does not depend on $a \geqslant a_0$).

If $\varphi \in \Phi_a$, then, by (33),

$$\begin{gathered} \varphi(x) \geqslant h^n, \quad x \in T_a; \qquad \varphi(x) \geqslant 0, \quad x \in T_{a+1}; \\ h^n \leqslant \varphi(0) \leqslant \|\varphi\|_{C^M(R^n,a)} \leqslant c(a_0, M, h)^n. \end{gathered} \tag{44}$$

We consider a standard function $\varphi_a \in \Phi_a$ given by

$$\varphi_a(x) \equiv \varphi_a^{(n)}(x) = \prod_{j=1}^n \nu\left(\frac{x_j}{a}\right), \tag{45}$$

where $0 \leqslant \nu(t) \in C_0^\infty(R^1)$ is a fixed function such that $\nu(t) = 1$ for $|t| \leqslant 1/2$, and $\nu(t) = 0$ for $|t| \geqslant 3/4$.

For $f \in Z^*(R^n, a)$ define the norm

$$\|f\|^{(3)}_{Z^*(R^n,a)} = \|f\|_{Z_{\varphi_a}(R^n)} = \|f\varphi_a\|_{Z(R^n)}. \tag{46}$$

By Theorem 3, this is one of the equivalent norms in $Z^*(R^n, a)$.

THEOREM 4. *Suppose that* 1(d) *and* 3(b) *hold and that* $a_0 > 0$. *Then for* $f \in Z^*(R^n, a)$ *with* $a \geqslant a_0$:

(a) *The norms* $\|f\|_{Z_\varphi(R^n)}$ *for different* $\varphi \in C^M(R^n, a)$ *with* $\varphi(x) > 0$ *on* T_a *are equivalent* (*in particular, they are equivalent to the norm* $\|f\|^{(3)}_{Z^*(R^n,a)}$) *and with constants not depending on* $a \geqslant a_0$ *when* $\varphi \in \Phi_a$.

(b) *For* $\varphi \in \Phi_a$ *these norms* (*in particular, the norm* $\|f\|^{(3)}_{Z^*(R^n,a)}$) *are equivalent to the norm* $\|f\|^{(1)}_{Z^*(R^n,a)}$, *with constants not depending on* $a \geqslant a_0$.

PROOF. 1. Let us prove (a). Note that a function $\varphi \in \Phi_a$ satisfies the conditions of Lemma 4 by (44). Substitution of $\varphi, \psi \in \Phi_a$ in (29) and (35) leads to constants $c(\varphi, \psi)$ and $c(\psi, \varphi)$ not depending on $a \geqslant a_0$, which is what was required.

2. We prove (b). Consider functions $\varphi, \psi \in \Phi_a$ such that $\varphi = \psi = 1$ on T_a, $\operatorname{supp}\varphi \subset \Pi \equiv T_{3a/2}$ and $\psi = 1$ on Π. Arguing as in the proof of Theorem 3, we get the estimate

$$\|f\varphi\|_{Z(R^n)}/\|\mathscr{P}_\varphi\| \leqslant \|f\|_{Z(T_{3a/2})} \leqslant \|f\psi\|_{Z(R^n)}. \tag{47}$$

But

$$\|\mathscr{P}_\varphi\| \leqslant B\|\varphi\|_{C^M(R^n)} \leqslant B\|\varphi\|_{C^M(R^n,a)} \leqslant Bc(a_0, M, 1)^n,$$

by property 3(b) (see also (44)). Therefore, (b) follows from (47) and (a).

REMARK 4. Under the conditions of Theorem 4 we introduce the norm $\|\cdot\|^{(3)}_{Z^*}(R^n, a)$ in the space $Z^*(R^n, a)$. Then for any $\varphi \in C_0^M(R^n)$ the operator $\mathscr{P}_\varphi$: $Z^*(R^n, a) \to Z(R^n)$ satisfies for all $a \geqslant a_0$ the estimate

$$\|\mathscr{P}_\varphi\| \leqslant c(M, n, a_0)B^2A|L_a(\varphi)| \cdot \|\varphi\|_{C^M(R^n)}, \tag{48}$$

where A, B, and M are the constants in 1(d) and 3(b), and $L_a(\varphi)$ is defined in (32). For a proof use (29) and (39), with $\psi = \varphi_a(x)$ (see (45)).

5°. Estimates of the norms of periodic extensions of compactly supported functions.

LEMMA 5. *Assume conditions* 1(c) *and* 2(a) (*or* 2(c), *or* 3(a)). *Then for* $a > 0$ *any compactly supported function* $f(x) \in Z(R^n)$ *has periodic extension satisfying* $Q_af \in Z^*(R^n, a)$ (*see* (26)), *and for* $j = 1$ (*respectively, for* $j = 2$, *or for* $j = 3$)

$$\|Q_af\|^{(j)}_{Z^*(R^n,a)} \leqslant c_j\left(\sum_{l\in L_j(f,a)} \|E_{la}\|\right)\|f\|_{Z(R^n)}, \tag{49}$$

where $c_1 = c_2 = 1$, $c_3 = \|\mathscr{P}_{\varphi_a}\|$, *and the sets* $L_1(f, a) = L_3(f, a) = L(f, a, T_{3a/2})$ *and* $L_2(f, a) = L(f, a, T_a)$ *are finite. Here*

$$L(f, a, \Pi) = \{l \in \mathbf{Z}^n \colon \Pi \cap \operatorname{supp} E_{la}f \neq \varnothing\}. \tag{50}$$

PROOF. Clearly, Q_af is an a-periodic function, and we need only estimate the norm $\|Q_af\|_{Z(\Pi)}$ for $\Pi \subset R^n$. Since $L(f, a, \Pi)$ is a finite set,

$$g_a(x) = \sum_{l\in L(f,a,\Pi)} (E_{la}f)(x) \in Z(R^n). \tag{51}$$

But $Q_a f = g_a|_{\Pi}$, and therefore,

$$\|Q_a f\|_{Z(\Pi)} \leqslant \|g_a\|_{Z(R^n)} \leqslant \Big(\sum_{l \in L(f,a,\Pi)} \|E_{la}\|\Big)\|f\|_{Z(R^n)}. \tag{52}$$

Hence, $Q_a f \in Z^*(R^n, a)$. Setting $\Pi = T_{3a/2}$ (or $\Pi = T_a$) here, we arrive at (49) for $j = 1$ (respectively, $j = 2$). Under condition 3(a), for $\Pi = T_{3a/2}$

$$\varphi_a Q_a f \equiv \varphi_a g_a \in Z(R^n) \quad \text{and} \quad \|\varphi_a Q_a f\|_{Z(R^n)} \leqslant \|\mathscr{P}_{\varphi_a}\| \cdot \|g_a\|_{Z(R^n)}.$$

Using the second estimate in (52), we get (49) from this with $j = 3$.

REMARK 5. *If $a(f)$ is taken sufficiently large, then for all $a \geqslant a(f)$*

$$\|Q_a f\|^{(j)}_{Z^*(R^n,a)} \leqslant \|f\|_{Z(R^n)} \qquad (j = 1, 2, 3), \tag{53}$$

and for $j = 3$ there is even equality.

For $j = 1$ or 2 this follows at once from (49), since $L_1(f, a) = L_2(f, a) = \{0\}$ for sufficiently large $a \geqslant a(f)$. For $j = 3$ note that

$$f \equiv \varphi_a Q_a f, \qquad a \geqslant a(f), \tag{54}$$

and so

$$\|f\|_{Z(R^n)} = \|\varphi_a Q_a f\|_{Z(R^n)} = \|Q_a f\|^{(3)}_{Z^*(R^n,a)}.$$

CHAPTER II

THE CONNECTION BETWEEN THE NORMS OF OPERATORS IN THE PERIODIC AND NONPERIODIC CASES

§2.1. A general theorem on estimating the norms of operators

1°. The concept of a norming operator and a norming family of operators. Let Z and W be normed linear spaces with norms $\|\cdot\|_Z$ and $\|\cdot\|_W$, and let Z_a^* and W_a^* be some families of normed linear spaces with norms $\|\cdot\|_{Z_a^*}$ and $\|\cdot\|_{W_a^*}$ (for example, the spaces $Z(R^n)$ and $W(R^m)$ and their periodic analogues $Z^*(R^n, a)$ and $W^*(R^m, a)$ considered in Chapter 1, with one of the equivalent norms introduced there).

DEFINITION 1. An operator $\mathscr{P}$: $Z_a^* \to Z$ is said to be *norming* if the expression $\|\mathscr{P} f\|_Z$ is equivalent to the norm $\|f\|_{Z_a^*}$.

This means that

$$\|\mathscr{P}\| = \sup_{0 \neq f \in Z_a^*} \frac{\|\mathscr{P} f\|_Z}{\|f\|_{Z_a^*}} < \infty \tag{1}$$

and

$$\|f\|_{Z_a^*} \leqslant C(\mathscr{P})\|\mathscr{P} f\|_Z \quad \forall f \in Z_a^*, \tag{2}$$

where $C(\mathscr{P})$ does not depend on f. If a norming operator $\mathscr{P}$ is linear, then the expression $\|\mathscr{P} f\|_Z$ is an equivalent norm in Z_a^*, and condition (2) is equivalent to the existence and boundedness of the inverse operator $\mathscr{P}^{-1}$: $\operatorname{Im} \mathscr{P} \to Z_a^*$.

DEFINITION 2. A family $\{\mathscr{P}_a\}_{a>0}$ of norming operators is said to be *norming* if

$$\|\{\mathscr{P}_a\}\| = \overline{\lim_{a\to\infty}} \|\mathscr{P}_a\| < \infty, \qquad C(\{\mathscr{P}_a\}) = \overline{\lim_{a\to\infty}} C(\mathscr{P}_a) < \infty. \tag{3}$$

Denote by $\mathscr{V}$ and $\{\mathscr{V}_a\}$, respectively, a norming operator and a norming family of operators from W_a^* to W.

2°. A theorem on estimating norms. Assume all the conditions and notation in 1°. We consider operators

$$S: Z \to W, \qquad S_a^*: Z_a^* \to W_a^*. \tag{4}$$

Let $\|S\|$ and $\|S_a^*\|$ be their norms, and $S^{\mathscr{P}}$ the restriction of S to the range $\operatorname{Im}\mathscr{P} \subset Z$ of the norming operator $\mathscr{P}$. Let $\|S^{\mathscr{P}}\| \leqslant \|S\|$ denote the norm of this restriction when $\operatorname{Im}\mathscr{P}$ is provided with the norm $\|\cdot\|_Z$.

THEOREM 5. 1. *Let $\mathscr{P}$ and $\mathscr{V}$ be norming operators. For*

$$\|S^{\mathscr{P}}\| < \infty \Rightarrow \|S_a^*\| < \infty \tag{5}$$

or

$$\|S_a^*\| < \infty \Rightarrow \|S^{\mathscr{P}}\| < \infty, \tag{6}$$

it is necessary and sufficient that $\|U\| < \infty$, where

$$U = \mathscr{V} S_a^* - S\mathscr{P}, \tag{7}$$

which is an operator from Z_a^* to W. Moreover,

$$\|S_a^*\| \leqslant C(\mathscr{V})(\|U\| + \|\mathscr{P}\| \cdot \|S^{\mathscr{P}}\|), \tag{8}$$

$$\|S^{\mathscr{P}}\| \leqslant C(\mathscr{P})\big(\|U\| + \|\mathscr{V}\| \cdot \|S_a^*\|\big). \tag{9}$$

2. *Let $\{\mathscr{P}_a\}_{a>0}$ and $\{\mathscr{V}_a\}_{a>0}$ be norming families of operators. For*

$$\varlimsup_{a\to\infty} \|S^{\mathscr{P}_a}\| < \infty \Rightarrow \varlimsup_{a\to\infty} \|S_a^*\| < \infty \tag{10}$$

or

$$\varlimsup_{a\to\infty} \|S_a^*\| < \infty \Rightarrow \varlimsup_{a\to\infty} \|S^{\mathscr{P}_a}\| < \infty, \tag{11}$$

it is necessary and sufficient that $\varlimsup_{a\to\infty}\|U_a\| < \infty$, where

$$U_a = \mathscr{V}_a S_a^* - S\mathscr{P}_a. \tag{12}$$

Moreover,

$$\varlimsup_{a\to\infty} \|S_a^*\| \leqslant C(\{\mathscr{V}_a\})\Big(\varlimsup_{a\to\infty} \|U_a\| + \|\{\mathscr{P}_a\}\| \cdot \varlimsup_{a\to\infty} \|S^{\mathscr{P}_a}\|\Big), \tag{13}$$

$$\varlimsup_{a\to\infty} \|S^{\mathscr{P}_a}\| \leqslant C(\{\mathscr{P}_a\})\Big(\varlimsup_{a\to\infty} \|U_a\| + \|\{\mathscr{V}_a\}\| \cdot \varlimsup_{a\to\infty} \|S_a^*\|\Big). \tag{14}$$

COROLLARY. *If under the conditions of the second part of Theorem* 5

$$Z = \bigcup_{a \geqslant b} \operatorname{Im}\mathscr{P}_a \tag{15}$$

for any $b > 0$, then for

$$\|S\| < \infty \Rightarrow \varlimsup_{a\to\infty} \|S_a^*\| < \infty \tag{10$'$}$$

or

$$\varlimsup_{a\to\infty} \|S_a^*\| < \infty \Rightarrow \|S\| < \infty, \tag{11$'$}$$

it is necessary and sufficient that (12) *hold. Moreover,* $\lim_{a\to\infty}\|S^{\mathscr{P}_a}\|$ *can be replaced by* $\|S\|$ *in* (13) *and* (14).

PROOF. It suffices to show that

$$\|S\| = \varlimsup_{a\to\infty} \|S^{\mathscr{P}_a}\|. \tag{16}$$

Indeed, for any $b > 0$

$$\|S\| = \sup_{0\neq f\in Z} \frac{\|Sf\|_W}{\|f\|_Z} = \sup_{a\geqslant b}\ \sup_{0\neq f\in \operatorname{Im}\mathscr{P}_a} \frac{\|Sf\|_W}{\|f\|_Z} = \sup_{a\geqslant b} \|S^{\mathscr{P}_a}\|.$$

Passing to the limit as $b \to +\infty$, we obtain (16).

PROOF OF THE THEOREM. 1. We show that the condition $\|U\| < \infty$ implies (5). Suppose that there exist an operator $\mathscr{V}$: $\operatorname{Im} S_a^* \to W$ such that

$$\|f\|_{W_a^*} \leqslant C(\mathscr{V})\|\mathscr{V}f\|_W \quad \forall f\in \operatorname{Im} S_a^* \subset W_a^*$$

and an operator $\mathscr{P}$: $Z_a^* \to Z$ with $\|\mathscr{P}\| < \infty$ such that $\|U\| < \infty$. Then the relations (5) and (8) are valid. Indeed, for $f \in Z_a^*$

$$\mathscr{V}S_a^*f = Uf + S\mathscr{P}f \in W.$$

From this we get

$$\|S_a^*f\|_{W_a^*} \leqslant C(\mathscr{V})\|\mathscr{V}S_a^*f\|_W \leqslant C(\mathscr{V})(\|U\| + \|S^{\mathscr{P}}\|\cdot\|\mathscr{P}\|)\|f\|_{Z_a^*},$$

which gives us (8), hence also (5).

2. We show that the condition $\|U\| < \infty$ implies (6). Suppose that there exist an operator $\mathscr{V}$: $\operatorname{Im} S_a^* \to W$ such that

$$\|\mathscr{V}\| = \sup_{0\neq f\in \operatorname{Im} S_a^*} \left[\|\mathscr{V}f\|_W/\|f\|_{W_a^*}\right] < \infty$$

and an operator $\mathscr{P}$: $Z_a^* \to Z$ with $C(\mathscr{P}) < \infty$ such that $\|U\| < \infty$. Then (6) and (9) hold. Indeed, for $f \in Z^*(R^n, a)$

$$S\mathscr{P}f = \mathscr{V}S_a^*f - Uf \in W.$$

Let $g = \mathscr{P}f \in \operatorname{Im}\mathscr{P}$. Then

$$\left\|S_g^{\mathscr{P}}\right\|_W \leqslant (\|\mathscr{V}\|\cdot\|S_a^*\| + \|U\|)\|f\|_{Z_a^*}.$$

But by (2) this implies that for any $g \in \operatorname{Im}\mathscr{P}$

$$\left\|S_g^{\mathscr{P}}\right\|_W \leqslant C(\mathscr{P})(\|U\| + \|\mathscr{V}\|\cdot\|S_a^*\|)\|g\|_Z,$$

i.e., (9) is valid, and hence also (6).

3. Let us show that each of the relations (5) and (6) implies that $\|U\|$ is finite. It suffices to show that if the norms $\|S^{\mathscr{P}}\|$ and $\|S_a^*\|$ are both finite, then $\|U\| < \infty$. But this follows at once from (7) with the finiteness of $\|\mathscr{V}\|$ and $\|\mathscr{P}\|$ taken into account.

4. The second assertion of the theorem for norming families $\{\mathscr{P}_a\}_{a>0}$ and $\{\mathscr{V}_a\}_{a>0}$ of operators follows from the first assertion by passing to the limit as $a \to \infty$. The theorem is proved.

§2.2. Estimates of the norms of operators commuting with multipliers

1°. A theorem on estimating the norms of operators commuting with multipliers. Suppose that $Z \equiv Z(R^n)$ and $W \equiv W(R^m)$ are normed linear spaces, $Z_a^* \equiv Z^*(R^n, a)$ and $W_a^* = W^*(R^m, a)$ are a-periodic analogues of them, for example, in the sense of Chapter I, and $\Phi_a = \{\varphi(x)\}$ is some set of norming multipliers from $Z^*(R^n, a)$ to $Z(R^n)$, i.e., functions $\varphi(x)$ such that the corresponding operators $\mathscr{P}_\varphi$: $Z^*(R^n, a) \to Z(R^n)$ of multiplication by $\varphi(x)$ are norming in the sense of §2.1. Similarly, $\Psi_a = \{\psi(x)\}$ is a set of norming multipliers from $W^*(R^m, a)$ to $W(R^m)$. Consider the operators S and S_a^* in (4). For $\varphi \in \Phi_a$ and $\psi \in \Psi_a$ denote by $\|S_a^*\|_{\varphi\to\psi}$ the norm of the operator S_a^* when $\|\mathscr{P}_\varphi f\|_{Z(R^n)}$ and $\|\mathscr{P}_\psi f\|_{W(R^m)}$, respectively, are chosen as equivalent norms in the spaces $Z^*(R^n, a)$ and $W^*(R^m a)$.

THEOREM 6. 1. *If for $a > 0$ there exist $\varphi \in \Phi_a$ and $\psi \in \Psi_a$ such that*

$$\mathscr{P}_\psi S_a^* = S\mathscr{P}_\varphi \tag{17}$$

on $Z^*(R^n, a)$, then

$$\|S_a^*\|_{\varphi\to\psi} = \|S^{\mathscr{P}_\varphi}\| \leqslant \|S\|. \tag{18}$$

2. *If for any sufficiently large $a > 0$ there exist $\varphi = \varphi_a \in \Phi$ and $\psi = \psi_a \in \Psi_a$ such that $\{\mathscr{P}_{\varphi_a}\}$ and $\{\mathscr{P}_{\psi_a}\}$ are norming families of operators and the equality* (17) *holds, then*

$$\overline{\lim_{a\to\infty}} \|S_a^*\|_{\varphi_a\to\psi_a} = \overline{\lim_{a\to\infty}} \|S^{\mathscr{P}_{\varphi_a}}\| \leqslant \|S\|. \tag{19}$$

A proof of the theorem follows at once from Theorem 5. Indeed, in this case for $\mathscr{P} = \mathscr{P}_\varphi$ and for $\mathscr{V} = \mathscr{P}_\psi$ we have $\|\mathscr{P}\| = C(\mathscr{P}) = \|\mathscr{V}\| = C(\mathscr{V}) = 1$, and $U = 0$, so that (8) and (9) yield (18), and (13) and (14) yield (19).

2°. The interconnection of periodic and nonperiodic imbedding theorems. We consider here an example of the application of Theorem 6 when it is possible to establish a two-sided estimate for $\|S_a^*\|$ and $\|S\|$. Suppose that $n = m$, and that $Z(R^n)$ and $W(R^m)$ satisfy requirements 1(c) and 3(a) in Chapter I. We define their periodic analogues $Z^*(R^n, a)$ and $W^*(R^n, a)$ according to Chapter I, and choose as norms the standard norms $\|\cdot\|^{(3)}_{Z^*(R^n,a)}$ and $\|\cdot\|^{(3)}_{W^*(R^n,a)}$ in (I.46) determined by one and the same function $\varphi_a(x)$ in (I.45).

THEOREM 7. *For there to be an imbedding $Z(R^n) \hookrightarrow W(R^n)$ it is necessary, and if $Z(R^n)$ consists of compactly supported functions then also sufficient, that for any $a > 0$ there is an imbedding $Z^*(R^n, a) \hookrightarrow W^*(R^n, a)$, and $\overline{\lim}_{a\to\infty}\|S_a^*\|^{(3)} < \infty$. Moreover,*

$$\|S_a^*\|^{(3)} \leqslant \|S\|, \qquad \|S\| = \overline{\lim_{a\to\infty}} \|S_a^*\|^{(3)}, \tag{20}$$

where S and S_a^* are the corresponding imbedding operators, and $\|S_a^*\|^{(3)}$ is the norm of S_a^* for the indicated choice of norms in $Z^*(R^n, a)$ and $W^*(R^n, a)$.

PROOF. 1. Let us show that

$$(Z(R^n) \hookrightarrow W(R^m)) \Rightarrow \left(Z^*(R^n, a) \hookrightarrow W^*(R^n, a),\ \|S_a^*\|^{(3)} \leqslant \|S\|\right).$$

The inclusion $Z^*(R^n, a) \subset W^*(R^n, a)$ obviously follows from the inclusion $Z(R^n) \subset W(R^n)$ and the method of constructing the periodic spaces. Hence, the identity operator $S_a^* f = f$ is an imbedding operator S_a^*: $Z^*(R^n, a) \to W^*(R^n, a)$. Since $Sg = g$, it follows that $\mathscr{P}_{\varphi_a} S_a^* f = S\mathscr{P}_{\varphi_a} f$ for any $f \in Z^*(R^n, a)$. Therefore, $\|S_a^*\|^{(3)} = \|S_a^*\|_{\varphi_a\to\varphi_a} \leqslant \|S\|$ by Theorem 6.

2. We show that if $Z(R^n)$ consists of compactly supported functions, then

$$\left(Z^*(R^n, a) \hookrightarrow W^*(R^n, a),\ \overline{\lim_{a\to\infty}} \|S_a^*\|^{(3)} < \infty\right) \Rightarrow (Z(R^n) \hookrightarrow W(R^n)),$$

and

$$\|S\| = \overline{\lim_{a\to\infty}} \|S_a^*\|^{(3)}. \tag{21}$$

Since the functions $g \in Z(R^n)$ are compactly supported, $Q_a g \in Z^*(R^n, a)$ for $a > 0$, and for sufficiently large $a(g)$ we have $g = \varphi_a Q_a g$ for all $a \geqslant a(g)$ (see Lemma 5, Remark 5, and (I.54)). This gives us that

$$Z(R^n) = \bigcup_{a>b} \operatorname{Im} \mathscr{P}_{\varphi_a}^Z \quad \forall b > 0, \tag{22}$$

while $\bigcup_{a>b} \operatorname{Im} \mathscr{P}_{\varphi_a}^W \subset W(R^n)$. Here $\operatorname{Im} \mathscr{P}_{\varphi_a}^Z$ and $\operatorname{Im} \mathscr{P}_{\varphi_a}^W$ are the ranges of $Z^*(R^n, a)$ and $W^*(R^n, a)$ under the mapping $\mathscr{P}_{\varphi_a}$.

The inclusion $Z^*(R^n, a) \subset W^*(R^n, a)$ implies the inclusion

$$\operatorname{Im} \mathscr{P}_{\varphi_a}^Z \subset \operatorname{Im} \mathscr{P}_{\varphi_a}^W,$$

and hence also the inclusion $Z(R^n) \subset W(R^n)$. By (22) and the corollary to Theorem 5,

$$\|S\| = \overline{\lim_{a\to\infty}} \|S^{\mathscr{P}_{\varphi_a}}\|.$$

It remains to note that $\|S_a^*\|^{(3)} = \|S_a^*\|_{\varphi_a\to\varphi_a}$ and apply Theorem 6 (equality (19)). As a result we get (21) and the (topological) imbedding $Z(R^n) \hookrightarrow W(R^n)$.

REMARK 6. *If $W(R^n)$ is a Banach space, then the sufficiency part of Theorem 7 extends to the case where $Z(R^n)$ coincides with the closure in the norm $\|\cdot\|_{Z(R^n)}$ of the space $Z_0(R^n)$ of all compactly supported functions in $Z(R^n)$ (more generally, when $Z(R^n)$ is contained in the completion of this closure with respect to $W(R^n)$).*

Indeed, if condition 3(a) holds, then $Z(R^n)$ and $Z_0(R^n)$ generate the same space $\check{Z}(R^n)$ (since for any $f \in \check{Z}(R^n)$ and any $\Pi \subset R^n$ there exist functions $g \in Z_0(R^n)$ with $g = f$ on Π, for example, $g = \varphi f \in Z_0(R^n)$, where $\varphi \in C_0^\infty(R^n)$ and $\varphi = 1$ on Π; see also Remark 1), and hence they generate the same space $Z^*(R^n, a)$, with the same norm $\|\cdot\|^{(3)}_{Z^*(R^n,a)}$. Therefore, applying the sufficiency part of Theorem 7 with $Z_0(R^n)$ instead of $Z(R^n)$, we get the imbedding $Z_0(R^n) \hookrightarrow W(R^n)$, and

$$\|S_0\| = \overline{\lim_{a\to\infty}} \|S_a^*\|^{(3)},$$

where S_0 is the corresponding imbedding operator. Extending it by continuity to $Z(R^n) = \overline{Z_0(R^n)}$, we get (21). The further extension to the relative completion is obvious (see [19], Chapter 1).

§2.3. Estimates of the norms of local operators

1°. A local operator $\check{S}$ and its restrictions S and S_a^*. Let $Z(R^n)$ and $W(R^m)$ be normed linear spaces with norms $\|\cdot\|_{Z(R^n)}$ and $\|\cdot\|_{W(R^m)}$, and let $Z^*(R^n, a)$ and $W^*(R^m, a)$ be their periodic analogues defined according to §1.1. We consider an operator $\check{S}\colon \check{Z}(R^n) \to \check{W}(R^m)$.

DEFINITION 3. Let $\delta \geqslant 0$. We say that the operator $\check{S}$ is *local with parameter* δ (briefly, $\check{S} \in \mathfrak{A}_\delta$) if for any parallelepiped $\Pi \subset R^l$, $l = \max(m, n)$, and any $f_1, f_2 \in \check{Z}(R^n)$ the equality $f_1 = f_2$ on $\Pi \cap R^n$ implies the equality

$$\check{S}f_1 = \check{S}f_2 \quad \text{on } \Pi_{(\delta)} \cap R^m. \tag{23}$$

(Here $\Pi_{(\delta)} = \{x \in \Pi\colon \rho(x, \partial\Pi) > \delta\}$.)

Imbedding operators, trace operators, differential operators, convolution operators with compactly supported kernel, etc., are examples of local operators.

Below we consider local operators $\check{S}$ such that

$$S = \check{S}|_{Z(R^n)}: Z(R^n) \to W(R^m), \tag{24}$$

and for any $a > 0$

$$S_a^* = \check{S}|_{Z^*(R^n,a)}: Z^*(R^n, a) \to W^*(R^m, a). \tag{25}$$

The condition (24) holds, for example, if $Z(R^n)$ consists of compactly supported functions, while $W(R^m)$ contains all compactly supported functions in $\check{W}(R^m)$. This follows at once from the fact that a local operator carries compactly supported functions into compactly supported functions. The condition (25) holds if and only if $\check{S}$ carries a-periodic functions into a-periodic functions. For example, if $m \leqslant n$, then the property follows from the invariance of $\check{S}$ under translations in R^m, i.e., from the condition that $E_{y'}\check{S} = \check{S}E_{(y',0)}$ for any $y' \in R^m$, where $y = (y', 0) \in R^n$. Indeed, $\check{S}f = \check{S}E_{(al',0)}f = E_{al'}\check{S}f$ for an a-periodic function $f(x)$ and any $l' \in \mathbf{Z}^m$, and this means that the function $\check{S}f$ is a-periodic.

2°. Estimates of the norms of the S_a^* in terms of the norm of S. Introducing the various equivalent norms defined in Chapter I in the spaces $Z^*(R^n, a)$ and $W^*(R^m, a)$, we write

$$\|S_a^*\|^{(j)} = \sup_{0 \neq f \in Z^*(R^n,a)} \frac{\|S_a^* f\|^{(j)}_{W^*(R^m,a)}}{\|f\|^{(j)}_{Z^*(R^n,a)}}, \qquad j = 1, 2, 3. \tag{26}$$

THEOREM 8. *Suppose that* $a_0 > 0$, $\delta \geqslant 0$ *and* $\check{S} \in \mathfrak{A}_\delta$.

1. *If* $Z(R^n)$ *and* $W(R^m)$ *satisfy* 1(a) *and* 2(a), *then for* $a > 0$

$$\|S_a^*\|^{(j)} \leqslant c_j \|S\|, \qquad c_j = c_j(a, \delta), \tag{27}$$

for $j = 1$. *Moreover, if* $\delta = 0$, *then* $c_1 = 1$; *if* 1(d) *and* 2(b) *hold, then* $c_1(a, \delta) \leqslant \tilde{c}_1(a_0, \delta)$ *for all* $a \geqslant a_0$.

2. *If* 1(b) *and* 2(c) *hold, then* (27) *is valid for* $j = 2$ *and* $a > 0$. *Moreover, if* $\delta = 0$, *then* $c_2 = 1$; *if* 1(d) *and* 2(d) *hold, then* $c_2(a, \delta) \leqslant \tilde{c}_2(a_0, \delta)$ *for all* $a \geqslant a_0$.

3. *If* 1(c) *and* 3(a) *hold, then* (27) *is valid for* $j = 3$ *and* $a > 0$, *and if* 1(d) *and* 3(b) *hold, then* $c_3(a, \delta) \leqslant \tilde{c}_3(a_0, \delta)$ *for all* $a \geqslant a_0$.

PROOF. 1. Suppose that $\Pi^m \subset R^m$, $\rho(\Pi^m) > a$, $\Pi^l \subset R^l$, $l = \max(m, n)$ and $(\Pi^l_{(\delta)} \cap R^m) \supset \Pi^m$. We let $\Pi^n = \Pi^l \cap R^n$ (note that $\rho(\Pi^n) > a$) and introduce the norms $\|\cdot\|_{Z(\Pi^n)}$ and $\|\cdot\|_{W(\Pi^m)}$ in $Z^*(R^n, a)$ and $W^*(R^m, a)$. Let us show that

$$\|S_a^*\|_{\Pi^n \to \Pi^m} \leqslant \|S\|, \tag{28}$$

where the left-hand side is the corresponding norm of S_a^*. To do this we let $\mathbf{P}_a = \{\mathscr{P}\}$, where the $\mathscr{P}$: $Z^*(R^n, a) \to Z(R^n)$ are bounded extension operators from Π^n to R^n (which are nonlinear in general; see Chapter I). Then

$$C(\mathscr{P}) \leqslant 1 \quad \forall \mathscr{P} \in \mathbf{P}_a, \qquad \inf_{\mathscr{P} \in \mathbf{P}_a} \|\mathscr{P}\| = 1, \tag{29}$$

i.e., the $\mathscr{P} \in \mathbf{P}_a$ are norming operators in the sense of §2.1.

Let $\mathscr{V}$: $W^*(R^m, a) \to W(\Pi^m)$ be the operator of restriction to Π^m. Clearly, $\|\mathscr{V}\| = C(\mathscr{V}) = 1$. Finally, $\mathscr{V}_1$: $W(R^m) \to W(\Pi^m)$ is also a restriction operator, and $\|\mathscr{V}_\lambda\| \leqslant 1$. We let $Z \equiv Z(R^n)$, $W \equiv W(\Pi^m)$, $Z_a^* \equiv Z^*(R^n, a)$ and $W_a^* \equiv W^*(R^m, a)$, and we apply Theorem 5 to the operators $S_1 = \mathscr{V}_1 S\colon Z \to W$ and $S_a^*\colon Z_a^* \to W_a^*$. From the condition that $\mathscr{P}f = f|_{\Pi^n}$ for any $\mathscr{P} \in \mathbf{P}_a$ and $f \in Z_a^*$ it follows by the local property that $\check{S}\mathscr{P}f = \check{S}f|_{\Pi^m}$, i.e.,

$$U = S_1\mathscr{P} - \mathscr{V}S_a^* = 0. \tag{30}$$

But then, by (8) and the facts that $\|S_1^{\mathscr{P}}\| \leqslant \|S_1\| = \|\mathscr{V}_1 S\|$ and $C(\mathscr{V}) = 1$, we get that

$$\|S_a^*\|_{\Pi^n \to \Pi^m} \leqslant \|\mathscr{P}\| \cdot \|\mathscr{V}_1 S\| \leqslant \|\mathscr{P}\| \cdot \|S\|,$$

from which, taking the infimum over $\mathscr{P} \in \mathbf{P}_a$ and using (29), we get (28).

2. Here let $\Pi^m = T^m_{3a/2}$ and $\Pi^n = T^n_{3a/2+2\delta}$. Then (27) for $j = 1$ follows from (28) by Theorem 1. If $\delta = 0$, then (27) with $c_1 = 1$ follows at once from (28). But if 1(d) and 2(b) hold, then $c_1(a, \delta) \leqslant \tilde{c}_1(a_0, \delta)$ for $a \geqslant a_0$, by Theorem 2.

3. The second part of the theorem can be proved similarly, and the third part follows from the first part and Theorems 3 and 4.

3°. An estimate of the norm of S in terms of the limit inferior of the norms of the S_a^*. To obtain a converse assertion we subject the space $W(R^m)$ to an additional requirement. Let

$$\|f\|_{W_1(R^m)} = \sup_{\Pi^m \subset R^m} \|f\|_{W(\Pi^m)} \leqslant \|f\|_{W(R^m)} \tag{31}$$

and suppose that for some $d \geqslant 1$

$$\|f\|_{W(R^m)} \leqslant d\|f\|_{W_1(R^m)} \quad \forall f \in W(R^m). \tag{32}$$

This requirement means that the norms $\|\cdot\|_{W_1}$ and $\|\cdot\|_W$ are equivalent. As a rule, it holds for concrete spaces with $d = 1$ (for more details see Chapter IV), i.e.,

$$\|f\|_{W_1(R^m)} = \|f\|_{W(R^m)}.$$

THEOREM 9. *Suppose that $\delta \geqslant 0$, $\check{S} \in \mathfrak{A}_\delta$, and $Z(R^n)$ consists of compactly supported functions.*

1. *If $Z(R^n)$ and $W(R^m)$ satisfy* 1(a) *and* 2(a) *and $W(R^m)$ also satisfies* (32), *then*

$$\|S\| \leqslant C \varliminf_{a \to \infty} \|S_a^*\|^{(j)} \tag{33}$$

for $j = 1$ and $C = d$.

2. *If* 1(b), 2(c), *and* (32) *hold, then* (33) *holds for $j = 2$ and $C = d$.*
3. *If* 1(c) *and* 3(a) *hold, then* (33) *holds for $j = 3$ and $C = 1$.*

PROOF. We give a straightforward derivation of (33) by using the specific nature of local operators. Citing the general Theorem 5 would enable us only to obtain (33) with the limit superior on the right-hand side.

1. Suppose that $g \in Z(R^n)$, $\Pi^m \subset R^m$, $\Pi^n \subset R^n$, and $\Pi^n_{(\delta)} \cap R^m \supset \Pi^m$. Since g is compactly supported, $g = Q_a g$ on Π^n for all $a \geqslant a(g, \Pi^n)$, whence $Sg = S_a^* Q_a g$ on Π^m, because $\check{S}$ is a local operator. Now the condition that $T^m_{3a/2} \supset \Pi^m$ for $a \geqslant a_1(\Pi^m)$ and the monotonicity of the norms $\|\cdot\|_{W(\Pi)}$ give us that

$$\begin{aligned}\|Sg\|_{W(\Pi^m)} &= \|S_a^* Q_a g\|_{W(\Pi^m)} \leqslant \|S_a^* Q_a g\|_{W(T^m_{3a/2})} \\ &= \|S_a^* Q_a g\|^{(1)}_{W^*(R^m, a)} \leqslant \|S_a^*\|^{(1)} \cdot \|Q_a g\|^{(1)}_{Z^*(R^n, a)}.\end{aligned}$$

Using (I.53) and letting a go to ∞, we get from this that

$$\|Sg\|_{W(\Pi^m)} \leqslant \varliminf_{a\to\infty} \|S_a^*\|^{(1)} \cdot \|g\|_{Z(R^n)}.$$

Passing to the supremum over the parallelepipeds $\Pi^m \subset R^m$ and taking (32) into account, we obtain (33) for $j = 1$. The second part of the theorem can be proved in exactly the same way.

2. Let $g \in Z(R^n)$. By compactness of the support, $g = Q_a g$ on $T^n_{a+2\delta}$ for all $a \geqslant a(g, \delta)$, and hence, by the property of being local, $Sg = S_a^* Q_a g$ on T_a^m. Since the operator is local, Sg is also compactly supported, and $\operatorname{supp} Sg \subset T_a^m$ for $a \geqslant a_1(g, \delta)$. Therefore,

$$(Sg)(x') \equiv \varphi_a(x')(S_a^* Q_a g)(x'), \qquad a \geqslant a_1(g, \delta).$$

Here $\varphi_a(x')$ is an m-dimensional standard function (I.45), and $\varphi_a(x') = 1$ on T_a^m. Hence, for all $a \geqslant a_1(g, \delta)$

$$\|Sg\|_{W(R^m)} = \|S_a^* Q_a g\|^{(3)}_{W^*(R^m, a)} \leqslant \|S_a^*\|^{(3)} \|Q_a g\|^{(3)}_{Z^*(R^n, a)}.$$

Using (I.53), we now get

$$\|Sg\|_{W(R^m)} \leqslant \varliminf_{a\to\infty} \|S_a^*\|^{(3)} \|g\|_{Z(R^n)},$$

which is what was required.

REMARK 7. It follows from Theorems 8 and 9 that if $\delta = 0$ and $d = 1$ in (32), then

$$\lim_{a\to\infty} \|S_a^*\|^{(j)} = \sup_{a>0} \|S_a^*\|^{(j)} = \|S\|, \qquad j = 1, 2. \tag{34}$$

§2.4. Recovery of a local operator from its restrictions

Everywhere in this section we assume that the spaces $Z(R^n)$ and $W(R^m)$ satisfy 1(a) and 2(a) in every case, so that $Z^*(R^n, a)$ and $W^*(R^m, a)$ are normable spaces (see §1.1). In §2.3 we assumed that the local operator $\check{S}$: $\check{Z}(R^n) \to \check{W}(R^m)$ was already given, and we studied the interconnection between its restrictions to $Z(R^n)$ and to $Z^*(R^n, a)$. It often happens, however, that only the restriction $S = \check{S}|_{Z(R^n)}$ or $S_a^* = \check{S}|_{Z^*(R^n, a)}$, $a > 0$, is given, and it is necessary to construct $\check{S}$ before studying the interconnection between the restrictions.

1°. Extension of an operator S.

LEMMA 6. (a) *A local operator S: $Z(R^n) \to W(R^m)$ admits a unique extension to a local operator $\check{S}$: $\check{Z}(R^n) \to \check{W}(R^m)$. Moreover, the parameters of the local property for the operators S and $\check{S}$ coincide*: $\delta(\check{S}) = \delta(S)$.

(b) *The extended operator $\check{S}$ has the properties of*

1) *linearity, and*

2) *invariance under translations in R^m ($m \leqslant n$)*

if and only if S has the corresponding properties.

PROOF. 1. We consider $f \in \check{Z}(R^n)$ and for any $\Pi^m \subset R^m$ let

$$\check{S}f = Sg|_{\Pi^m}, \tag{35}$$

where $g \in Z(R^n)$, $g = f|_{\Pi^n}$, $\Pi^n = \Pi^l \cap R^n$, $l = \max(m, n)$, and $\Pi^l_{(\delta)} \cap R^m \supset \Pi^m$; $\delta = \delta(S)$ is the local property parameter of S. This gives a function $\check{S}f \in \check{W}(R^m)$

(since $Sg \in W(R^m)$). To show that the definition (35) is unambiguous we must verify that the values of $\check{S}f$ on Π^m are independent of the choice of Π^l and the extension g. This follows easily from the fact that S is a local operator. Moreover, if $f \in Z(R^n)$, then $\check{S}f = Sf$. We show that S is a local operator and $\delta(\check{S}) = \delta(S)$. Indeed, for any $\Pi^m \subset R^m$ we have that if $f_1 = f_2|_{\Pi^n}$, where $f_1, f_2 \in \check{Z}(R^n)$, $\Pi^n = \Pi^l \cap R^n$, and $\Pi^l_{(\delta)} \cap R^m \supset \Pi^m$, $\delta = \delta(S)$, then $\check{S}f_1 = \check{S}f_2 = Sg$ for $g \in Z(R^n)$, $g = f_1 = f_2|_{\Pi^n}$. Hence, $\check{S} \in \mathfrak{A}_\delta$, $\delta \leqslant \delta(S)$. But obviously, $\delta(\check{S}) \geqslant \delta(S)$, so $\delta(\check{S}) = \delta(S)$.

2. We prove the uniqueness of a local operator $\check{S}$ whose restriction to $Z(R^n)$ coincides with S.

Suppose that $\check{S}_1$: $\check{Z}(R^n) \to \check{W}(R^m)$ is another local operator with $\delta(\check{S}_1) \geqslant 0$ and $\check{S}_1|_{Z(R^n)} = S$. Let $\delta = \max(\delta(\check{S}), \delta(\check{S}_1))$. For any $f \in \check{Z}(R^n)$ and any $\Pi^m \subset R^m$ choose $\Pi^l \subset R^l$ such that $\Pi^l_{(\delta)} \cap R^m \supset \Pi^m$. Since $\check{S}$ and $\check{S}_1$ are local operators and $\check{S}|_{Z(R^n)} = \check{S}_1|_{Z(R^n)} = S$, we get for any function $g \in Z(R^n)$ with $g = f|_{\Pi^n}$, $\Pi^n = \Pi^l \cap R^n$, that $\check{S}f = \check{S}_1 f = Sg|_{\Pi^m}$. By the arbitrariness of $\Pi^m \subset R^m$ and f, this means that $\check{S} = \check{S}_1$.

3. We now prove that for $m \leqslant n$ the invariance of S under translations in R^m implies the analogous property for $\check{S}$ (the converse is obvious).

Let $y' \in R^m$, $y = (y', 0) \in R^n$ and $\Pi^m \subset R^m$. We consider $\Pi^n \subset R^n$ with $\Pi^n_{(\delta)} \cap R^m \supset \Pi^m$ and choose Π^n_1 such that $\Pi^n \subset \Pi^n_1$ and $\Pi^n + (y', 0) \subset \Pi^n_1$. If $\Pi^m_1 = (\Pi^n_1)_{(\delta)} \cap R^m$, then $\Pi^m \subset \Pi^m_1$ and $\Pi^m + y' \subset \Pi^m_1$. Suppose that $f \in \check{Z}(R^n)$ and $g \in Z(R^n)$, $g = f|_{\Pi^n_1}$. Then $\check{S}f = Sg|_{\Pi^m_1}$; in particular, $\check{S}f = Sg|_{\Pi^m + y'}$. But then $E_{y'}\check{S}f = E_{y'}Sg|_{\Pi^m}$, and so

$$E_{y'}\check{S}f = SE_{(y',0)}g|_{\Pi^m}. \tag{36}$$

Further, $g = f$ on Π^n_1; in particular, $g = f$ on $\Pi^n + (y', 0)$. Therefore, $E_{(y',0)}f = E_{(y',0)}g \in Z(R^n)$ on Π^n, which, by the definition (35) of $\check{S}$, implies that $\check{S}E_{(y',0)}f = SE_{(y',0)}g|_{\Pi^m}$. This and (36) give us that $\check{S}$ is invariant under translations in R^m.

4. Let us prove that $\check{S}$ is linear if S is linear (the converse is obvious). Indeed, if $f = \lambda_1 f_1 + \lambda_2 f_2$, $f_1, f_2 \in \check{Z}(R^n)$, then for any $\Pi^m \subset R^m$ and $\Pi^l \subset R^l$, $l = \max(m, n)$, with $\Pi^l_{(\delta)} \cap R^m \supset \Pi^m$, and any $g_1, g_2 \in Z(R^n)$ with $g_1 = f_1|_{\Pi^n}$ and $g_2 = f_2|_{\Pi^n}$, where $\Pi^n = \Pi^l \cap R^n$, we have that $g = \lambda_1 g_1 + \lambda_2 g_2 \in Z(R^n)$ and $g = f|_{\Pi^n}$, i.e., by (35),

$$\check{S}f = Sg = \lambda_1 Sg_1 + \lambda_2 Sg_2 = (\lambda_1 \check{S}f_1 + \lambda_2 \check{S}f_2)|_{\Pi^m}.$$

REMARK 8. If the constructed operator $\check{S}$ carries a-periodic functions into a-periodic functions (for example, $m \leqslant n$ and $\check{S}$ is invariant under translations in R^m), then, writing $S^*_a = \check{S}|_{Z^*(R^n, a)}, a > 0$, we can apply all the results in §2.3 to the operators S and S^*_a.

2°. Extension of the family $\{S^*_a\}_{a > a_0}$. Suppose that $a_0 > 0$ and that we are given a family of bounded operators

$$S^*_a \colon Z^*(R^n, a) \to W^*(R^m, a), \qquad a \geqslant a_0. \tag{37}$$

DEFINITION 4. Let $\delta \geqslant 0$. The family $\{S^*_a\}_{a \geqslant a_0}$ is *jointly local with parameter* δ if for any parallelepiped $\Pi \subset R^l$, $l = \max(m, n)$, and any $f_1 \in Z^*(R^n, a)$ and $f_2 \in Z^*(R^n, a')$, $a, a' \geqslant a_0$, the equality $f_1 = f_2$ on $\Pi \cap R^n$ implies the equality $S^*_a f_1 = S^*_{a'} f_2$ on $\Pi_{(\delta)} \cap R^m$.

The restrictions $S^*_a = \check{S}|_{Z^*(R^n, a)}$ of an operator $\check{S} \in \mathfrak{A}_\delta$ obviously have this property.

LEMMA 7. *Suppose that $Z(R^n)$ consists of compactly supported functions.*

(a) *A family $\{S_a^*\}_{a \geqslant a_0}$ of operators* (37) *that is jointly local with parameter $\delta \geqslant 0$ admits a unique extension to a local operator $\check{S}$: $\check{Z}(R^n) \to \check{W}(R^m)$. Moreover, $\check{S} \in \mathfrak{A}_\delta$.*

(b) *The extended operator $\check{S}$ has the properties of*

1) *linearity, and*

2) *invariance under translations in R^n*

if and only if the operators S_a^* have the corresponding properties.

PROOF. We confine ourselves to the determination of $\check{S}$ and to the proof that it is well defined. The proofs of all the properties are analogous to those given in 1°.

For $f \in \check{Z}(R^n)$ and any $\Pi^m \subset R^m$ let

$$\check{S}f = \lim_{a \to \infty} S_a^* Q_a g|_{\Pi^m} \quad \left(\text{limit in } \|\cdot\|_{W(\Pi^m)}\right). \tag{38}$$

Here $g \in Z(R^n)$, $g = f|_{\Pi^n}$ and $\Pi^n = \Pi^l \cap R^n$, where $\Pi^l \subset R^l$, $l = \max(m, n)$, is such that $\Pi^l_{(\delta)} \cap R^m \supset \Pi^m$. The operator Q_a was introduced in (I.26).

We first show that the limit on the right-hand side in (38) exists. Since $g \in Z(R^n)$ is compactly supported, we have $Q_a g = Q_{a'} g = g|_{\Pi^n}$ for any $a, a' \geqslant a(\Pi^n, g)$, and $Q_a g \in Z^*(R^n, a)$ and $Q_{a'} g \in Z^*(R^n, a')$ by Lemma 5. Hence, $S_a^* Q_a g = S_{a'}^* Q_{a'} g|_{\Pi^m}$, i.e., $S_a^* Q_a g|_{\Pi^m}$ does not depend on $a \geqslant a(\Pi^n, g)$; in particular, the indicated limit exists:

$$\check{S}f = S_a^* Q_a g|_{\Pi^m}, \qquad a \geqslant a(\Pi^n, g). \tag{39}$$

By using the property of being local it is easy to show that the limit does not depend on the choice of Π^l and the extension g. Note further that if $f \in Z^*(R^n, a)$, then $\check{S}f = S_a^* f$, i.e., $\check{S}$ is really an extension of the family $\{S_a^*\}$.

REMARK 9. If the restriction $S = \check{S}|_{Z(R^n)}$ of the local operator $\check{S}$ constructed above satisfies condition (24), i.e., $\operatorname{Im} S \subset W(R^m)$, then the results in §2.3 are applicable to the operators S and S_a^*.

For example, this is so if $W(R^m)$ satisfies 3(a). Indeed, since S is a local operator, the function $Sg \in \check{W}(R^m)$ is compactly supported whenever $g \in Z(R^n)$ is, and hence $Sg \in W(R^m)$ (see Remark 1 in §1.2).

CHAPTER III

TRACE AND EXTENSION OPERATORS

§3.1. Periodic and nonperiodic trace operators

Let $1 \leqslant m < n$ and $x = (x_1, \ldots, x_n) = (u, v) \in R^n$, where $u = (x_1, \ldots, x_m) \in R^m$ and $v = (x_{m+1}, \ldots, x_n) \in R^{n-m}$, and suppose that the spaces $Z(R^n)$ and $W(R^m)$ satisfy 1(c) and 3(a). The spaces $\check{Z}(R^n)$, $\overline{Z}(R^n)$, $Z^*(R^n, a)$, $\check{W}(R^m)$, $\overline{W}(R^m)$ and $W^*(R^m, a)$ are defined as in §1.1.

DEFINITION 5. A linear operator $\check{S}$: $\check{Z}(R^n) \to \check{W}(R^m)$ is called a *trace operator* if:

1) $\check{S} \in \mathfrak{A}_0$;

2) $\check{S}$ is invariant under translations in R^m; and

3) there exists an integer $L = L(\check{S}) \geqslant 0$ such that, for all $\psi(u) \in C_0^\infty(R^m)$ and $\mu(v) \in C_0^\infty(R^{n-m})$ for which $\mu(0) = 1$ and $(D^\alpha \mu)\ (0) = 0$ for any integer vector $\alpha = (\alpha_1, \ldots, \alpha_{n-m}) \geqslant 0$ with $0 < |\alpha| = \alpha_1 + \cdots + \alpha_{n-m} \leqslant L$,

$$\mathscr{P}_\psi \check{S} = \check{S} \mathscr{P}_\varphi, \qquad \varphi(x) = \psi(u)\mu(v). \tag{1}$$

For example, if $(\check{S}f)(u) = (D_v^\beta f)(u,0)$ for a sufficiently smooth function f, then

$$\psi(u)\big(D_v^\beta f\big)(u,0) = \big(D^\beta[\psi(u)\mu(v)f(u,v)]\big)\big|_{v=0}$$

for $L \geqslant |\beta|$ and the indicated $\psi(u)$ and $\mu(v)$; and this coincides with (1).

There are many different definitions of trace operators in concrete spaces (in the sense of L_p, in the sense of the theory of generalized functions, as the closure of an operator given initially on the smooth functions, and so on). In this article it is irrelevant just what definition is taken as a basis; the only important thing is that the requirements 1)–3) be satisfied.

LEMMA 8. (a) *If* $\check{S}$: $\check{Z}(R^n) \to \check{W}(R^m)$ *is a trace operator and* $S = \check{S}|_{Z(R^n)}$: $Z(R^n) \to W(R^m)$ *is its restriction, then* S *is a trace operator.*

(b) *If* $\check{S}$: $Z(R^n) \to W(R^m)$ *is a trace operator, then it admits a unique extension to a trace operator* $\check{S}$: $\check{Z}(R^n) \to \check{W}(R^m)$.

PROOF. The first assertion is obvious (note that $\mathscr{P}_\varphi f \in Z(R^n)$ for $f \in Z(R^n)$, by 3(a)). Next, by Lemma 6, there exists a unique extension $\check{S}$ of S that is linear, local ($\delta = 0$), and invariant under translations in R^m. We prove (1). Suppose that $f \in Z(R^n)$ and $\Pi^m \subset R^m$, and let $\Pi^n \subset R^n$ be such that $\Pi^n \cap R^m \supset \Pi^m$. Then for any $g \in Z(R^n)$ with $g = f$ on Π^n we have

$$\mathscr{P}_\psi \check{S}f = \mathscr{P}_\psi Sg = S\mathscr{P}_\varphi g = \check{S}\mathscr{P}_\varphi f,$$

because $\mathscr{P}_\varphi g \in Z(R^n)$ and $\mathscr{P}_\varphi g = \mathscr{P}_\varphi f$ on Π^n.

We have understood the trace property for S: $Z(R^n) \to W(R^n)$ just as in Definition 5.

The equality (1) does not make sense for the operators S_a^*: $Z^*(R^n, a) \to W^*(R^m, a)$, because $\mathscr{P}_\varphi f$ is a nonperiodic function. In this case we call S_a^* a *trace operator* if (1) is replaced in Definition 5 by the equality

$$\mathscr{P}_{\psi^*} S_a^* = S_{aP\varphi^*}^*, \qquad \varphi^*(x) = \psi^*(u)\mu^*(v), \tag{2}$$

where $\psi^* = Q_a^{(m)}\psi$, $\mu^* = Q_a^{(n-m)}\mu$ (see (I.26)), and $\operatorname{supp}\mu \subset T_a^{n-m}$ (under this assumption $\mu^*(0) = 1$ and $(D^\alpha\mu^*)(0) = 0$, $0 < |\alpha| \leqslant L$).

LEMMA 9. (a) *If* $\check{S}$: $\check{Z}(R^n) \to W(R^m)$ *is a trace operator, then for any* $a > 0$ *its restriction satisfies*

$$S_a^* = \check{S}|_{Z^*(R^n,a)}: Z^*(R^n, a) \to W^*(R^m, a)$$

and is a trace operator, and the family $\{S_a^*\}_{a>0}$ *is jointly local with* $\delta = 0$.

(b) *If* $Z(R^n)$ *consists of compactly supported functions and* $\{S_a^*\}_{a>0}$ *is a family of trace operators* S_a^*: $Z^*(R^n, a) \to W^*(R^m, a)$ *that is jointly local with* $\delta = 0$, *then it admits a unique extension to a trace operator* $\check{S}$: $\check{Z}(R^n) \to \check{W}(R^m)$.

PROOF. 1. In the first part of the lemma proof is required only for property (2), which can be rewritten in the form

$$\big(Q_a^{(m)}\mathscr{P}_\psi S_a^* f\big)(u) = \big(S_a^* Q_a^{(n)}\mathscr{P}_\varphi f\big)(u), \qquad f \in Z^*(R^n, a), \tag{3}$$

and, by periodicity, it suffices to prove this equality on T_a^m.

It follows from (1) that $\check{S}\mathscr{P}_\varphi f = \mathscr{P}_\psi \check{S}f = \mathscr{P}_\psi S_a^* f$ for $f \in Z^*(R^n, a)$ (according to Remark 1). Therefore,

$$Q_a^{(m)}\check{S}\mathscr{P}_\varphi f = Q_a^{(m)}\mathscr{P}_\psi S_a^* f. \tag{4}$$

2. We prove that for compactly supported functions $g \in Z(R^n)$

$$Q_a^{(m)}g \in \check{Z}(R^n) \quad \text{and} \quad Q_a^{(m)}\check{S}g = \check{S}Q_a^{(m)}g. \tag{5}$$

Let $\Pi = \Pi^m \times \Pi^{n-m} \subset R^n$ be an arbitrary parallelepiped, and let $\operatorname{supp} g \subset \Pi_0 = \Pi_0^m \times \Pi_0^{n-m}$. If $L = \{l \in \mathbf{Z}^m\colon (\Pi_0^m + al) \cap \Pi^m \neq \varnothing\}$ (L is a finite set), then $Q_a^{(m)}g = \sum_{l \in L} E_{(al,0)}g \in Z(\Pi)$ on Π, by 1(c). Consequently, $Q_a^{(m)}g \in \check{Z}(R^n)$. Further, since $\check{S} \in \mathfrak{A}_0$, on Π^m we have that

$$\check{S}Q_a^{(m)}g = \check{S}\Big(\sum_{l \in L} E_{(al,0)}g\Big) = \sum_{l \in L} E_{al}\check{S}g = \sum_{l \in \mathbf{Z}^m} E_{al}\check{S}g = Q_a^{(m)}\check{S}g.$$

The arbitrariness of $\Pi^m \subset R^m$ gives us (5).

3. By (4) and (5),

$$\check{S}Q_a^{(m)}\mathscr{P}_\varphi f = Q_a^{(m)}\mathscr{P}_\psi S_a^* f. \tag{6}$$

On $R^m \times T_a^{n-m}$ we have $\mathscr{P}_\varphi f = Q_a^{(n-m)}\mathscr{P}_\varphi f$ (since $\operatorname{supp}\mu \subset T_a^{n-m}$); consequently,

$$Q_a^{(m)}\mathscr{P}_\varphi f = Q_a^{(n)}\mathscr{P}_\varphi f \in Z^*(R^n, a)$$

on T_a^n, and

$$\check{S}Q_a^{(m)}\mathscr{P}_\varphi f = \check{S}Q_a^{(n)}\mathscr{P}_\varphi f = S_a^* Q_a^{(n)}\mathscr{P}_\varphi f$$

on T_a^m by the property of being local. This and (6) give us (3).

4. We prove the second part of the lemma. According to Lemma 7, there exists a unique extension $\check{S}$ of the family $\{S_a^*\}_{a>0}$ that is linear, local with $\delta = 0$, and invariant under translations in R^m. It remains to establish the equality (1) by starting from (3). For $f \in \check{Z}(R^n)$ and any $\Pi = \Pi^m \times \Pi^{n-m} \subset R^n$ we consider a compactly supported function $g \in Z(R^n)$ with $g = f$ on Π; then, by (2.39),

$$\mathscr{P}_\psi \check{S}f = \mathscr{P}_\psi S_a^* Q_a^{(n)} g \quad \text{on } \Pi^m, \qquad a \geqslant a(\Pi, g). \tag{7}$$

Note that $\mathscr{P}_\varphi g \in Z(R^n)$ is a compactly supported function and $\mathscr{P}_\varphi g = \mathscr{P}_\varphi f$ on Π; hence, according to (2.39),

$$\check{S}\mathscr{P}_\varphi f = S_a^* Q_a^{(n)}\mathscr{P}_\varphi g \quad \text{on } \Pi^m, \qquad a \geqslant a(\Pi, g). \tag{8}$$

But, writing $h = Q_a^{(n)}g \in Z^*(R^n, a)$, we get that $Q_a^{(n)}\mathscr{P}_\varphi g = Q_a^{(n)}\mathscr{P}_\varphi h$ on Π^n for all $a \geqslant a(\Pi, \varphi)$ because $\mathscr{P}_\varphi g$ and $\mathscr{P}_\varphi h$ are compactly supported, and, by the property of being local and (3), this gives us that

$$S_a^* Q_a^{(n)}\mathscr{P}_\varphi g = S_a^* Q_a^{(n)}\mathscr{P}_\varphi h = Q_a^{(m)}\mathscr{P}_\psi S_a^* h = Q_a^{(m)}\mathscr{P}_\psi S_a^* Q_a^{(n)} g$$

on Π^m for all $a \geqslant a(\Pi, \varphi)$.

It remains to observe that

$$Q_a^{(m)}\mathscr{P}_\psi S_a^* Q_a^{(n)} g = \mathscr{P}_\psi S_a^* Q_a^{(n)} g$$

on Π^m for $a \geqslant a(\Pi^m, \psi)$ because ψ is compactly supported; therefore, $\mathscr{P}_\psi \check{S}f = \check{S}\mathscr{P}_\varphi f$ on Π^m by (8) and (7), and this gives us (1) in view of the arbitrariness of $\Pi^m \subset R^m$.

Lemmas 8 and 9 enable us to apply all the results in §§2.4 and 2.3 to the trace operators S and S_a^*. The following theorem is a result.

THEOREM 10. *Suppose that* $1 \leqslant m < n$ *and that* $Z(R^n)$ *and* $W(R^m)$ *satisfy* 1(c) *and* 3(a).

(a) *For any* $a > 0$ *a trace operator*

$$S\colon Z(R^n) \to W(R^m) \tag{9}$$

uniquely determines a trace operator

$$S_a^*: Z^*(R^n, a) \to W^*(R^m, a); \tag{10}$$

moreover, the family $\{S_a^*\}_{a>0}$ *of operators is jointly local with* $\delta = 0$, *and*

$$\|S_a^*\|^{(3)} \leqslant \|S\|. \tag{11}$$

(b) *If* $Z(R^n)$ *consists of compactly supported functions, then under the condition of being jointly local with* $\delta = 0$ *a family* $\{S_a^*\}_{a>0}$ *of trace operators of the form* (10) *uniquely determines a trace operator of the form* (9), *and*

$$\lim_{a\to\infty} \|S_a^*\|^{(3)} = \sup_{a>0}\|S_a^*\|^{(3)} = \|S\|. \tag{12}$$

PROOF. 1. Part (a) of the theorem follows from Lemmas 8(b) and 9(a). To obtain (11) we must take into account that, according to property (1) of a trace, the operators S and S_a^* satisfy the equality $\mathscr{P}_{\varphi_a^{(m)}}S_a^* = S\mathscr{P}_{\varphi_a^{(n)}}$, where $\varphi_a^{(n)}(x)$ and $\varphi_a^{(m)}(x)$ are standard functions (I.45) (in R^n and R^m, respectively), and then apply Theorem 6, in which the estimate (II.18) coincides with (11).

2. To prove Theorem 10(b) we use Lemma 9(b) to construct a trace operator $\check{S}$, and let $S = \check{S}|_{Z(R^n)}$. Note that the functions in $Z(R^n)$ are compactly supported, and the operator S (since it is local) carries them into compactly supported functions in $\check{W}(R^m)$, i.e., into functions in $W(R^m)$ (see Remark 1). Hence, S is a trace operator of the form (9). The remaining assertions in (b) follow from Theorem 9 and inequality (11).

REMARK 10. If $W(R^m)$ is a Banach space and

$$\varlimsup_{a\to\infty} \|S_a^*\|^{(3)} < \infty,$$

then Theorem 10(b) extends to the case where $Z(R^n)$ is the closure of the set $Z_0(R^n)$ of compactly supported functions in $Z(R^n)$ with respect to the norm $\|\cdot\|_{Z(R^n)}$.

Indeed, in this case (according to the arguments in Remark 6) Theorem 10(b), applied with $Z_0(R^n)$ instead of $Z(R^n)$, determines a bounded trace operator

$$S_0 = \check{S}|_{Z_0(R^n)}: Z_0(R^n) \to W(R^m), \qquad \|S_0\| = \lim_{a\to\infty} \|S_a^*\|^{(3)}.$$

We now consider the trace operator

$$S = \check{S}|_{Z(R^n)}: Z(R^n) \to \check{W}(R^m)$$

and the operator

$$S_1: Z(R^n) \to W(R^m), \qquad \|S_1\| = \lim_{a\to\infty} \|S_a^*\|^{(3)},$$

obtained by extending S_0 by continuity. To prove that they coincide it suffices to verify that $S_1 \in \mathfrak{A}_0$ (two extensions of S_0 that are both local operators coincide; see the second part of the proof of Lemma 6). Thus, the required assertion follows from Lemma 10, which is also of independent interest.

LEMMA 10. *Suppose that* $W(R^m)$ *is a Banach space,* $Z(R^n)$ *coincides with the closure of* $Z_0(R^n)$ *in the norm* $\|\cdot\|_{Z(R^n)}$ *and satisfies* 3(a), $S_0 \in \mathfrak{A}_\delta$ *is a bounded linear operator from* $Z_0(R^n)$ *to* $W(R^m)$, *and* S_1 *is the extension by continuity to* $Z(R^n)$. *Then* $S_1 \in \mathfrak{A}_\delta$.

PROOF. Suppose that $\Pi^m \subset R^m$, $\Pi^l \subset R^l$, $l = \max(m, n)$, $\Pi^l_{(\delta)} \cap R^m \supset \Pi^m$ and $\Pi^n = \Pi^l \cap R^n$. We prove that if $f \in Z(R^n)$ and $f = 0|_{\Pi^n}$, then $S_1 f = 0$ on Π^m. For any $\varepsilon > 0$ consider $\Pi^l_{(\varepsilon)}$ and $\Pi^n_{(\varepsilon)} = \Pi^l_{(\varepsilon)} \cap R^n$, and choose $\varphi(x) \in C_0^\infty(R^n)$ such that $\varphi = 1|_{\Pi^n_{(\varepsilon)}}$ and $\operatorname{supp}\varphi \subset \Pi^n$. Next consider a sequence $\{f_\nu\}_1^\infty$ such that $f_\nu \in Z_0(R^n)$ and $\lim_{\nu\to\infty} f_\nu = f$ in $Z(R^n)$. Then $\lim_{\nu\to\infty} S_0 f_\nu = S_1 f$ in $W(R^m)$. We have that

$$f_\nu - \varphi f_\nu \in Z_0(R^n), \qquad \lim_{\nu\to\infty} (f_\nu - \varphi f_\nu) = f - \varphi f = f$$

in $Z(R^n)$, so that $\lim_{\nu\to\infty} S_0(f_\nu - \varphi f_\nu) = S_1 f$ in $W(R^m)$. But $f_\nu - \varphi f_\nu = 0|_{\Pi^n_{(\varepsilon)}}$, and hence, $S_0(f_\nu - \varphi f_\nu) = 0$ on $\Pi^m_{(\varepsilon)} \subset (\Pi^l_{(\varepsilon)})_{(\delta)} \cap R^m$, $\nu = 1, 2, \ldots$.

Thus,

$$\begin{aligned} \|S_1 f\|_{W(\Pi^m_{(\varepsilon)})} &= \|S_1 f - S_0(f_\nu - \varphi f_\nu)\|_{W(\Pi^m_{(\varepsilon)})} \\ &\leqslant \|S_1 f - S_0(f_\nu - \varphi f_\nu)\|_{W(R^m)} \to 0 \qquad (\nu \to \infty), \end{aligned}$$

i.e., $S_1 f = 0|_{\Pi^m_{(\varepsilon)}}$. Since $\varepsilon > 0$ is arbitrary, $S_1 f = 0|_{\Pi^m}$, and this proves the relation $S_1 \in \mathfrak{A}_\delta$.

§3.2. Periodic and nonperiodic extension operators

1°. The concept of an extension operator. Let $1 \leqslant m < n$ and $x = (x_1, \ldots, x_n) = (u, v) \in R^n$, where $u = (x_1, \ldots, x_m) \in R^m$ and $v = (x_{m+1}, \ldots, x_n)$; the spaces $Z(R^n)$ and $W_k(R^m)$, $k = 1, \ldots, N$, are assumed to satisfy 1(c) and 3(a); $\mathbf{W}(R^m) = \Pi_1^N W_k(R^m)$ is the direct product of the corresponding normed spaces, with

$$\|f\|_{\mathbf{W}(R^m)} = \max_{1\leqslant k\leqslant N} \|f_k\|_{W_k(R^m)}$$

for $f = \{f_1, \ldots, f_N\} \in \mathbf{W}(R^m)$. Similarly,

$$\check{\mathbf{W}}(R^m) = \prod_{k=1}^N \check{W}_k(R^m) \quad \text{and} \quad \mathbf{W}^*(R^m, a) = \prod_{k=1}^N W_k^*(R^m, a).$$

We consider N-tuples of trace operators $\check{\mathbf{S}} = \{\check{S}_k\}_1^N$, where the $\check{S}_k$: $\check{Z}(R^n) \to \check{W}_k(R^m)$, $k = 1, \ldots, N$, are trace operators in the sense of Definition 5, and $\mathbf{S}_a^* = \{S_{a,k}^*\}_1^N$, where the $S_{a,k}^*$: $Z^*(R^n, a) \to W_k^*(R^m, a)$ are periodic trace operators. In abbreviated notation,

$$\check{\mathbf{S}}\colon \check{Z}(R^n) \to \check{\mathbf{W}}(R^m); \qquad \mathbf{S}_a^*\colon Z^*(R^n, a) \to \mathbf{W}^*(R^m, a). \tag{13}$$

DEFINITION 6. An operator $\mathbf{T}$: $\mathbf{W}(R^m) \to \check{Z}(R^n)$ assigning a function $g = \mathbf{T}f \in \check{Z}(R^n)$ to an N-tuple of functions $f = \{f_k\}_1^N \in \mathbf{W}(R^m)$ is called an *extension operator corresponding to the trace operator* $\check{\mathbf{S}}$ *in* (13) *if* $\check{\mathbf{S}}\mathbf{T} = \mathbf{I}$ (*the identity operator in* $\mathbf{W}(R^m)$), *i.e.*,

$$\check{S}_k \mathbf{T} f = f_k \qquad (k = 1, \ldots, N) \quad \forall f = \{f_k\}_{k=1}^N \in \mathbf{W}(R^m). \tag{14}$$

If there exists an extension operator $\mathbf{T}$ for the trace operator $\check{\mathbf{S}}$, then obviously

$$\operatorname{Im}\check{\mathbf{S}} = \prod_{k=1}^N \operatorname{Im}\check{S}_k \supset \mathbf{W}(R^m).$$

We introduce

$$\|\mathbf{T}\| = \sup_{0\neq f\in\mathbf{W}(R^n)} \frac{\|\mathbf{T}f\|_{\check{Z}(R^n)}}{\|f\|_{\mathbf{W}(R^m)}}.$$

If $\|\mathbf{T}\| < \infty$, then $\mathbf{T}$: $\mathbf{W}(R^m) \to \bar{Z}(R^n)$ is a bounded operator. Similarly, the operator $\mathbf{T}_a^*$: $\mathbf{W}^*(R^m, a) \to Z^*(R^n, a)$ is a periodic extension operator corresponding to the trace operator $\mathbf{S}_a^*$ in (13) if $\mathbf{S}_a^*\mathbf{T}_a^* = \mathbf{I}_a^*$ (the identity operator in $\mathbf{W}^*(R^m, a)$), i.e.,

$$S_{a,k}^*\mathbf{T}_{af}^* = f_k, \qquad k = 1,\ldots,N, \quad \forall f = \{f_k\} \in \mathbf{W}^*(R^m, a). \tag{15}$$

If there exists an extension operator $\mathbf{T}_a^*$ for the operator $\mathbf{S}_a^*$, then $\operatorname{Im}\mathbf{S}_a^* = \mathbf{W}^*(R^m, a)$.

We remark that there is not so rigid a connection between extension operators in the periodic and nonperiodic cases as there is between trace operators (see Theorem 10), since infinitely many extension operators, including nonlocal ones, can correspond to a given trace operator.

2°. Construction of periodic extension operators.

THEOREM 11. *Suppose that* $1 \leqslant m < n$, $Z(R^n)$ *and* $W_k(R^m)$ $(k = 1,\ldots,N)$ *satisfy* 1(c) *and* 3(a), *and a trace operator* $\check{\mathbf{S}}$ *in* (13) *and an extension operator* $\mathbf{T}$ *in* (14) *are defined. Then for any* $a > 0$ *there exists a periodic extension operator* $\mathbf{T}_a^*$ *in* (15) *for the corresponding periodic trace operator* $\mathbf{S}_a^*$ *in* (13), *and*

$$\|\mathbf{T}_a^*\|^{(3)} \leqslant c(a)\|\mathbf{T}\|. \tag{16}$$

If 1(d) *and* 3(b) *hold, then for any* $a_0 > 0$ *the inequality* $c(a) \leqslant \bar{c}(a_0)$ *holds for all* $a \geqslant a_0$. *The operators* $\mathbf{T}_a^*$ *are also linear if* $\mathbf{T}$ *is linear. Here* $\|\mathbf{T}_a^*\|^{(3)}$ *is the norm of the extension operator when the norms* $\|\cdot\|^{(3)}$ *are introduced in* $Z^*(R^n, a)$ *and* $\mathbf{W}^*(R^m, a)$ (*i.e., in* $W_k^*(R^m, a)$, $k = 1,\ldots,N$).

PROOF. 1. Suppose that $a_0 > 0$ and $a \geqslant a_0$. Consider a function

$$\psi(u) \in C_0^\infty(T_{3a/2}^m), \qquad \psi(u) = 1 \quad \text{on } T_{a/2}^m, \qquad Q_a^{(m)}\psi(u) \equiv 1, \tag{17}$$

such that for $M \geqslant 0$

$$\|\psi\|_{C^M(R^n)} \leqslant d_1(M, m, a_0), \tag{18}$$

along with a function

$$\chi(u) \in C_0^\infty(T_{3a/2+2}^m), \qquad \chi(u) = 1 \quad \text{on } T_{3a/2}^m, \tag{19}$$

with $\|\chi\|_{C^M(R^m)} \leqslant d_2(M, m, a_0)$. Note that $\chi(u)\psi(u) \equiv \psi(u)$, and, therefore,

$$Q_a^{(m)}(\chi\psi) \equiv 1. \tag{20}$$

Such functions are not hard to construct (a construction of ψ can be obtained from Lemma 2). Consider further a function

$$\begin{gathered}\mu(v) \in C_0^\infty(T_a^{n-m}), \qquad \mu(v) = 1 \quad \text{on } T_{a/2}^{n-m},\\ \|\mu\|_{C^M(R^{n-m})} \leqslant d_3(M, n-m, a_0).\end{gathered} \tag{21}$$

Then for the function $\varphi(x) = \chi(u)\mu(v) \in C_0^\infty(R^n)$ we get that

$$\|\varphi\|_{C^M(R^n)} \leqslant d_4(M, n, m, a_0). \tag{22}$$

2. We now construct an extension operator $\mathbf{T}_a^*$. For $f = \{f_k\}_1^N \in \mathbf{W}^*(R^m, a)$ let $\mathbf{P}_\psi f = \{\mathscr{P}_\psi f_1,\ldots,\mathscr{P}_\psi f_N\}$. Then $\mathbf{P}_\psi f \in \mathbf{W}(R^m)$ is a vector of compactly supported functions. Further, $\mathscr{P}_\varepsilon g = \varphi g \in Z_0(R^n)$ for $g \in \check{Z}(R^n)$. We define

$$\mathbf{T}_a^* f = Q_a^{(n)}\mathscr{P}_\varphi\mathbf{T}\mathbf{P}_\psi f, \qquad f \in \mathbf{W}^*(R^m, a). \tag{23}$$

Obviously, $\mathbf{T}_a^* f \in Z^*(R^n, a)$, and if $\|\mathbf{T}\| < \infty$, then, by Lemma 5 and (I.23),

$$\|\mathbf{T}_a^* f\|^{(3)}_{Z^*(R^n,a)} \leqslant c\Big(\sum_{l\in L(\varphi,a)} \|E_{la}\|\Big)\|\mathscr{P}_\varphi \mathbf{T}\mathbf{P}_\psi f\|_{Z(R^n)}$$

$$\leqslant c\Big(\sum_{l\in L(\varphi,a)} \|E_{la}\|\Big)\|\mathscr{P}_\varphi\|\cdot\|\mathbf{T}\|\cdot\|\mathbf{P}_\psi\|\cdot\|f\|^{(3)}_{\mathbf{W}^*(R^m,a)}.$$

Here $L(\varphi, a) = \{l \in \mathbf{Z}^n \colon T^n_{3a/2} \cap \operatorname{supp} E_{la}\varphi \neq \varnothing\}$ is a finite set,

$$\|\mathbf{P}_\psi\| = \max_{l\leqslant k\leqslant N} \|\mathscr{P}_\psi\|_{W_k^*(R^m,a)\to W_k(R^m)},$$

and the norm $\|\cdot\|^{(3)}$ is introduced in $W_k^*(R^m, a)$. This estimate coincides with (16). If 1(d) and 3(b) hold, then by using the inequality $|L(\varphi, a)| \leqslant c_1$, (22), and then (I.48) (with the corresponding notation) and (18) to estimate $\|\mathbf{P}_\psi\|$, we get in (16) that $c(a) \leqslant \bar{c}(a_0)$ for all $a \geqslant a_0$.

3. Let us proceed to the proof that $\mathbf{T}_a^*$ is an extension operator, i.e., $\mathbf{S}_a^*\mathbf{T}_a^* = \mathbf{I}_a^*$. Note first that if $g \in Z_0(R^n)$ and $\operatorname{supp} g \subset R^m \times T_a^{n-m}$, then

$$S^*_{a,k} Q_a^{(n)} g = Q_a^{(m)} S_k g, \qquad k = 1, \dots, N. \tag{24}$$

By periodicity, it suffices to prove this equality on $\mathbf{T}_a^m$. We have that $Q_a^{(n)} g = Q_a^{(m)} g$ on T_a^n, and from this, since $S^*_{a,k}$ and its extension $\check{S}_k$ are local operators and (5) holds, we have that

$$S^*_{a,k} Q_a^{(n)} g = \check{S}_k Q_a^{(m)} g = Q_a^{(m)} S_k g$$

on T_a^m, which gives us (24). Then, by (23), (24), (1), (14), and (20),

$$S^*_{a,k}\mathbf{T}_a^* f = S^*_{a,k} Q_a^{(n)} \mathscr{P}_\varphi \mathbf{T}\mathbf{P}_\psi f = Q_a^{(m)} S_k \mathscr{P}_\varphi \mathbf{T}\mathbf{P}_\psi f$$

$$= Q_a^{(m)} \mathscr{P}_\chi \check{S}_k \mathbf{T}\mathbf{P}_\psi f = Q_a^{(m)} \mathscr{P}_\chi \mathscr{P}_\psi f_k = f_k,$$

which coincides with the definition (15) of an extension. The theorem is proved.

REMARK 11. *The operators* $\mathbf{T}_a^*$ *thus constructed converge to the operator* $\mathbf{T}$ *as* $a \to \infty$, *in the following sense: for any* $f = \{f_k\}_1^N \in \mathbf{W}_0(R^m)$ ($\mathbf{W}_0(R^m)$ *is the set of all compactly supported functions in* $\mathbf{W}(R^m)$) *and any* $\Pi \subset R^n$

$$\mathbf{T}_a^* \mathbf{Q}_a^{(m)} f = \mathbf{T} f \quad \text{on } \Pi \quad \forall a \geqslant a(\Pi, f). \tag{25}$$

PROOF. For $f \in \mathbf{W}_0(R^m)$ we get that $\mathbf{Q}_a^{(m)} f = \{Q_a^{(m)} f_k\}_{k=1}^N \in \mathbf{W}^*(R^m, a)$, so that is makes sense to write $\mathbf{T}_a^* \mathbf{Q}_a^{(m)} f \in Z^*(R^n, a)$. By (23),

$$\mathbf{T}_a^* \mathbf{Q}_a^{(m)} f = Q_a^{(n)} \mathscr{P}_\varphi \mathbf{T}\mathbf{P}_\psi \mathbf{Q}_a^{(m)} f.$$

For sufficiently large $a(f)$ and for all $a \geqslant a(f)$

$$f \equiv \mathbf{P}_\psi f \equiv \mathbf{P}_\psi \mathbf{Q}_a^{(m)} f.$$

The first equality follows from the fact that f is compactly supported and the condition that $\psi(u) = 1$ on $T^m_{a/2}$. The second takes into account also the condition that $\operatorname{supp}\psi \subset T^m_{3a/2}$. Hence,

$$\mathbf{T}_a^* \mathbf{Q}_a^{(m)} f = Q_a^{(n)} \mathscr{P}_\varphi \mathbf{T} f \quad \forall a \geqslant a(f).$$

But for any $\Pi \subset R^n$ and all $a \geqslant a(\Pi, f)$

$$\mathbf{T} f = \mathscr{P}_\varphi \mathbf{T} f = \mathbf{Q}_a^{(n)} \mathscr{P}_\varphi \mathbf{T} f \quad \text{on } \Pi.$$

The first equality follows from the condition that $\varphi(x) = 1$ on $T^n_{a/2}$ (see (17) and (21)), and the second takes into account also the condition that $\operatorname{supp}\varphi \subset T^n_{3a/2}$. Thus, (25) holds.

3°. Construction of a nonperiodic extension operator.

THEOREM 12. *Suppose that* $1 \leqslant m < n$, $Z(R^n)$ *and* $W_k(R^m)$ *satisfy* 1(c) *and* 3(a), *with* $W_k(R^m)$ *consisting of compactly supported functions for* $k = 1, \ldots, N$, *and* $\check{\mathbf{S}}$ *is a continuous trace operator* (13). *If the extension operators* $\mathbf{T}^*_a$ *in* (15) *are given, and the limit relation*

$$\lim_{a\to\infty} \mathbf{T}^*_a \mathbf{Q}^{(m)}_a f \quad \text{in } \check{Z}(R^n) \tag{26}$$

is valid for any $f = \{f_k\}_1^N \in \mathbf{W}(R^m)$, *then there exists an extension operator* $\mathbf{T}$: $\mathbf{W}(R^m) \to \check{Z}(R^n)$ *corresponding to the trace operator* $\check{\mathbf{S}}$, *and*

$$\|\mathbf{T}\| \leqslant \varlimsup_{a\to\infty} \|\mathbf{T}^*_a\|^{(3)}. \tag{27}$$

If the operators $\mathbf{T}^*_a$ *are linear* (*invariant under translations in* R^m), *then* $\mathbf{T}$ *is also linear* (*invariant under translations in* R^m).

PROOF. 1. Let $\mathbf{T}$: $\mathbf{W}(R^m) \to \check{Z}(R^n)$ be the operator acting according to the rule

$$\mathbf{T}f = \lim_{a\to\infty} \mathbf{T}^*_a \mathbf{Q}^{(m)}_a f \quad \text{in } \check{Z}(R^n). \tag{28}$$

We show that $\mathbf{T}$ is an extension operator. Indeed, since the operator $\check{\mathbf{S}}$ in (13) is continuous,

$$\check{S}_k \mathbf{T} f = \lim_{a\to\infty} \check{\mathbf{S}}_k \mathbf{T}^*_a \mathbf{Q}^{(m)}_a f = \lim_{a\to\infty} Q^{(m)}_a f_k = f_k \quad \text{in } \check{W}_k(R^m).$$

2. Let us obtain (27). To do this, note that, along with (28), we have

$$\mathbf{T}f = \lim_{a\to\infty} \varphi_a \mathbf{T}^*_a \mathbf{Q}^{(m)}_a f \quad \text{in } \check{Z}(R^n), \tag{29}$$

where $\varphi_a(x)$ is a function as in (I.45). But

$$\begin{aligned}\|\varphi_a \mathbf{T}^*_a \mathbf{Q}^{(m)}_a f\|_{Z(R^n)} &= \|\mathbf{T}^*_a \mathbf{Q}^{(m)}_a f\|^{(3)}_{Z^*(R^n,a)} \\ &\leqslant \|\mathbf{T}^*_a\|^{(3)} \|\mathbf{Q}^{(m)}_a f\|^{(3)}_{\mathbf{W}^*(R^m,a)} \leqslant \|\mathbf{T}^*_a\|^{(3)} \|f\|_{\mathbf{W}(R^n)}\end{aligned}$$

for sufficiently large $a \geqslant a(f)$, by (1.53). Thus,

$$\varlimsup_{a\to\infty} \|\varphi_a \mathbf{T}^*_a \mathbf{Q}^{(m)}_a f\|_{Z(R^n)} \leqslant \Big(\varlimsup_{a\to\infty} \|\mathbf{T}^*_a\|^{(3)}\Big) \|f\|_{\mathbf{W}(R^m)}.$$

From this and (29) and the definition of the relative completion in §1.1,

$$\|\mathbf{T}f\|_{\check{Z}(R^n)} \leqslant \Big(\varlimsup_{a\to\infty} \|\mathbf{T}^*_a\|^{(3)}\Big) \|f\|_{\mathbf{W}(R^m)},$$

which gives us (27).

3. The linearity of $\mathbf{T}$ (when the $\mathbf{T}^*_a$, $a > 0$, are linear) is obvious, and invariance under translations in R^m follows from (28) and the continuity of the translation operator E_y in $\check{Z}(R^n)$, which follows from 1(c).

REMARK 12. The existence of extension operators $\mathbf{T}^*_a$ with the property (26) is necessary for the existence of the extension operator $\mathbf{T}$: $\mathbf{W}(R^m) \to \check{Z}(R^n)$, as follows from Remark 11. Moreover, the condition $\overline{\lim}_{a\to\infty}\|\mathbf{T}_a^*\|^{(3)} < \infty$ (under the requirements 1(d) and 3(b) is necessary for the norm of $\mathbf{T}$ to be finite), according to Theorem 11. In this case $\mathbf{T}$ acts from $\mathbf{W}(R^m)$ to $\overline{Z}(R^n)$.

CHAPTER IV

REMARKS. SUPPLEMENTS. EXAMPLES

§4.1. Remarks of a general nature

1°. The "intrinsically" defined norm. In studying concrete function spaces it is usually more convenient to deal with the "intrinsically" defined norm $\|\cdot\|^{+}_{Z(\Pi)}$ (without the infimum over extensions), which is equivalent to the norm $\|\cdot\|_{Z(\Pi)}$ defined by (1.1). These norms have been thoroughly studied in many cases.

Note that $\|\cdot\|^{+}_{Z(\Pi)} \sim \|\cdot\|_{Z(\Pi)}$ if and only if:

1) the restriction operator S_Π: $Z(R^n) \to Z^+(\Pi)$ is bounded ($f \in Z^+(\Pi)$ if $\|f\|^{+}_{Z(\Pi)} < \infty$), and

2) there exists a bounded extension operator T_Π: $Z^+(\Pi) \to Z(R^n)$.

The necessity of these conditions is obvious, since they hold by definition for $Z(\Pi)$. Let us prove their sufficiency. For $f \in Z(\Pi)$ there exists a function $g \in Z(R^n)$ such that $f = S_\Pi g$ and $\|g\|_{Z(R^n)} \leqslant 2\|f\|_{Z(\Pi)}$. Then

$$\|f\|^{+}_{Z(\Pi)} = \|S_\Pi g\|^{+}_{Z(\Pi)} \leqslant 2\|S_\Pi\|\|f\|_{Z(\Pi)}.$$

On the other hand,

$$\|f\|_{Z(\Pi)} \leqslant \|T_\Pi f\|_{Z(R^n)} \leqslant \|T_\Pi\|\|f\|^{+}_{Z(\Pi)}.$$

2°. On quasinormed spaces. All the results in this paper are valid for the case of quasinormed spaces, in which the triangle inequality has the form $\|f+g\| \leqslant d(\|f\| + \|g\|)$, $d > 1$. All the proofs use the triangle inequality for finite sums and carry over to this case without changes (only the constants may change). The exception is Lemma 4 and Theorem 4 (which is based on it), where the triangle inequality is used for the sum of a series, and it does not hold in the case of quasinormed spaces. Instead of the triangle inequality we can use the estimate $\|\sum_l f_l\| \leqslant C(\sum_l \|f_l\|^\rho)^{1/\rho}$, where $(2d)^\rho = 2$, which is valid for quasinormed spaces (see [20], 3.10). As a result we get the validity of the same statements, except that the norm in $C^M(R^n, a)$ must be introduced differently:

$$\|\varphi\|_{C^M(R^n,a)} = \left(\sum_{l\in\mathbf{Z}^n} \|\varphi\|^\rho_{C^M(T_{a+1}+al)}\right)^{1/\rho}.$$

This quasinorm is finite for the functions $\varphi \in S(R^n)$, for any $\rho > 0$.

3°. On seminormed spaces. What is presented in Chapters I–III can be applied also to the case where $Z(R^n)$ and hence $W(R^m)$ are seminormed spaces. Let $N(R^n) = \{f \in Z(R^n)\colon \|f\|_{Z(R^n)} = 0\}$. For seminormed spaces a function f is regarded as compactly supported if there exists a function $g \in Z(R^n)$ with compact support such that $f - g \in N(R^n)$.

A typical situation for seminormed spaces is the case when the operator $\mathscr{P}_\varphi$ acts from $Z(R^n)$ to $Z(R^n)$ for $\varphi \in C_0^\infty(R^n)$ but is not bounded. Therefore, in their periodic analogues it makes sense to consider only the seminorms $\|\cdot\|^{(j)}_{Z^*(R^n,a)}$, $j = 1, 2$ (see Example 2° in §4.4 for more details).

4°. On the norm $\|\cdot\|_{Z_1(R^n)}$. For $f(x) \in Z(R^n)$ consider the norm

$$\|f\|_{Z_1(R^n)} = \sup_{\Pi \subset (R^n)} \|f\|_{Z(\Pi)}.$$

Obviously, $\|f\|_{Z_1(R^n)} \leqslant \|f\|_{Z(R^n)}$. On the other hand, note that in the definition of $\|\cdot\|_{Z(\Pi)}$ in (1.1) the two norms $\|\cdot\|_{Z(R^n)}$ and $\|\cdot\|_{Z_1(R^n)}$ give the same result, namely,

$$\|f\|_{Z(\pi)} = \inf_{\substack{g \in Z(R^n) \\ g = f|_\Pi}} \|g\|_{Z_1(R^n)} \quad \forall \Pi \subset R^n. \tag{1}$$

Indeed, the right-hand side of (1) is no larger than $\|f\|_{Z(\Pi)}$, and on the other hand, $\|f\|_{Z(\Pi)} = \|g\|_{Z(\Pi)} \leqslant \|g\|_{Z_1(R^n)}$ for any $g \in Z(R^n)$ with $g = f$ on Π; taking the infimum over such g, we get from this that $\|f\|_{Z(\Pi)}$ is no larger than the right-hand side of (1).

Let $\overline{Z}_1(R^n)$ denote the completion of $Z(R^n)$ with the norm $\|\cdot\|_{Z_1(R^n)}$ relative to $\check{Z}(R^n)$. Then

$$\|f\|_{\overline{Z}_1(R^n)} = \|f\|_{Z_1(R^n)}, \qquad f \in \overline{Z}_1(R^n).$$

The inequality $\|f\|_{\overline{Z}_1(R^n)} \leqslant \|f\|_{Z_1(R^n)}$ (the converse is obvious) can be proved by the same argument as used in the proof of (I.3). Thus,

$$\overline{Z}_1(R^n) = \{f \in \check{Z}(R^n) \colon \|f\|_{Z_1(R^n)} < \infty\}.$$

All the definitions, assertions, and proofs in this article carry over word-for-word to the case where the norms $\|\cdot\|_{Z_1(R^n)}$ and $\|\cdot\|_{W_1(R^m)}$ are used in them everywhere in place of the norms $\|\cdot\|_{Z(R^n)}$ and $\|\cdot\|_{W(R^m)}$ (this does not apply to §2.1 nor to 1° in §2.2, where a more general situation is considered). Moreover, in Theorem 9 and Remark 7 the necessity of the requirement (2.32) drops out, and (2.33) is valid with the constant $c = 1$. In the main text we used the more simply constructed norm $\|\cdot\|_{Z(R^n)}$. However, as a rule, these two norms are equivalent or even coincide for concrete spaces (see condition (2.32)).

We say that the space $Z(R^n)$ satisfies requirement 3(c) *if in* 3(b)

$$\|\mathscr{P}_\varphi\| \leqslant \|\varphi\|_{C(R^n)} + B_1 \sum_{0<|\alpha|\leqslant M} \|D^\alpha \varphi\|_{C(R^n)}. \tag{2}$$

LEMMA 11. *Suppose that the compactly supported functions are dense in $Z(R^n)$. If* 3(b) *holds, then*

$$\|f\|_{Z(R^n)} \leqslant B\|f\|_{Z_1(R^n)}. \tag{3}$$

If 3(c) *holds, then*

$$\|f\|_{Z(R^n)} = \|f\|_{Z_1(R^n)}. \tag{4}$$

PROOF. Suppose that $f \in Z(R^n)$ is a compactly supported function, $\Pi \subset R^n$, and $\operatorname{supp} f \subset \Pi$. Consider an arbitrary function $g \in Z(R^n)$ such that $g = f$ on Π, and construct a function $\varphi_\Pi(x) \in C_0^M(\Pi)$ such that $0 \leqslant \varphi_\Pi(x) \leqslant 1$ and $\varphi_\Pi = 1$ on $\operatorname{supp} f$. Then $f \equiv \varphi_\Pi g$ and $\|f\|_{Z(R^n)} \leqslant \|\mathscr{P}_{\varphi_\Pi}\| \cdot \|g\|_{Z(R^n)}$. Taking the infimum over g, we get that

$$\|f\|_{Z(R^n)} \leqslant \|\mathscr{P}_{\varphi_\Pi}\| \cdot \|f\|_{Z(\Pi)} \leqslant \|\mathscr{P}_{\varphi_\Pi}\| \cdot \|f\|_{Z_1(R^n)}.$$

If we now expand Π to R^n, then we can ensure that

$$\sum_{0<|\alpha|\leqslant M} \|D^\alpha \varphi_\Pi\|_{C(R^n)} \to 0,$$

while $\|\varphi_\Pi\|_{C^M(R^n)} \to 1$. By 3(b), this gives us (3), and (4) follows from 3(c) and the fact that $\|f\|_{Z_1(R^n)} \leqslant \|f\|_{Z(R^n)}$. The compactly supported functions are dense in $Z(R^n)$ in the norm $\|\cdot\|_{Z(R^n)}$, hence also in the norm $\|\cdot\|_{Z_1(R^n)}$. Therefore, (3) and (4) are valid for any $f \in Z(R^n)$.

We indicate one more property (useful below) of the norm $\|\cdot\|_{Z_1(R^n)}$.

LEMMA 12. *If* $Z(R^n)$ *satisfies* 3(c), *then*

$$\lim_{s\to\infty} \|f(x)\varphi(x/s)\|_{Z(R^n)} = \|f\|_{Z_1(R^n)}, \tag{5}$$

where $\varphi(x) \in C_0^\infty(R^n)$, $0 \leqslant \varphi(x) \leqslant 1$, $\varphi(x) = 1$ *for* $x \in T_1$, *and* $\varphi(x) = 0$ *for* $x \in T_2$.

PROOF. Given any $\Pi \subset R^n$, $f(x) = f(x)\varphi(x/s)$ on Π when $s \geqslant s(\Pi)$; therefore, $\|f\|_{Z(\Pi)} \leqslant \|f(x)\varphi(x/s)\|_{Z(R^n)}$, whence

$$\|f\|_{Z(\Pi)} \leqslant \varliminf_{s\to\infty} \|f(x)\varphi(x/s)\|_{Z(R^n)}.$$

Taking the supremum over $\Pi \subset R^n$, we get that

$$\|f\|_{Z_1(R^n)} \leqslant \varliminf_{s\to\infty} \|f(x)\varphi(x/s)\|_{Z(R^n)}.$$

On the other hand, for any s and any $g(x) \in Z(R^n)$ with $g = f$ on T_{2s} we have that $f(x)\varphi(x/s) \equiv g(x)\varphi(x/s)$, and so, by 3(c),

$$\left\|f(x)\varphi\left(\frac{x}{s}\right)\right\|_{Z(R^n)} \leqslant \left(\left\|\varphi\left(\frac{x}{s}\right)\right\|_{C(R^n)} + B_1 \sum_{0<|\alpha|\leqslant M} \left\|D^\alpha\left[\varphi\left(\frac{x}{s}\right)\right]\right\|_{C(R^n)}\right)\|g\|_{Z(R^n)}. \tag{6}$$

When the infimum over such g is taken, this gives us (6) with $\|f\|_{Z(T_{2s})} \leqslant \|f\|_{Z_1(R^n)}$ instead of $\|g\|_{Z(R^n)}$. Passing here to the limit as $s \to \infty$, we see that

$$\varlimsup_{s\to\infty} \|f(x)\varphi(x/s)\|_{Z(R^n)} \leqslant \|f\|_{Z_1(R^n)},$$

which implies (5).

5°. On the periodic spaces generated by compactly supported functions and those generated by functions not compactly supported. Suppose that $Z(R^n)$ is not a space consisting of compactly supported functions. Then consider along with $Z(R^n)$ the space $Z_0(R^n)$ consisting of the compactly supported functions belonging to $Z(R^n)$, with the same norm. Let $Z^*(R^n, a)$ and $Z_0^*(R^n, a)$ be the spaces of periodic functions generated by them (see §1.1).

If 3(a) *holds, then* $Z_0^*(R^n, a) = Z^*(R^n, a)$ (*see* Remark 6), *and the norms* $\|\cdot\|^{(3)}_{Z_0^*(R^n,a)}$ *and* $\|\cdot\|^{(3)}_{Z^*(R^n,a)}$ *coincide.*

If 3(c) *holds, then for any* $\Pi \subset R^n$

$$\|\cdot\|_{Z(\Pi)} = \|\cdot\|_{Z_0(\Pi)}, \tag{7}$$

where these norms are generated by the spaces $Z(R^n)$ *and* $Z_0(R^n)$.

Indeed, it is obvious that $\|f\|_{Z(\Pi)} \leqslant \|f\|_{Z_0(\Pi)}$. On the other hand, according to (1) and Lemma 12, for any $\varepsilon > 0$ there exist a $g \in Z(R^n)$ with $g = f$ on Π and an s such that

$$\|f\|_{Z(\Pi)} \geqslant \|g\|_{Z_1(R^n)} - \varepsilon \geqslant \|g(x)\varphi(x/s)\|_{Z(R^n)} - 2\varepsilon \geqslant \|f\|_{Z_0(\Pi)} - 2\varepsilon.$$

Consequently, $\|f\|_{Z(\Pi)} \geqslant \|f\|_{Z_0(\Pi)}$.

Thus, if 3(c) holds, then the norms $\|\cdot\|^{(j)}_{Z_0^*(R^n,a)}$ and $\|\cdot\|^{(j)}_{Z^*(R^n,a)}$ coincide ($j = 1, 2$).

It follows from the theorems in Chapters II and III that the most complete results on the connection between the norms of periodic and nonperiodic operators are obtained when the spaces $Z_0(R^n)$ and $Z_0^*(R^n, a)$ are considered.

We now assume that $Z^*(R^n, a) = Z_0^*(R^n, a)$, and that

$$\|f\|^{(j)}_{Z^*(R^n,a)} \sim \|f\|^{(j)}_{Z_0^*(R^n,a)}$$

for some j. Along with the norm

$$\|S\| = \sup_{0\neq j\in Z(R^n)} \left(\|Sf\|_{W(R^m)}/\|f\|_{Z(R^n)}\right)$$

consider the norm

$$\|S\|^{(0)} = \sup_{0\neq j\in Z_0(R^n)} \left(\|Sf\|_{W(R^m)}/\|f\|_{Z(R^n)}\right) \leqslant \|S\|.$$

Estimates of the norms $\|S_a^*\|^{(j)}$ in terms of $\|S\|^{(0)}$ (and, a fortiori, in terms of $\|S\|$) were essentially established in Chapter II, along with, conversely, estimates of $\|S\|^{(0)}$ in terms of $\|S_a^*\|^{(j)}$. The question of getting estimates of $\|S\|$ in terms of $\|S_a^*\|^{(j)}$ is thus determined by the possibility of estimating $\|S\|$ in terms of $\|S\|^{(0)}$. The same applies to the case of the extension operators studied in Chapter III.

We remark that if the functions in $Z_0(R^n)$ are dense in $Z(R^n)$ and the operator S (which is nonlinear in general) is continuous, then $\|S\| = \|S\|^{(0)}$. If there is a sequence of functions $\psi_s \in C_0^\infty(R^n)$ such that for any $f \in Z(R^n)$

$$\lim_{s\to\infty} \|f\psi_s\|_{Z(R^n)} \sim \|f\|_{Z(R^n)}, \qquad \lim_{s\to\infty} \|S(f\psi_s)\|_{W(R^m)} \sim \|Sf\|_{W(R^m)},$$

then $\|S\| \leqslant c\|S\|^{(0)}$ (if the equality sign can be written here in place of the equivalence sign, then $\|S\| = \|S\|^{(0)}$).

§4.2. Comments

1°. On the space $\tilde{Z}(R^n)$ (Comment to §1.1). In considering operators that are not local it can be expedient to modify the scheme in Chapter I by replacing $\check{Z}(R^n)$ by a narrower space $\tilde{Z}(R^n)$ which contains $Z(R^n)$ and $Z^*(R^n, a)$. We assume that $Z(R^n)$ satisfies 1(c) and use the notation

$$\tilde{Z}(R^n) = \left\{ f \in \check{Z}(R^n) \colon \|f\|_{\tilde{Z}(\Pi)} = \sup_{y\in R^n} \|E_y f\|_{Z(\Pi)} < \infty \ \forall \Pi \subset R^n \right\}.$$

Then $Z(R^n) \hookrightarrow \tilde{Z}(R^n)$ and $Z^*(R^n, a) \hookrightarrow \tilde{Z}(R^n)$, $a > 0$. Under the conditions in Theorems 1 and 2 the norms $\|\cdot\|_{\tilde{Z}(\Pi)}$ are also equivalent norms in the spaces $Z^*(R^n, a)$.

2°. On periodic partitions of unity (Comment to Theorem 4). In Theorem 3 it is noted that the condition that $\varphi(x) > 0$ on T_a can be replaced by the requirement that $Q_a\varphi(x) \equiv 1$, i.e., $\varphi(x)$ is a periodic partition of unity. An analogous remark is true for Theorem 4.

Let $\tilde{\Phi}_a = \tilde{\Phi}_a(n, a_0, M)$ be the set of functions of the form (I.42), where for all $a \geqslant a_0 > 0$

1) $\sum_{l=-\infty}^{\infty} \mu_a(t + al) = 1$,

2) $\|\mu_a\|_{C^M(R^1,a)} \leqslant C(a_0, M)$, and

3) for any $\varepsilon > 0$ there exists a set $L(\varepsilon, a_0, M) \subset \mathbf{Z}^1$, $|L| < \infty$, such that

$$\sum_{l \in Z^1 - L} \|\mu_a(\cdot + la)\|_{C^M(T^1_{a+1})} < \varepsilon.$$

For example, if $\mu_1(t)$ satisfies conditions 1)–3) for $a = 1$, then $\mu_a(t) = \mu_1(t/a)$ satisfies 1)–3) for any $a \geqslant a_0$.

Theorem 4 is valid for all $\varphi(x) \in \tilde{\Phi}_a$ (instead of $\varphi(x) \in \Phi_a$).

3°. On existence of a limit for norms of periodic imbedding operators (Comment to Theorem 7). It was shown in Theorem 7 that $\|S\| = \overline{\lim}_{a \to \infty} \|S^*_a\|^{(3)}$ for imbedding operators. This proof did not use the property of the imbedding operator to be local. The equality $\|S\| = \lim_{a \to \infty} \|S^*_a\|^{(3)}$ can be obtained as in Theorem 10 by taking into account the fact that $S \in \mathfrak{A}_0$.

4°. Analogue of Theorem 9 for nonlocal operators. Suppose that $Z(R^n)$ and $W(R^m)$ satisfy 1(a) and 2(a), $Z(R^n)$ consists of compactly supported functions, and the S^*_a: $Z^*(R^n, a) \to W^*(R^m, a)$ are operators converging to the operator S: $Z(R^n) \to \check{W}(R^m)$ in the following sense:

$$Sg = \lim_{a \to \infty} S^*_a Q_a g \quad \text{in } \check{W}(R^m) \quad \forall g \in Z(R^n). \tag{8}$$

Then for $j = 1$

$$\|S\|_{Z(R^n) \to \overline{W}_1(R^m)} \leqslant \overline{\lim_{a \to \infty}} \|S^*_a\|^{(j)}. \tag{9}$$

If 1(b) and 2(c) hold, then the estimate holds also for $j = 2$, and if 1(c) and 3(a) hold, then also for $j = 3$, with $\overline{W}(R^m)$ instead of $\overline{W}_1(R^m)$ (see §4.1).

If (II.32) holds, then $\overline{W}_1(R^m)$ can also be replaced by $\overline{W}(R^m)$ for $j = 1$ and 2, with the constant d introduced into the estimate.

For linear operators S this assertion extends to the case where $Z(R^n)$ is the closure of some set of compactly supported functions on which (8) holds. The proof is analogous to that of Theorem 12. If the operators S^*_a are linear (invariant under translations in R^m for $m \leqslant n$), then so is the S in (8).

5°. Recovery of a weakly local operator from its restrictions (Comment to §2.4.). In getting rid of the requirement that $\check{S}$ be local (be in $\mathfrak{A}_\delta$) it is more convenient to replace $\check{Z}(R^n)$ by $\tilde{Z}(R^n)$ (see 1°) and consider an operator $\tilde{S}$: $\tilde{Z}(R^n) \to \check{W}(R^m)$. We say that $\tilde{S}$ is weakly local (briefly, $\tilde{S} \in \mathfrak{A}_\infty$) if $\lim_{k \to \infty} \tilde{S}g_k = \tilde{S}f$ in $\check{W}(R^m)$ for any $f \in \tilde{Z}(R^n)$ and any sequence $\{g_k\}_1^\infty$ that is bounded in $\tilde{Z}(R^n)$ (i.e., $\sup_k \|g_k\|_{\tilde{Z}(\Pi)} < \infty$ for any $\Pi \subset R^n$) and such that that $g_k = f$ on T_k $(k = 1, 2, \dots)$. This condition clearly holds for $\tilde{S} \in \mathfrak{A}_\delta$, $\delta \geqslant 0$.

Let $S = \tilde{S}|_{Z(R^n)}$ and $S^*_a = \tilde{S}|_{Z^*(R^n, a)}$.

(a) For an operator S: $Z(R^n) \to \check{W}(R^m)$, any $f \in \tilde{Z}(R^n)$, and any sequence $\{g_k\}_1^\infty$, $g_k \in Z(R^n)$ that is bounded in $\tilde{Z}(R^n)$ and such that $g_k = f$ on T_k the limit $\lim_{k \to \infty} Sg_k$ exists in $\check{W}(R^m)$ and does not depend on the choice of $\{g_k\}_1^\infty$. This property is also sufficient for S to admit an extension $\tilde{S} \in \mathfrak{A}_\infty$. Moreover, the extended operator in $\mathfrak{A}_\infty$ is unique and has the properties of linearity and invariance under translations in R^m $(m \leqslant n)$ whenever S has these properties. If $S \in \mathfrak{A}_\delta$, then also $\tilde{S} \in \mathfrak{A}_\delta$.

(b) For a family $\{S^*_a\}_{a>0}$, any $f \in \tilde{Z}(R^n)$, and any sequence of compactly supported functions $g_k \in Z(R^n)$ that is bounded in $\tilde{Z}(R^n)$ and satisfies $g_k = f$ on T_k

the following double limit exists and does not depend on the choice of $\{g_k\}$:

$$\lim_{k\to\infty}\left(\lim_{a\to\infty} S_a^* Q_a g_k\right) \quad \text{in } \check{W}(R^m).$$

This property is also sufficient for the family $\{S_a^*\}_{a>0}$ to admit an extension $\tilde{S} \in \mathfrak{A}_\infty$. Moreover, the extended operator in $\mathfrak{A}_\infty$ is unique and has the properties of linearity and invariance under translations in R^m $(m \leqslant n)$ whenever the S_a^* have these properties. If the S_a^* are jointly local with parameter $\delta \geqslant 0$, then $\tilde{S} \in \mathfrak{A}_\delta$.

We remark that if $\tilde{S} \in \mathfrak{A}_\infty$, then its restrictions are connected by the relation (8), so that the estimates (9) are true for them.

6°. On thinning out a family $\{S_a^*\}_{a>a_0}$ (Comment to 2° in §2.4). Everything in that subsection is valid if instead of the family $\{S_a^*\}_{a>0}$ we consider a subfamily $\{S_a^*\}_{a\in\mathscr{A}}$, where $\mathscr{A}$ is a set of numbers $a \geqslant a_0$ (not necessarily all of them) having limit point $(+\infty)$. For example, we can consider $\{S_{a^\nu}^*\}_{\nu=0}^\infty$, $a > 1$. In this case the spaces $Z^*(R^n, a^\nu)$ increase with increasing ν, and the condition of being jointly local reduces to the requirement that $S_{a^\nu}^*|_{Z^*(R^n, a^\mu)} = S_{a^\mu}^*$ for any $\nu > \mu$.

§4.3. On interrelation of estimates for entire functions and trigonometric polynomials

1°. Spaces of entire functions and trigonometric polynomials. Suppose that $m \leqslant n$, $Z(R^n) \subset S'(R^n)$ and $W(R^m) \subset S'(R^m)$ are normed (quasinormed) linear spaces satisfying 1(d) and 3(b), with norms $\|\cdot\|_{Z(R^n)}$ and $\|\cdot\|_{W(R^m)}$, and G^n (G^m) is a bounded open region in R^n (R^m). We fix $\varepsilon > 0$ and consider values $a > 0$ large enough that the sets

$$L_{G^n} = \left\{l \in \mathbf{Z}^n \colon 2\pi l/a \in G^n_{(\varepsilon/a)}\right\}, \qquad L_{G^m} = \left\{l' \in \mathbf{Z}^m \colon 2\pi l'/a \in G^m_{(\varepsilon/a)}\right\}$$

are nonempty, where $G^n_{(\varepsilon/a)} = \{x \in G^n \colon \rho(x, \partial G^n) > \varepsilon/a\}$.

Consider the subspace $\mathfrak{M}(R^n) = \{f \in Z(R^n) \colon \operatorname{supp}\mathscr{F}f \subset G^n\}$ of $Z(R^n)$, with the norm $\|\cdot\|_{Z(R^n)}$. Here $\mathscr{F}$ is the Fourier transformation in R^n (generalized, in general). Clearly, $\mathfrak{M}(R^n)$ consists of entire functions of exponential type. The spaces $Z(R^n)$ and $W(R^m)$ determine periodic analogues $Z^*(R^n, a)$ and $W^*(R^m, a)$ (see Chapter I) which are normed (quasinormed) linear spaces. Let $\|\cdot\|_{Z^*(R^n,a)}$ and $\|\cdot\|_{W^*(R^m,a)}$ be norms in these spaces defined in Chapter I; for example, the norms $\|\cdot\|^{(3)}$. Consider the subspace

$$\mathfrak{M}^*(R^n, a) = \left\{f \in Z^*(R^n, a) \colon f = \sum_{l\in L_{G^n}} f_l e^{i(2\pi l/a)x}\right\}$$

of $Z^*(R^n, a)$ consisting of trigonometric polynomials, with the norm $\|\cdot\|_{Z^*(R^n,a)}$.

Similarly, we introduce

$$\mathfrak{N}(R^m) = \{f \in W(R^m) \colon \operatorname{supp}\mathscr{F}f \subset G^m\}$$

with the norm $\|\cdot\|_{W(R^m)}$, and

$$\mathfrak{N}^*(R^m, a) = \left\{f \in W^*(R^m, a) \colon f = \sum_{l'\in L_{G^m}} f_{l'} e^{2\pi l' u/a}\right\}$$

with the norm $\|\cdot\|_{W^*(R^m,a)}$. Here $x = (x_1,\dots,x_n) = (u, v) \in R^n$, $u = (x_1,\dots,x_m) \in R^m$, $lx = l_1x_1 + \cdots + l_nx_n$ and $l'u = l_1x_1 + \cdots + l_mx_m$.

To study the interconnection between estimates for entire functions and for trigonometric polynomials it is necessary to introduce multipliers connecting these spaces. We fix $0 < h < 1$, $\varepsilon = \varepsilon(h) > 0$, and $a_0 > 0$, and let $\Phi_a = \Phi_a^n = \{\varphi(x)\}$ be a system of functions $\varphi(x)$ with the following properties:

1) $\mathscr{F}\varphi(\xi) \in C_0^\infty(K_{\varepsilon/a}^n)$, where $K_{\varepsilon/a}^n = \{\xi \in R^n\colon |\xi_i| < \varepsilon/a\}$.

2) $\varphi(x) \geqslant h^n > 0$ for $x \in T_{2a}^n$, and $\|\varphi\|_{C^M(R^n,a)} \leqslant c_0$.

Here $a \geqslant a_0$, and $c_0 = c_0(M)$ does not depend on a, and $M = 1, 2, \ldots$.

Functions of this type are not hard to construct. For example, we can take $\mu_1(t) \in S(R^1)$ such that $\mu_1(0) = 1$, $\mu_1(t) \geqslant h$ for $|t| < 1$, and $\mathscr{F}\mu_1 \in C_0^\infty(-\varepsilon, \varepsilon)$ (this can always be achieved by taking $\varepsilon = \varepsilon(h)$ sufficiently large), and then define $\varphi(x) = \Pi_{j=1}^n \mu_1(x_j/a)$. Further, let $\Psi_a = \Psi_a^m = \{\psi(u)\}$ and $\mathrm{M}_a = \mathrm{M}_a^{n-m} = \{\mu(v)\}$ be systems with analogous properties. Note that $\varphi(x) = \psi(u)\mu(v) \in \Phi_a$ if $\psi \in \Psi_a$ and $\mu \in \mathrm{M}_a$.

An important property of the systems constructed is that the operators

$$\mathscr{P}_\varphi\colon \mathfrak{M}^*(R^n, a) \to \mathfrak{M}(R^n), \qquad \varphi \in \Phi_a,$$
$$\mathscr{P}_\psi\colon \mathfrak{N}^*(R^m, a) \to \mathfrak{N}(R^m), \qquad \psi \in \Psi_a,$$

are norming operators (in the sense of §2.1), while $\{\mathscr{P}_{\varphi_a}\}_{a>a_0}$, $\varphi_a \in \Phi_a$, and $\{\mathscr{P}_{\psi_a}\}_{a>a_0}$, $\psi_a \in \Psi_a$, are norming families of operators as $a \to \infty$. To see this, observe first that $\operatorname{Im}\mathscr{P}_\varphi = \mathscr{P}_\varphi(\mathfrak{M}^*(R^n, a)) \subset \mathfrak{M}(R^n)$ and $\operatorname{Im}\mathscr{P}_\psi \subset \mathfrak{N}(R^m)$. Indeed, if $f \in \mathfrak{M}^*(R^n, a)$ and $\varphi(x) \in \Phi_a$, then

$$\operatorname{supp}\mathscr{F}(f\varphi) = \operatorname{supp}(\mathscr{F}f * \mathscr{F}\varphi) \subset G^n.$$

Moreover, $f\varphi \in Z(R^n)$ by Theorem 4, so $f\varphi \in \mathfrak{M}(R^n)$. Similarly, $g\psi \in \mathfrak{N}(R^m)$ if $g \in \mathfrak{N}^*(R^m, a)$ and $\psi \in \Psi_a$. Further, by Theorem 4, the norm $\|f\|_{Z_\varphi(R^n)}$, $\varphi \in \Phi_a$, is one of the equivalent norms in $Z^*(R^n, a)$, with constants not depending on $a \geqslant a_0$.

Thus, $\|\cdot\|_{Z_\varphi(R^n)}$, $\varphi \in \Phi_a$, is an equivalent norm in $\mathfrak{M}^*(R^n, a)$, and with constants in the equivalence condition that do not depend on $a \geqslant a_0$ nor G^n. Similarly, $\|\cdot\|_{W_\psi(R^m)}$, $\psi \in \Psi_a$, is an equivalent norm in $\mathfrak{N}^*(R^m, a)$, with constants not depending on $a \geqslant a_0$ nor G^m.

2°. Estimates for entire functions and trigonometric polynomials. Suppose that it is known that the operators

$$S\colon \mathfrak{M}(R^n) \to \mathfrak{N}(R^m), \qquad S_a^*\colon \mathfrak{M}^*(R^n, a) \to \mathfrak{N}^*(R^m, a)$$

are such that there exist $\varphi_a \in \Phi_a$ and $\psi_a \in \Psi_a$ for which

$$\mathscr{P}_{\psi_a} S_a^* = S\mathscr{P}_{\varphi_a} \tag{10}$$

(as an operator from $\mathfrak{M}^*(R^n, a)$ into $\mathfrak{N}(R^m)$). We give examples.

1) Suppose that $n = m$, $G^n = G^m$, and S and S_a^* are imbedding operators (the identities). Then (10) holds for any $\varphi_a \in \Phi_a$ and $\psi_a = \varphi_a$.

2) Suppose that $m < n$, G^m is the projection $G^n|_{R^m}$ of G^n onto R^m, and S and S_a^* are trace operators of a function (or its derivative), i.e., $(Sf)(u) = (D_v^\alpha f)(u, 0)$, and the same formula for S_a^*. Then (10) follows for any function $\psi_a \in \Psi_a$ and for $\varphi_a(x) = \psi_a(u)\mu_a(v) \in \Phi_a$, where $\mu_a \in M_a$ is such that $\mu_a(0) = 1$ and $D_v^\alpha \mu_a(0) = 0$.

For operators S and S_a^* with property (10) we can use Theorem 6, according to which

$$\|S_a^*\|_{\varphi_a \to \psi_a} \leqslant \|S\|, \qquad a > 0.$$

If we pass to the original norms in $\mathfrak{M}^*(R^n, a)$ and $\mathfrak{N}^*(R^m, a)$, we get

$$\|S_a^*\| \leqslant c(a_0, h)\|S\| \quad \forall a \geqslant a_0, \tag{11}$$

where the constant is independent not only of a, but also of G^n.

We shall consider some examples of application of (11).

1. Suppose that $m \leqslant n$, $Z(R^n) = L_p(R^n)$, $W(R^m) = L_q(R^m)$, $0 < p \leqslant q \leqslant \infty$, with the corresponding norms (quasinorms), $G^n = \{\xi \in R^n\colon |\xi_j| \leqslant \lambda_j, j = 1,\dots,n\}$, and $G^m = G^n|_{R^m}$, $\lambda_j \geqslant 1$. Then $\mathfrak{M}(R^n)$ and $\mathfrak{N}(R^m)$ are spaces of entire functions of exponential type λ_j in x_j, with the norms $\|\cdot\|_{L_p(R^n)}$ and $\|\cdot\|_{L_q(R^m)}$, and $\mathfrak{M}^*(R^n, a)$ and $\mathfrak{N}(R^m, a)$ are spaces of trigonometric polynomials of degree at most $\lambda_j - \varepsilon/a$ (i.e., at most λ_j) in x_j, with the norms $\|\cdot\|_{L_p^*(R^n, a)}$ and $\|\cdot\|_{L_q^*(R^m,a)}$, where

$$\|f\|_{L_p^*(R^n,a)} \equiv \|f\|^{(2)}_{\mathfrak{M}^*(R^n,a)} = \|f\|_{L_p(T_a^n)}.$$

By Theorems 4 and 2, these norms are equivalent to $\|\cdot\|_{Z_\varphi(R^n)}$, $\varphi \in \Phi_a$, or, respectively, to $\|\cdot\|_{W_\psi(R^m)}$, $\psi \in \Psi_a$ (see 1°). By an inequality of Nikol'skiĭ [1], for $g \in \mathfrak{M}(R^n)$

$$\|g\|_{L_q(R^m)} \leqslant c \prod_{j=1}^{n} \lambda_j^{1/p-1/q} \prod_{j=m+1}^{n} \lambda_j^{1/q} \|g\|_{L_p(R^n)}. \tag{12}$$

This inequality gives an estimate of the norm of the corresponding trace operators S. According to (11), it implies that for $f(x) \in \mathfrak{M}^*(R^n, a)$

$$\|f\|_{L_q^*(R^m,a)} \leqslant c_1(a_0, h) \prod_{j=1}^{n} \lambda_j^{1/p-1/q} \prod_{j=m+1}^{n} \lambda_j^{1/q} \|f\|_{L_p^*(R^n,a)}$$

(the constants do not depend on $\lambda_j \geqslant 1$). This estimate is of course known. But the same method can also be used to get periodic analogues of some new inequalities for entire functions.

2. Suppose that $m = n$, $Z(R^n) = L_p(R^n)$ and $W(R^n) = L_q(R^n)$, $1 < p \leqslant q < \infty$. Denote by $Q \subset R^n$ the set consisting of the vector $(1/p - 1/q,\dots,1/p - 1/q)$ and the vectors obtained from it by replacing some coordinates by zeros, and let $G(Q; \gamma)$, $\gamma > 0$, be the "hyperbolic cross"

$$G = G(Q; \gamma) = \left\{\xi \in R^n\colon \prod_{i=1}^{n} |\xi_i|^{l_i} \leqslant \gamma;\ l = (l_1,\dots,l_n) \in Q\right\}.$$

Let $N = \operatorname{meas} G$. It is clear that the "cross" G expands and N increases as γ increases. As above, we introduce the spaces $\mathfrak{M}(R^n)$ and $\mathfrak{N}(R^n)$ of entire functions with spectra supported in G and with norms $\|\cdot\|_{L_p(R^n)}$ and $\|\cdot\|_{L_q(R^n)}$, and the corresponding spaces $\mathfrak{M}^*$ and $\mathfrak{N}^*$ of trigonometric polynomials. For $f \in \mathfrak{M}^*(R^n, a)$ we have an inequality obtained by Magaril-Il'yaev [4] (see also [21]–[24]):

$$\|g\|_{L_q(R^n)} \leqslant c(n, p, q)(N \ln^{-(n+1)} N)^{1/p-1/q} \|g\|_{L_p(R^n)},$$

which is sharp with respect to order as $N \to \infty$. Note that for $m = n$ and for a rectangle G with $\operatorname{meas} G = N$ the inequality (12) of Nikol'skiĭ gives the larger order $cN^{1/p-1/q}$. Using (11) for the corresponding imbedding operator, for $f \in \mathfrak{M}^*(R^n, a)$ we get

$$\|f\|_{L_q^*(R^n,a)} \leqslant c_1(N \ln^{-(n+1)} N)^{1/p-1/q} \|f\|_{L_p^*(R^n,a)} \quad \forall a \geqslant a_0$$

with a constant c_1 not depending on a nor N.

3. Suppose now that $m \leqslant n$, and $Z(R^n)$ and $W(R^m)$ are symmetric Banach spaces (see [19], Chapter II). They satisfy 1(d) and 3(b), and, moreover, $\|f\varphi\| \leqslant \|f\|$ if $|\varphi(x)| \leqslant 1$; by Remark 3, this estimate ensures 2(d). Therefore, by Theorems 2 and 4, the norms $\|\cdot\|_{Z_\varphi(R^n)}$, $\varphi \in \Phi_a$, and $\|\cdot\|_{W_\psi(R^m)}$, $\psi \in \Psi_a$, are equivalent to the "natural norms" in the periodic spaces $Z^*(R^n, a)$ and $W^*(R^m, a)$:

$$\|f\|_{Z^*(R^n,a)} = \|\chi_{T_a} f\|_{Z(R^n)} = \|f\|_{Z(T_a)} = \|f\|^{(2)}_{Z^*(R^n,a)}$$

(and, similarly, to the norm $\|\cdot\|_{W^*(R^m,a)}$).

Hence, we have the estimate (11), which permits us to carry the inequalities with different metrices and different dimensions in symmetric spaces obtained for entire functions by Berkolaĭko and Ovchinnikov in [5] over to trigonometric polynomials.

3°. Converses of the estimates. Assume all the notation and requirements in 1° and 2°.(3) We shall assume in addition that the following conditions hold:

(a) The function $\varphi_a \in \Phi_a$ in (10) has the form $\varphi_a(x) = \varphi(x/a)$, where $\varphi(0) = 1$ and $\varphi(x) \in \Phi_1$.

(b) The operator S: $\mathfrak{M}(R^n) \to \mathfrak{N}(R^m)$ is such that, for any $g \in \mathfrak{M}(R^n)$ and any sequence $\{g_\nu\} \subset \mathfrak{M}(R^n)$: $g_\nu \to g$ $(\nu \to \infty)$ uniformly along with any derivatives on any compact subset of R^n, we have the relation

$$Sg_\nu \to Sg \in \mathfrak{N}(R^m) \qquad (\nu \to \infty) \quad \text{in } \check{W}(R^m)$$

(i.e., in $W(\Pi)$ for any $\Pi \subset R^n$).

THEOREM 13. *Under the above assumptions and with the above notation*

$$\|S\| \leqslant c \varliminf_{a \to \infty} \|S_a^*\|,$$

where c does not depend on G^n or G^m.

We first prove the following assertion.

LEMMA 13. *Suppose that $Z(R^n)$ satisfies* 1(d) *and* 3(b), *and let $\varphi(x) \in S(R^n)$. Then*

$$\varlimsup_{a \to \infty} \|Q_a(\varphi f)\|^{(3)}_{Z^*(R^n,a)} \leqslant d\|\varphi\|_{C^M(R^n)}\|f\|_{Z(R^n)},$$

where d does not depend on φ nor $f \in Z(R^n)$.

PROOF OF LEMMA 13. 1. Let us first prove that under the conditions of the lemma

$$\|Q_a(\varphi f)\|^{(3)}_{Z^*(R^n,a)} \leqslant c_1\|\varphi\|_{C^M(R^n,a)}\|f\|_{Z(R^n)}$$

for all $a \geqslant a_0 > 0$, with a constant c_1 not depending on a, f, nor φ. This is also of independent interest.

Choose a function $\varphi_1(x) \in C_0^M(T_{a+1})$ such that $\|\varphi_1\|_{C^M(R^n)} \leqslant c_2$ (c_2 does not depend on a) and $\varphi_1(x) = 1$ on T_a. Then, by Theorem 4 and properties 3(b) and 1(d),

$$\|Q_a(\varphi f)\|^{(3)}_{Z^*(R^n,a)} \leqslant c_3\|\varphi_1 Q_a(\varphi f)\|_{Z(R^n)} \leqslant c_3 \sum_{l \in \mathbf{Z}^n} \|\varphi_1 E_{al}(\varphi f)\|_{Z(R^n)}$$

$$\leqslant c_3 AB\|f\|_{Z(R^n)} \sum_{l \in \mathbf{Z}^n} \|\varphi_1 E_{al}\varphi\|_{C^M(R^n)}.$$

(3) *Added in proof.* See also the authors' article [30].

But

$$\|\varphi_1 E_{al}\varphi\|_{C^M(R^n)} \leqslant c_4\|E_{al}\varphi\|_{C^M(T_{a+1})} = c_4\|\varphi\|_{C^M(T_{a+1}+al)}.$$

The last two estimates give the required inequality.

2. To obtain the lemma it remains to show that if $\varphi(x) \in S(R^n)$, then

$$\lim_{a\to\infty} \|\varphi\|_{C^M(R^n,a)} = \|\varphi\|_{C^M(R^n)}.$$

Indeed,

$$\|\varphi\|_{C^M(R^n,a)} = \|\varphi\|_{C^M(T_{a+1})} + \sum_{0\neq l\in\mathbf{Z}^n} \|\varphi\|_{C^M(T_{a+1}+al)}.$$

The first term converges to $\|\varphi\|_{C^M(R^n)}$, and the second to zero as $a \to \infty$ for $\varphi \in S(R^n)$. The lemma is proved.

REMARK 13. A more detailed analysis shows that if 3(c) holds, then we can take $d = A$ in the statement of the lemma, where A is the constant in 1(d).

PROOF OF THEOREM 13. 1. Suppose that $\lambda(x) \in S(R^n)$ and $\mathscr{F}\lambda \in C_0^\infty(K_1)$, where $K_\varepsilon = \{\xi \in R^n\colon |\xi_j| < \varepsilon,\ j = 1,\dots,n\}$, $\varepsilon > 0$, and $|\lambda(x)| \leqslant \lambda(0) = 1$, $x \in R^n$. For $s = 1,2,\dots$ we set $\lambda_s(x) = \lambda(x/s)$; then $\mathscr{F}\lambda_s \in C_0^\infty(K_{1/s})$. Note that for any integer $M \geqslant 0$

$$\lim_{s\to\infty} \|\lambda_s\|_{C^M(R^n)} = \|\lambda\|_{C(R^n)} = 1$$

and that, uniformly on any compact set,

$$\lim_{s\to\infty} \lambda_s(x) = 1, \qquad \lim_{s\to\infty} D^\alpha\lambda_s(x) = 0 \quad \forall\alpha = (\alpha_1,\dots,\alpha_n).$$

2. Let $f \in \mathfrak{M}(R^n)$. Since G^n is an open set, there exists a $\delta = \delta(f) > 0$ such that $\operatorname{supp}\mathscr{F}f \subset G^n_\delta$. But then for sufficiently large s, $s \geqslant s(f)$, we have $\operatorname{supp}\mathscr{F}(\lambda_s f) \subset G^n_{\delta/2}$, i.e., $\lambda_s f \in \mathfrak{M}(R^n)$. Moreover, since f is an entire function in $S'(R^n)$ and $\lambda_s \in S(R^n)$, it follows that $\lambda_s f \in S(R^n)$. The Poisson summation formula ([2], Theorem 2.4, Corollary 6) is applicable for $Q_a(\lambda_s f) \in Z^*(R^n, a)$ (see the lemma), and according to it the Fourier coefficients $(Q_a(\lambda_s f))_l$ in the trigonometric system $\{\exp(i2\pi lx/a)\}_{l\in\mathbf{Z}^n}$ coincide with $\mathscr{F}(\lambda_s f)(2\pi l/a)$. Hence, $Q_a(\lambda_s f) \in \mathfrak{M}(R^n, a)$, i.e., it is a trigonometric polynomial. Note that the condition $\lambda_s f \in S(R^n)$ implies that $Q_a(\lambda_s f) \to \lambda_s f\ (a \to \infty)$ and also (by assumption (a))

$$\varphi_a Q_a(\lambda_s f) \to \lambda_s f \qquad (a \to \infty)$$

together with any derivatives uniformly on any compact set.

3. From the results of the second step of the proof and property (b) it follows that for any $\Pi \subset R^m$

$$\begin{aligned}\|S(\lambda_s f)\|_{W(\Pi)} &= \lim_{a\to\infty} \|S\mathscr{P}_{\varphi_a} Q_a(\lambda_s f)\|_{W(\Pi)} \\ &\leqslant \varliminf_{a\to\infty} \|S\mathscr{P}_{\varphi_a} Q_a(\lambda_s f)\|_{W(R^m)}.\end{aligned} \tag{13}$$

Using (10), we get

$$\begin{aligned}\|S(\lambda_s f)\|_{W(\Pi)} &\leqslant \varliminf_{a\to\infty} \|\mathscr{P}_{\psi_a} S_a^* Q_a(\lambda_s f)\|_{W(R^m)} \\ &\leqslant \varliminf_{a\to\infty} \left(\|S_a^*\|_{\varphi_a\to\psi_a} \|Q_a(\lambda_s f)\|_{Z^*_{\varphi_a}(R^n,a)}\right).\end{aligned}$$

Theorem 4 and the lemma now give us that

$$\|S(\lambda_s f)\|_{W(\Pi)} \leqslant c \varliminf_{a\to\infty} \|S_a^*\|_{\varphi_a\to\psi_a}\|\lambda_s\|_{C^M(R^n)}\|f\|_{Z(R^n)},$$

where c does not depend on s, G^m, G^n, f, nor Π. Let $s \to \infty$ here. Taking into account the results in step 1 and their consequence the relation that $\lambda_s f \to f$ $(s \to \infty)$ together with any derivatives uniformly on any compact set, we get that

$$\begin{aligned}\|Sf\|_{W(\Pi)} &= \lim_{s\to\infty} \|S(\lambda_s f)\|_{W(\Pi)} \\ &\leqslant c \varliminf_{a\to\infty} \|S_a^*\|_{\varphi_a\to\psi_a}\|f\|_{Z(R^n)}.\end{aligned}$$

Now take the supremum over $\Pi \subset R^m$. This gives us that

$$\|Sf\|_{W_1(R^m)} \leqslant c \varliminf_{a\to\infty} \|S_a^*\|_{\varphi_a\to\psi_a}\|f\|_{Z(R^n)}.$$

Since the norms $\|\cdot\|_{W(R^m)}$ and $\|\cdot\|_{W_1(R^m)}$ are equivalent under condition 3(b) (see (11)), while φ_a and ψ_a are norming multipliers as $a \to \infty$, we arrive at the statement of the theorem.

4°. On an inequality of Bernstein type. If (10) does not hold under the conditions in 1°–3°, then instead of Theorem 6 it is necessary to use the general Theorem 5 in estimating the operator

$$U_a = S\mathscr{P}_\varphi - \mathscr{P}_\psi S_a^*\colon \mathfrak{M}^*(R^n, a) \to \mathfrak{N}(R^m). \tag{14}$$

For example, consider the case $m = n$, with S and S_a^* the differentiation operators $Sg \equiv D^\alpha g$, $g \in \mathfrak{M}(R^n)$, and $S_a^* f \equiv (D^\alpha)_a^* f = D^\alpha f$, $f \in \mathfrak{M}^*(R^n, a)$, $\alpha = (\alpha_1, \dots, \alpha_n)$. The question of the interconnection of the norms of these operators coincides with the question of the interconnection between the Bernstein inequalities in a certain space of entire functions and in the corresponding space of trigonometric polynomials.

THEOREM 14. 1) *If under the conditions of* 2° *it is known that* $\|D^\gamma\| < \infty$ *for all* γ *with* $0 \leqslant \gamma \leqslant \beta$, *then*

$$\|(D^\beta)_a^*\| \leqslant \sum_{0\leqslant\gamma\leqslant\beta} c_\gamma(a)\|D^\gamma\|, \tag{15}$$

where $c_\gamma(a) \leqslant \bar{c}_\gamma(a_0)$ *for all* $a \geqslant a_0 > 0$ *and* $\bar{c}_\gamma(a_0)$ *does not depend on* G^n *nor* G^m; *moreover*, $\lim_{a\to\infty} c_\gamma(a) = 0$ *for* $0 \leqslant \gamma < \beta$.(4)

2) *If under the conditions in* 3° *it is known that* $\varlimsup_{a\to\infty}\|(D^\gamma)_a^*\| < \infty$ *for all* γ *with* $0 \leqslant \gamma < \beta$, *then*

$$\|D^\beta\| \leqslant d_\beta \varliminf_{a\to\infty} \|(D^\beta)_a^*\|, \tag{16}$$

where d_β *does not depend on* G^m *nor* G^n.

PROOF. 1. Suppose that $0 < \beta$, $S = D^\beta$ and $S_a^* = (D^\beta)_a^*$. We choose a function $\varphi(x) \in \Phi_a$ of the form $\varphi(x) = \varphi_1(x/a)$, $\varphi_1 \in \Phi_1$, let $\psi = \varphi$, and consider the

(4)In this subsection $\gamma \leqslant \beta$ means that $\gamma_i \leqslant \beta_i$ $(i = 1, \dots, n)$, and $\gamma < \beta$ means that $\gamma \leqslant \beta$ and $\gamma \neq \beta$.

operator U_a in (14), which in this case has the form

$$U_a f = D^\beta(\varphi f) - \varphi(D^\beta)^*_a f = \sum_{0 \leqslant \gamma < \beta} c_\gamma (D^{\beta-\gamma}\varphi)(D^\gamma)^*_a f, \quad f \in \mathfrak{M}^*(R^n, a).$$

Using the estimates (I.29) and (I.35) of Lemma 4 with $Z(R^n)$ replaced in them by $W(R^m)$, φ by $D^{\beta-\gamma}\varphi$, and ψ by the function φ_a in (I.45), we conclude from this that

$$\|U_a f\|_{W(R^n)} \leqslant \sum_{0 \leqslant \gamma < \beta} d_\gamma \|D^{\beta-\gamma}\varphi\|_{C^M(R^n,a)} \|\varphi_a (D^\gamma)^*_a f\|_{W(R^n)},$$

where d_γ does not depend on $a \geqslant a_0$.

Taking into account the definition of $\|\cdot\|^{(3)}_{W^*(R^n,a)}$ (see (I.46)), we get the estimate

$$\|\varphi_a (D^\gamma)^*_a f\|_{W(R^n)} \leqslant \|(D^\gamma)^*_a\|^{(3)} \|f\|^{(3)}_{Z^*(R^n,a)}.$$

Consequently,

$$\|U_a\| \leqslant \sum_{0 \leqslant \gamma < \beta} d_\gamma \|D^{\beta-\gamma}\varphi\|_{C^M(R^n,a)} \|(D^\gamma)^*_a\|^{(3)}. \tag{17}$$

2. Note that for $\beta = 0$ inequality (15) is valid because of (11), since the identity operators $S = D^0$ and $S^*_a = (D^0)^*_a$ satisfy (10) for any $\varphi \in \Phi_a$.

Making the induction hypothesis that (15) is true for any γ with $0 \leqslant \gamma < \beta$, we shall obtain it for $\gamma = \beta$.

Indeed, under our assumption it follows from (17) that

$$\|U_a\| \leqslant \sum_{0 \leqslant \delta < \beta} e_\delta(a) \|D^\delta\|, \tag{18}$$

where

$$e_\delta(a) = c_\delta(a) \sum_{\delta \leqslant \gamma < \beta} d_\gamma \|D^{\beta-\gamma}\varphi\|_{C^M(R^n,a)}.$$

Since $\varphi \in \Phi_a$, we have that $e_\delta(a) \leqslant \bar{e}_\delta(a_0)$, $a \geqslant a_0$, and it is not hard to see that $e_\delta(a) \to 0$ $(a \to \infty)$, i.e.,

$$\lim_{a \to \infty} \|U_a\| = 0. \tag{19}$$

But it then follows from Theorem 5 (see II.8)) and the estimates $C(\mathscr{P}_\varphi) \leqslant c(a_0)$ and $\|\mathscr{P}_\varphi\| \leqslant c(a_0)$ for $\varphi \in \Phi_a$ and $a \geqslant a_0$ (see Theorem 4) that

$$\|(D^\beta)^*_a\|_{\varphi \to \varphi} \leqslant d(a_0)(\|U_a\| + \|D^\beta\|),$$

which gives (15) when (18) and (19) are taken into account.

3. We prove the second part of the theorem. Note that for $\beta = 0$ inequality (16) is valid by Theorem 13. We use the arguments in the proof of Theorem 13 with $S = D^\beta$ up to formula (13). Further (with $\varphi = \varphi_a$),

$$D^\beta \mathscr{P}_\varphi Q_a(\lambda_s f) = \mathscr{P}_\varphi (D^\beta)^*_a Q_a(\lambda_s f) + U_a(Q_a(\lambda_s f)).$$

The first term can be estimated as in Theorem 13. To estimate the second we use (17) and the fact that

$$\lim_{a \to \infty} \|D^{\beta-\gamma}\varphi\|_{C^M(R^n,a)} \to 0 \qquad (a \to \infty)$$

for $\gamma < \beta$. As a result we get, by Lemma 13, that

$$\lim_{a \to \infty} \|U_a(Q_a(\lambda_s f))\|_{W(R^m)} = 0.$$

REMARK 14. The assumptions in 3° hold for all the examples considered in 2°. Thus, the typical situation in the problem of estimating the norms of entire functions and of trigonometric polynomials is the same as in the imbedding Theorem 7. Namely, an existing estimate for entire functions gives us at once the corresponding estimate for trigonometric polynomials; the converse is true if and only if the constants in the inequality for trigonometric polynomials of period a are bounded as $a \to \infty$.

We remark in conclusion that the condition that G^n and G^m be bounded was not used in the proofs. The same results are true for spaces of not necessarily entire functions with restrictions on the spectra.

§4.4. Examples

1°. The spaces $W_p^l(R^n)$ and $W_p^{l*}(R^n, a)$. As $Z(R^n)$ we consider the anisotropic Sobolev spaces $W_p^l(R^n)$, where $l = (l_1, \ldots, l_n)$, the l_i being nonnegative integers, and $1 \leqslant p \leqslant \infty$, with the finite norm

$$\|f\|_{W_p^l(R^n)} = \|f\|_{L_p(R^n)} + \sum_{i=1}^n \|D_i^{l_i} f\|_{L_p(R^n)}$$

(the $D_i^{l_i}$ are the generalized derivatives of order l_i with respect to the variables x_i).

The spaces $W_p^l(R^n)$ have the properties 1(d) with $A = 1$, 2(b) and 2(d) with $C(\Pi_1, \Pi_2) = 1$, and 3(c) with $M = \max_{0 \leqslant i \leqslant n} l_i$. According to §1.1, they determine periodic Banach spaces $W_p^{l*}(R^n, a)$ with norm $\|f\|^{(1)}_{W_p^{l*}(R^n, a)}$.

For parallelepipeds $\Pi \subset R^n$

$$\|f\|_{W_p^l(\Pi)} \sim \|f\|^{+}_{W_p^l(\Pi)} = \|f\|_{L_p(\Pi)} + \sum_{i=1}^n \|D_i^{l_i} f\|_{L_p(\Pi)}$$

(since the conditions of 1° in §4.1 hold; for $l = 0$ we have $\|f\|_{L_p(\Pi)} = \|f\|^{+}_{L_p(\Pi)}$). Moreover,

$$\|f\|_{(W_p^l)_1(R^n)} = \|f\|_{W_p^l(R^n)},$$

because

$$\|f\|_{W_p^l(R^n)} = \sup_{\Pi} \|f\|^{+}_{W_p^l(\Pi)} \leqslant \sup_{\Pi} \|f\|_{W_p^l(\Pi)} = \|f\|_{(W_p^l)_1(R^n)}$$

(it suffices to refer to Lemma 11 for $p < \infty$).

Finally, according to Lemma 12,

$$\lim_{s \to \infty} \|f(x)\varphi(x/s)\|_{W_p^l(R^n)} = \|f\|_{W_p^l(R^n)}, \tag{20}$$

where $\varphi(x) \in C_0^\infty(R^n)$, $0 \leqslant \varphi(x) \leqslant 1$, $\varphi = 1$ for $x \in T_1$, and $\varphi(x) = 0$ for $x \notin T_2$, and according to 5° in §4.1,

$$\|f\|_{(W_p^l)_0(\Pi)} = \|f\|_{W_p^l(\Pi)}.$$

The next assertion follows from Theorems 1–4.

LEMMA 14. 1. *The space $W_p^{l*}(R^n, a)$, $1 \leqslant p \leqslant \infty$, coincides with the space of all a-periodic functions which have generalized derivatives $D_i^{l_i} f$ in R^n and satisfy*

$$\|f\|_{W_p^{l*}(R^n, a)} = \|f\|^{+}_{W_p^l(T_a)} < \infty. \tag{21}$$

2. *Suppose that* $a_0 > 0$, $\lambda > 1$, *and* $1 \leqslant p \leqslant \infty$. *For* $f \in W_p^{l*}(R^n, a)$ *the norm* $\|f\|_{W_p^{l*}(R^n,a)}$ *is equivalent to*:

1) *the norms* $\|f\|_{W_p^l(\Pi)}$, *where* $\rho(\Pi) \geqslant a$, *and the equivalence constants do not depend on* $a \geqslant a_0$ *nor on* Π *if* $\rho(\Pi) \leqslant \lambda a$;

2) *the norms* $\|f\|^{+}_{W_p^l(\Pi)}$, *where* $\rho(\Pi) \geqslant a$, *and the equivalence constants do not depend on* $a \geqslant a_0$ *nor on* Π *if* $\rho(\Pi) \leqslant \lambda a$; *and*

3) *the norms* $\|f\|_{W^l_{p,\varphi}(R^n)} = \|f\varphi\|_{W_p^l(R^n)}$, *where* $\varphi(x) > 0$ *on* T_a, $\varphi \in C^M(R^n, a)$, *and the equivalence constants do not depend on* $a \geqslant a_0$ *for* $\varphi \in \Phi_a$.

In particular, the norm $\|f\|_{W_p^{l*}(R^n,a)}$ *is equivalent to the norms* $\|f\|^{(j)}_{W_p^{l*}(R^n,a)}$, $j = 1, 2, 3$, *and the equivalence constants do not depend on* $a \geqslant a_0$.

To prove the assertions involving the norm $\|\cdot\|^{+}_{W_p^l(\Pi)}$ it is necessary to take into account that there is an extension operator

$$T\colon \left(W_p^l\right)^{+}(\Pi) \to W_p^l(R^n)$$

for which $\|T\| \leqslant c_l(1 + a^{-1})$ (for example, this can be the Hestenes extension operator with subsequence multiplication by a suitable compactly supported function).

2°. The spaces $w_p^l(R^n)$ and $w_p^{l*}(R^n, a)$. Suppose now that $Z(R^n) = w_p^l(R^n)$, which is a seminormed space with the finite seminorm

$$\|f\|_{w_p^l(R^n)} = \sum_{i=1}^{n} \left\|D_i^{l_i} f\right\|_{L_p(R^n)}.$$

The spaces $w_p^l(R^n)$ also have properties 1(d) with $A = 1$, and 2(b) and 2(d) with $C(\Pi_1, \Pi_2) = 1$. The operator of multiplication by a function $\varphi \in C_0(R^n)$ acts from $w_p^l(R^n)$ to $w_p^l(R^n)$, but is not bounded.

According to §1.1, $w_p^l(R^n)$ determines a seminormed periodic space $w_p^{l*}(R^n, a)$ (see 3° in §4.1) with the seminorm $\|f\|^{(1)}_{w_p^{l*}(R^n,a)}$.

For parallelepipeds $\Pi \subset R^n$

$$\|f\|_{w_p^l(\Pi)} \simeq \|f\|^{+}_{w_p^l(\Pi)} = \sum_{i=1}^{n} \left\|D_i^{l_i} f\right\|_{L_p(\Pi)}$$

(because, according to Burenkov [28], the conditions in 1° of §4.1 hold).

Moreover, as for the spaces W_p^l,

$$\|f\|_{(w_p^l)_1(R^n)} = \|f\|_{w_p^l(R^n)}.$$

A lemma analogous to Lemma 14 is valid for the spaces $w_p^{l*}(R^n, a)$, with the exception of the assertions relating to the norm $\|\cdot\|_{w^l_{p,\varphi}(R^n)}$, which in this case is not equivalent to $\|\cdot\|_{w_p^{l*}(R^n,a)}$ for any function $\varphi \in C_0^\infty(R^n)$ (indeed, if $f(x) \equiv 1$, then $\|f\|_{w_p^{l*}(R^n,a)} = 0$ but $\|f\|_{w^l_{p,\varphi}(R^n)} \neq 0$ for $\varphi \not\equiv 0$).

3°. The spaces $B^l_{p\theta}(R^n)$ and $B^{l*}_{p\theta}(R^n, a)$. As another example of the spaces $Z(R^n)$ we consider the anisotropic Nikol'skiĭ-Besov spaces $B^l_{p\theta}(R^n)$, where $l = (l_1, \dots, l_n)$, $l_i > 0$ and $1 \leqslant p, \theta \leqslant \infty$, with finite norm

$$\|f\|_B = \|f\|_{L_p(R^n)} + \sum_{i=1}^{n} \left(\int_0^1 h^{-1-\theta l_i} \left\|\Delta^2_{h,i} D_i^{m_i} f\right\|^{\theta}_{L_p(R^n)} dh \right)^{1/\theta}$$

(here $\Delta^2_{h,i} f$ is the second-order difference with step h with respect to the variable x_i, and m_i is the greatest integer less than l_i).

The spaces $B^l_{p\theta}(R^n)$ have property 1(d) with $A = 1$. Property 2(b) holds, where $C(\Pi_1, \Pi_2)$ satisfies the estimates $C(\Pi_1, \Pi_2) \leqslant c_1(\alpha)$ (by Lemma 1) and $C(\Pi_1, \Pi_2) < c_2(\beta)$ (see Burenkov [9] for $p = \theta$, and Kuznetsov [13], [14] for $p \neq \theta$). For $1 \leqslant p < \infty$ property 2(c) holds if and only if $l_i - [l_i] \neq 1/p$, $i = 1, \dots, n$. (This follows from results of Nikol'skiĭ [6] ($\theta = \infty$), Yakovlev [7] ($l_i < 1$, $p = \theta$), Burenkov [8] ($p = \theta$), and Kuznetsov [11], [12] ($p \neq \theta$); counterexamples are given in Burenkov's paper [10].) Condition 3(c) is satisfied with positive integers $M > \max_{1 \leqslant i \leqslant n} l_i$ (this follows from the proof of Theorem 17.11 and Lemma 16.7 in Besov, Il'in and Nikol'skiĭ).

According to §1.1, the space $B^l_{p\theta}(R^n)$ determines a periodic Banach space $B^{l*}_{p\theta}(R^n, a)$ with norm $\|f\|^{(1)}_{B^{l*}_{p\theta}(R^n,a)}$.

For parallelepipeds $\Pi \subset R^n$

$$\|f\|_{B^l_{p\theta}(\Pi)} \sim \|f\|^{+}_{B^l_{p\theta}(\Pi)} = \|f\|_{L_p(\Pi)} + \sum_{i=1}^{n} \left(\int_0^1 h^{-1-\theta l_i} \|\Delta^2_{h,i} D^{m_i}_i f\|^{\theta}_{L_p(\Pi_{(2h)})} \, dh \right)^{1/\theta} \tag{22}$$

(since the conditions in 1° of §4.1 are satisfied). Moreover, just as for the spaces W^l_p,

$$\|f\|_{(B^l_{p\theta})_1(R^n)} = \|f\|_{B^l_{p\theta}(R^n)}, \qquad \lim_{s \to \infty} \|f(x)\varphi(x/s)\|_{B^l_{p\theta}(R^n)} = \|f\|_{B^l_{p\theta}(R^n)} \tag{23}$$

for the same φ as in (20), $1 \leqslant p, \theta \leqslant \infty$; and

$$\|f\|_{(B^l_{p\theta})_0(\Pi)} = \|f\|_{B^l_{p\theta}(\Pi)}, \qquad \Pi \subset R^n,$$

for $1 \leqslant p, \theta \leqslant \infty$.

LEMMA 15. 1. *The space $B^{l*}_{p\theta}(R^n, a)$, $1 \leqslant p, \theta \leqslant \infty$, coincides with the space of all a-periodic functions such that the following norm is finite*:

$$\|f\|_{B^*_a} = \|f\|_{L_p(T_a)} + \sum_{i=1}^{n} \left(\int_0^1 h^{-1-\theta l_i} \|\Delta^2_{h,i} D^{m_i}_i f\|^{\theta}_{L_p(T_a)} \, dh \right)^{1/\theta}. \tag{24}$$

2. *Suppose that $a_0 > 0$, $\lambda > 1$, and $1 \leqslant p, \theta \leqslant \infty$. For $f \in B^{l*}_{p\theta}(R^n, a)$ the norm $\|f\|_{B^*_a}$ is equivalent to*:

1) *the norms $\|f\|_{B^l_{p\theta}(\Pi)}$, where $\rho(\Pi) > a$, and the equivalence constants are independent of $a \geqslant a_0$ and of Π if $\rho(\Pi) \leqslant \lambda a$;*

2) *the norms $\|f\|_{B^l_{p\theta}(\Pi)}$ for $1 \leqslant p < \infty$, where $\rho(\Pi) = a$, if and only if $l_i - [l_i] \neq 1/p$, $i = 1, \dots, n$; in this case the equivalence constants are independent of $a \geqslant a_0$;*

3) *the norms $\|f\|^{+}_{B^l_{p\theta}(\Pi)}$, where $\rho(\Pi) > a$, and the equivalence constants are independent of $a \geqslant a_0$ and of Π if $\rho(\Pi) \leqslant \lambda a$;*

4) *the norms $\|f\|^{+}_{B^l_{p\theta}(\Pi)}$ for $1 \leqslant p < \infty$, where $\rho(\Pi) = a$, if and only if $l_i - [l_i] \neq 1/p$, $i = 1, \dots, n$; in this case the equivalence constants are independent of $a \geqslant a_0$; and*

5) *the norms $\|f\|_{B^l_{p\theta,\varphi}(R^n)} = \|f\varphi\|_{B^l_{p\theta}(R^n)}$, where $\varphi(x) > 0$ on T_a, $\varphi \in C^M(R^n, a)$, and the equivalence constants are independent of $a \geqslant a_0$ for $\varphi \in \Phi_a$.*

In particular, the norm $\|f\|_{B^*_a}$ is equivalent to the norms $\|f\|^{(j)}_{B^{l*}_{p\theta}(R^n,a)}$ for $j = 1, 2, 3$, and the equivalence constants are independent of $a \geqslant a_0$ (for $j = 2$ and $1 \leqslant p < \infty$ this is true if and only if $l_i - [l_i] \neq 1/p$, $i = 1, \dots, n$).

PROOF. First of all, there exists a bounded extension operator $T\colon (B^l_{p\theta}) + (\Pi) \to B^l_{p\theta}(R^n)$ such that $\|T\| \leqslant c(a_0)$ for $a \geqslant a_0$ (for example, as for the spaces W^l_p, the Hestenes extension operator with subsequent multiplication by a suitable compactly supported function). Hence, according to 1° in §4.1, in the second part of the lemma the norms 1) and 3), as well as the norms 2) and 4), are equivalent. Let us prove that for a-periodic functions the norms $\|f\|_{B_a^*}$ and $\|f\|^{(1)}_{B^{l*}_{p\theta}(R^n,a)} \sim \|f\|^{+}_{B^l_{p\theta}(T_{3a/2})}$ are equivalent, with constants independent of $a \geqslant a_0$. The norm $\|f\|_{B_a^*}$ is equivalent to the norm (22) with the integral from 0 to 1 replaced by the integral from 0 to $\min\{a_0/8, 1\}$, with constants independent of $a \geqslant a_0$; the latter norm obviously does not exceed $\|f\|^{+}_{B^l_{p\theta}(T_{3a/2})}$. On the other hand, since f is periodic,

$$\left\|\Delta^2_{h,i} D_i^{m_i} f\right\|_{L_p((T_{3a/2})_{(2/h)})} \leqslant \sum_{-1 \leqslant k \leqslant 1} \left\|\Delta^2_{h,i} D_i^{m_i} f\right\|_{L_p(T_a+ka)} = 3^n \left\|\Delta^2_{h,i} D_i^{m_i} f\right\|_{L_p(T_a)},$$

and so

$$\|f\|^{+}_{B^l_{p\theta}(T_{3a/2})} \leqslant 3^n \|f\|_{B_a^*}.$$

Thus, the first part of the lemma is proved, along with (by Theorems 2–4) all the assertions in the second part except for the necessity part in 2) and 4). To prove the necessity it suffices to construct a counterexample to assertion 4).

We prove that for $f \in B^{l*}_{p\theta}(R^n, a)$ the norms $\|f\|^{+}_{B^l_{p\theta}(T_a)}$ and $\|f\|^{+}_{B^l_{p\theta}(T_b)}$, where $b > a$, are not equivalent if $1 < p < \infty$, and $l_i - [l_i] = 1/p$ for some i. Suppose first that $n = 1$ and $l = 1/p$. Consider a function $g(x) \in B^l_{p\theta}(-3a/8, 3a/8)$, which is equal to 1 for $-3a/8 \leqslant x \leqslant -a/4$ and equal to -1 for $a/4 \leqslant x \leqslant 3a/8$. For $0 < \varepsilon \leqslant a/4$ we construct a function $f_\varepsilon(x) \in B^{l*}_{p\theta}(R^n, a)$ as follows: $f_\varepsilon(x) = \frac{1}{\varepsilon}(x + ak/2)$ for $|x + ak/2| \leqslant \varepsilon$; $f_\varepsilon(x) = 1$ for $\varepsilon \leqslant x + ak/2 \leqslant a/4$; $f_\varepsilon(x) = -1$ for $a/2 - \varepsilon \leqslant x + ak/2 \leqslant a/2$; and $f_\varepsilon(x) = g(x + ak/2)$ for $a/8 \leqslant x + ak/2 \leqslant 3a/8$ $(k = 2m + 1, m \in \mathbf{Z}^1)$. The computations show that

$$\|f_\varepsilon\|_{L_p(T_b)} \leqslant c, \quad \|f_\varepsilon\|^{+}_{B^{1/p}_{p\theta}(T_a)} \leqslant c; \qquad \|f_\varepsilon\|^{+}_{B^{1/p}_{p\theta}(T_b)} \to \infty \quad (\varepsilon \to 0), \tag{25}$$

where c does not depend on ε (a similar example in the nonperiodic case was constructed by Burenkov in [10]). If $l = m + 1/p$ $(m = 1, 2, \ldots)$, then the relations (25) with l instead of $1/p$ are satisfied for the function $\Phi_\varepsilon(x) \in B^{l*}_{p\theta}(R^n, a)$ defined for $-a/2 \leqslant x \leqslant a/2$ by

$$\Phi_\varepsilon(x) = \frac{1}{(m-1)!} \int_{-a/2}^{x} (x-t)^{m-1} f_\varepsilon(t)\, dt + \mathscr{P}_{m-1,\varepsilon}(x),$$

where $\mathscr{P}_{m-1,\varepsilon}(x)$ is the polynomial of degree at most $m - 1$ chosen so that $\Phi_\varepsilon^{(k)}(-a/2) = \Phi_\varepsilon^{(k)}(a/2)$, $k = 0, \ldots, m-1$ (it suffices to note that $\|\mathscr{P}_{m-1,\varepsilon}\|_{L_p(T_a)} \leqslant c_1 \|f\|_\varepsilon$ and $\Phi_\varepsilon(x) = f_\varepsilon(x)$). Finally, for $n > 1$ it suffices to consider the function $G_\varepsilon(x) = \Phi_\varepsilon(x_i)$, where i is such that $l_i - [l_i] = 1/p$.

REMARK 15. Just as in 2°, we can consider the seminormed spaces $b^l_{p\theta}(R^n)$ corresponding to $B^l_{p\theta}(R^n)$, and their periodic analogues $b^{l*}_{p\theta}(R^n, a)$. For these a lemma analogous to Lemma 15 is valid, except for the statement relating to the norm $\|\cdot\|_{b^l_{p\theta,\varphi}(R^n)}$. Property 2(b) with the estimate $C(\Pi_1, \Pi_2) \leqslant c_1(\beta)$ (Lemma 1 is not applicable in this case) and property 2(c) for $1 \leqslant p < \infty$ and $l_i - [l_i] \neq 1/p$ follow from the aforementioned papers of Burenkov and Kuznetsov, where inequalities are obtained not only for the norms $\|\cdot\|_{B^l_{p\theta}}$ but also for the corresponding seminorms $\|\cdot\|_{b^l_{p\theta}}$.

4°. The connection between imbedding and extension theorems for the spaces $W^l_p(R^n)$, $W^{l*}_p(R^n, a)$, and $B^l_{p\theta}(R^n)$, $B^{l*}_{p\theta}(R^n, a)$. Suppose that $1 \leqslant p, p' \leqslant \infty$, $1 \leqslant \theta$, $\theta' \leqslant \infty$, $l, l' \geqslant 0$ and $0 < m \leqslant n$. Let $X(R^n)$ be any one of the spaces $W^l_p(R^n)$ or

$B^l_{p\theta}(R^n)$, let $Y(R^m)$ be $W^{l'}_{p'}(R^m)$ or $B^{l'}_{p'\theta'}(R^m)$, respectively (for example, $X(R^n) = W^l_p(R^n)$, $Y(R^m) = B^{l'}_{p'\theta'}(R^m)$), and let $X^*(R^n, a)$ and $Y^*(R^m, a)$ be their periodic analogues with the norms defined by (21) and (24).

The next assertion follows easily from Theorems 7–9, the remarks made in 5° of §4.1, and the properties listed in 1° and 3° for these spaces.

For

$$X(R^n) \hookrightarrow Y(R^m) \tag{26}$$

to hold it is necessary and sufficient that

$$X^*(R^n, a) \hookrightarrow Y^*(R^m, a) \tag{27}$$

for any $a > 0$, and

$$\overline{\lim_{a\to\infty}}\ \|I^*_a\| < \infty. \tag{28}$$

In this case,

$$\|I^*_a\| \leqslant c_1(a)\|I\|,$$

and for $a \geqslant a_0 > 0$ we have $c_1(a) \leqslant c_2(a_0)$ and

$$\|I\| \leqslant c_3 \overline{\lim_{a\to\infty}}\ \|I^*_a\|.$$

(Here I and I^*_a are the corresponding imbedding operators; in this case $\|I\| = \|I\|^{(0)}$ according to 5° in §4.1, along with (20) and (23).)

If $0 < m < n$, then I and I^*_a are trace operators. In this case we have in mind the following definition of a trace operator $\check{S}$, which is suitable for $f \in L_1^{\text{loc}}(R^n)$: the expression $\check{S}f \equiv f|_{R^m} = \varphi$ means that $\varphi \in L_1^{\text{loc}}(R^m)$, while f is equivalent, in the sense of n-dimensional Lebesgue measure, to a function g such that $g(x_1,\dots,x_m,0,\dots,0) = \varphi(x_1,\dots,x_m)$, and

$$\lim_{x_{m+1},\dots,x_n\to 0} \|g(x_1,\dots,x_m,x_{m+1},\dots,x_n) - \varphi(x_1,\dots,x_m)\|_{L_1(\Pi\cap R^m)} = 0.$$

for any parallelepiped $\Pi \subset R^n$ such that $\Pi \cap R^m \neq \varnothing$. (Here, obviously, the corresponding trace operators S^*_a (restriction operators $\check{S}$) are jointly local.)

We remark that condition (27), valid for any $a > 0$ (without condition (28)), does not imply (26) in general. For example $L^*_p(R^n, a) \hookrightarrow L^*_{p'}(R^n, a)$ for $1 \leqslant p' < p \leqslant \infty$ and any $a > 0$, but the imbedding $L_p(R^n) \hookrightarrow L_{p'}(R^m)$ is not valid (in this case $\|I^*_a\| = a^{n(1/p'-1/p)} \to \infty$ as $a \to \infty$).

Theorem 11 implies the following assertion. *If there exists an extension operator (not necessarily linear)*

$$T\colon Y_0(R^m) \to X(R^n), \qquad Tf|_{R^m} = f, \tag{29}$$

where $Y_0(R^m)$ is the space of compactly supported functions in $Y(R^m)$ (and, all the more so, if there exists an extension operator T: $Y(R^m) \to X(R^m)$), then for any $a > 0$ there exists an extension operator

$$T^*_a\colon Y^*(R^m, a) \to X^*(R^n, a), \qquad T^*_a f|_{R^m} = f, \tag{30}$$

and

$$\|T^*_a\| \leqslant c_4(a)\|T\|,$$

where for any $a_0 > 0$ the inequality $c_4(a) \leqslant c_5(a_0)$ holds for all $a \geqslant a_0$.

Theorem 12 implies the converse. *If there exist extension operators* (30) *and we have the existence of*

$$\lim_{a\to\infty} T_a^* Q_a^{(m)} f \quad \text{in } \check{X}(R^n), \tag{31}$$

then there exists an extension operator

$$T\colon Y_0(R^m) \to X(R^n), \qquad Tf|_{R^m} = f, \tag{32}$$

and

$$\|T\| \leqslant c_6 \varlimsup_{a\to\infty} \|T_a^*\|. \tag{33}$$

If T is linear, then so are the T_a^*; conversely, if the T_a^* are linear, then so is T. The existence of extension operators with the property (31) is necessary for the existence of an extension operator T.

CHAPTER V

APPLICATION TO GENERAL ANISOTROPIC NIKOL'SKII - BESOV SPACES

§5.1. Definition and general properties of general Nikol'skiĭ-Besov spaces

1°. Definition and examples. Let $\mathscr{P} = \{P_t\}_0^\infty$ be a system of rectangular parallelepipeds

$$P_t = \{\xi = (\xi_1,\dots,\xi_n) \in R^n\colon |\xi_j - \xi_j^t| \leqslant \rho_{t,j}/2, j = 1,\dots,n\}, \tag{1}$$

where $\xi^t = (\xi_1^t,\dots,\xi_n^t)$ is the center of the parallelepiped, and $\rho_{t,j}$ is the length of the edge parallel to the ξ_j-axis. Everywhere below it is assumed that

$$1)\quad P_t \subset P_{t+1} \quad (t = 0,1,\dots), \qquad 2)\quad \bigcup_{t=0}^{\infty} P_t = R^n. \tag{2}$$

We associate with the covering $\mathscr{P} = \{P_t\}_0^\infty$ a system $s = \{s_t\}_0^\infty$ of positive numbers, and for $1 \leqslant p \leqslant \infty$ and $0 < \theta \leqslant \infty$ we define a general anisotropic Nikol'skiĭ-Besov space.

Definition 7. The space $B \equiv B_{p\theta}^{(\mathscr{P},s)}(R^n)$ is the collection of all functions $f(x) \in L_p(R^n)$ having an expansion

$$(q)\colon f(x) \underset{L_p(R^n)}{=} \sum_{t=0}^{\infty} q_t(x), \qquad q_t \in \mathfrak{M}_p(P_t, R^n), \tag{3}$$

such that

$$\left.\begin{aligned} &|f|_{(q)} = \left\{\sum_{t=0}^{\infty} \left(s_t\|q_t\|_{L_p(R^n)}\right)^\theta\right\}^{1/\theta} < \infty \qquad (0 < \theta < \infty),\\ \text{or}\quad &|f|_{(q)} = \sup_{t\geqslant 0}\left[s_t\|q_t\|_{L_p(R^n)}\right] < \infty \qquad (\theta = \infty).\end{aligned}\right\} \tag{4}$$

Let

$$\|f\|_B \equiv \|f\|_{B_{p\theta}^{(\mathscr{P},s)}(R^n)} = \inf_{(q)} |f|_{(q)}. \tag{5}$$

Here for a closed bounded region $G \subset R^n$ we have used the notation

$$\mathfrak{M}_p(G, R^n) = \{q(x) \in L_p(R^n)\colon \operatorname{supp} \mathscr{F} q \subseteq G\}, \tag{6}$$

where $\mathscr{F}$ is the Fourier transformation in R^n (generalized, in general).

By the Paley-Wiener theorem, the functions $q(x) \in \mathfrak{M}_p(G, R^n)$ are entire functions of exponential type, so that formula (3) gives an expansion for $f(x)$ as a series of entire functions. In (5) the infimum is taken over all such expansions satisfying (4).

Before discussing the general properties of the space B, we note that if $r = (r_1, \dots, r_n)$, $r_j > 0$, $1 \leqslant \theta$, $p \leqslant \infty$, and

$$P_t = \left\{\xi = (\xi_1, \dots, \xi_n) \in R^n \colon |\xi_j| \leqslant 2^{t/r_j}, j = 1, \dots, n\right\}, \qquad s_t = 2^t, \qquad t = 0, 1, \dots, \tag{7}$$

then the space $B_{p\theta}^{(\mathscr{P},s)}(R^n)$ coincides with the known anisotropic Nikol'skiĭ-Besov space $B_{p\theta}^r(R^n)$ considered in §4.4 (in this case Definition 7 coincides with the description of the space $B_{p\theta}^r$ in §5.6 of Nikol'skiĭ's book [1]). Definition 7 includes also spaces with conditions on the modulus of continuity of an L_p-function more general than the Hölder conditions on which the definition of the class $B_{p\theta}^r$ is based.

For example, suppose that $\omega(\tau) = \{\omega_1(\tau), \dots, \omega_n(\tau)\}$, $0 \leqslant \tau \leqslant 1$, where the functions $\omega_j(\tau)$ are continuous, $\omega_j(0) = 0$, $\omega_j(\tau)$ is increasing, but there exists a number $\beta_j > 0$ such that $\omega_j(\tau)\tau^{-\beta_j}$ is almost decreasing (i.e., $\omega_j(\tau)\tau^{-\beta_j} \leqslant d_j\omega_j(t)t^{-\beta_j}$ for any $0 \leqslant t \leqslant \tau \leqslant 1$, the $d_j \geqslant 1$ being fixed). The class $B_{p\theta}^{(\omega)}(R^n)$ is defined as the set of functions $f(x) \in L_p(R^n)$ with norm

$$\|f\|_B = \|f\|_{L_p} + \sum_{j=1}^{n} \left\{ \int_0^1 \left[\frac{\omega_{k_j}(f, \tau)_p}{\omega_j(\tau)}\right]^\theta \frac{d\omega_j(\tau)}{\omega_j(\tau)} \right\}^{1/\theta} < \infty \tag{8}$$

(the supremum of the expression in square brackets in (8) is taken instead of the integral when $\theta = \infty$). Here $\omega_{k_j}(f, \tau)_p$ is the modulus of continuity of order $k_j \geqslant 1$ for the function $f(x)$ in $L_p(R^n)$ with respect to the variable x_j, with step τ, and $0 < \beta_j < k_j$. If $\omega_j(\tau) = \tau^{r_j}$, $0 < r_j < k_j$, then $B_{p\theta}^{(\omega)}(R^n) = B_{p\theta}^r(R^n)$. In the general case it is possible to choose sequences $\{\lambda_j(t)\}_{t=0}^\infty$, $j = 1, \dots, n$, such that $\lambda_j(0) = 1$, $\lambda_j(t)/\lambda_j(t+1) \leqslant \mu_t < 1$ $(t = 0, 1, \dots)$, and such that the numbers

$$s_t = 1/\omega_j\big(1/\lambda_j(t)\big) \tag{9}$$

do not depend on $j = 1, \dots, n$ and satisfy the conditions $1 < \lambda_0 \leqslant s_{t+1}/s_t \leqslant \Lambda_0 < \infty$ $(t = 0, 1, \dots)$. For example, if $\omega_j(\tau) = \tau^{r_j}$, then we can take $\lambda_j(t) = 2^{t/r_j}$, and then $s_t = 2^t$. It can be proved by approximation theory methods that for the indicated choice of numbers $\lambda_j(t)$ an equivalent description of the space $B_{p\theta}^{(\omega)}(R^n)$ is obtained by setting

$$P_t = \left\{\xi = (\xi_1, \dots, \xi_n) \in R^n \colon |\xi_j| \leqslant \lambda_j(t), j = 1, \dots, n\right\} \tag{10}$$

in Definition 7 and defining the system $s = (s_t)_0^\infty$ by the conditions (9) (see, for example, [25], [15], and [26]). The spaces $B_{p\theta}^{(\mathscr{P},s)}(R^n)$ were defined in full generality by Gol'dman [26], [17]; a definition slightly less general was given by Kalyabin [16].

ASSERTION 1. *Suppose that*

$$\sup_{t \geqslant 0} s_t^{-1} < \infty \quad (0 < \theta \leqslant 1), \qquad \sum_{t=0}^{\infty} s_t^{-\theta'} < \infty \quad (1 < \theta \leqslant \infty), \tag{11}$$

where $1/\theta + 1/\theta' = 1$. *Then the following statements are true*:

(a) *For any* $f \in B$ *the expansion* (3) *converges to it absolutely in the* $L_p(R^n)$*-norm, and* $\|f\|_{L_p(R^n)} \leqslant c\|f\|_B$, *where c does not depend on* $f \in B$.

B *is a Banach* (*quasi-Banach if* $0 < \theta < 1$) *space, and the norm* (*quasinorm*) *in it is translation invariant, i.e.,* $\|E_y f\|_B = \|f\|_B$ *for any* $y \in R^n$.

Assertion 1 and the necessity of the conditions (11) for Definition 7 to make sense are stated in [17] (Theorem 1, Remark 4 to it, and §6).

2°. Boundedness of the operator of multiplication by a function in $C^M(R^n)$.

THEOREM 15. *Suppose that* $1 \leqslant p \leqslant \infty$, $0 < \theta \leqslant \infty$, *the conditions* (11) *hold, and there exist numbers* $\lambda_0 > 1$, $0 < \lambda < 1$, *and an integer* $L \geqslant 0$ *such that*

1) $$s_{t+1} \leqslant \lambda_0 s_t \qquad (t = 0, 1, \ldots), \tag{12}$$

2) $$P_t \subset P^\lambda_{t+L} \qquad (t = 0, 1, \ldots), \tag{13}$$

where

$$P_t^\lambda = \left\{\xi = (\xi_1, \ldots, \xi_n) \in R^n : |\xi_j - \xi_j^t| \leqslant \lambda \rho_{t,j}/2, j = 1, \ldots, n\right\} \tag{14}$$

is a "contraction" of the parallelepiped P_t *in* (1). *Then for some integer* $M = M(\lambda_0, \lambda, L) \geqslant 0$ *the operator* $\mathscr{P}_\varphi$ *of multiplication by* $\varphi \in C^M(R^n)$ *is bounded in the space* $B_{p\theta}^{(\mathscr{P}, s)}(R^n)$, *and* $\|\mathscr{P}_\varphi\| \leqslant B\|\varphi\|_{C^M(R)^n}$.

Before proceeding to the proof of the theorem, we obtain an auxiliary assertion. Fix λ, $0 < \lambda < 1$, and consider a function $0 \leqslant \mu(\tau) \in C_0^\infty(-1, 1)$ such that $\mu(\tau) = 1$ for $|\tau| \leqslant \lambda$. With its help we define the functions

$$\psi_t(\xi) = \prod_{j=1}^n \mu\left((\xi_j - \xi_j^t)/\rho_{t,j}\right) \in C_0^\infty(P_t), \qquad \psi_t(\xi) = 1, \xi \in P_t^\lambda, \tag{15}$$

$$\varphi_t(x) = \left(\mathscr{F}^{-1}\psi_t\right)(x) = (2\pi)^{-n/2} \int_{R^n} \psi_t(\xi) e^{ix\xi}\, d\xi \in L_1(R^n). \tag{16}$$

LEMMA 16. *For the functions* $\varphi_t(x)$

$$\|x^\alpha \varphi_t(x)\|_{L_1(R^n)} \leqslant c_\alpha \prod_{j=1}^n \rho_{t,j}^{-\alpha_j}, \qquad \alpha = (\alpha_1, \ldots, \alpha_n), \tag{17}$$

$$x^\alpha = \prod_{j=1}^n x_j^{\alpha_j}, \tag{18}$$

where the $\alpha_j \geqslant 0$ *are integers.*

PROOF. It is easy to see that

$$\varphi_t(x) = \prod_{j=1}^n e^{i\xi_j^t x_j} \rho_{t,j} \varphi^0(\rho_{t,j} x_j), \tag{19}$$

where

$$\varphi^0(u) = \frac{1}{\sqrt{2\pi}} \int_{-1}^1 \mu(\tau) e^{i\tau u}\, d\tau, \qquad u \in R^1. \tag{20}$$

Note that we have the estimates

$$|\varphi^0(u)| \leqslant c_0, \qquad |\varphi^0(u)| \leqslant c_l |u|^{-l} \quad (l = 1, 2, \ldots). \tag{21}$$

The first is obvious, and the second follows from (20) by integration l times by parts.

Let $J = \{j_1, \ldots, j_r\}$ denote a set of indices $1 \leqslant j_1 < \cdots < j_r \leqslant n$, where $0 \leqslant r = r(J) \leqslant n$; $\bar{J} = \{i_1, \ldots, i_{n-r}\}$ is the set of indices between 1 and n complementing J. If $J = \{1, \ldots, n\}$, then $\bar{J} = \varnothing$, and if $J = \varnothing$, then $\bar{J} = \{1, \ldots, n\}$. For any set J, any integer $l \geqslant 1$, and any $x \in R^n$ we have from (19) and (21) the estimate

$$|x^\alpha \varphi_t(x)| \leqslant c_{\alpha, l} \prod_{j \in J} \left(\rho_{t,j} |x_j^{\alpha_j}|\right) \prod_{i \in \bar{J}} \left(\rho_{t,i}^{1-l} x_i^{\alpha_i - l}\right), \tag{22}$$

which we shall use for $l < \max_{1 \leqslant i \leqslant n} \alpha_i + 1$ and for $x \in S_t(J) = \{x \in R^n \colon |x_j| \leqslant 1/\rho_{t,j},\, j \in J;\, |x_i| > 1/\rho_{t,i},\, i \in \bar{J}\}$. Then $R^n = \bigcup_J S_t(J)$, where the union (which is finite) is taken over all distinct sets J (including $\varnothing$ and the whole set $\{1, \ldots, n\}$). Therefore, it suffices to estimate the integral over any region $S_t(J)$, and this we do by using (22):

$$\int_{S_t(J)} |x^\alpha \varphi_t(x)|\,dx \leqslant c \prod_{j \in J} \left(\rho_{t,j} \int_{|\tau| \leqslant \rho_{t,j}^{-1}} \tau^{\alpha_j}\,d\tau\right) \prod_{i \in \bar{J}} \left(\rho_{t,i}^{1-l} \int_{|\tau| > \rho_{t,i}^{-1}} \tau^{\alpha_i - l}\,d\tau\right) \leqslant c \prod_{j=1}^{n} \rho_{t,j}^{-\alpha_j}.$$

The lemma is proved.

PROOF OF THEOREM 15. 1. Condition (11) for $0 < \theta \leqslant 1$ is equivalent to the condition that $\underline{\lim}_{t \to \infty} s_t > 0$. If also $\overline{\lim}_{t \to \infty} s_t < \infty$, then, as noted in [17], $B_{p\theta}^{(\mathscr{P}, s)}(R^n) = L_p(R^n)$, and in this case Theorem 15 is obvious. Therefore, without loss of generality it can be assumed that

$$\lim_{t \to \infty} s_t = \infty \quad (0 < \theta \leqslant 1), \qquad \sum_{t=0}^{\infty} s_t^{-\theta'} < \infty \quad (1 < \theta \leqslant \infty,\, 1/\theta + 1/\theta' = 1).$$

But then (after thinning out the covering $\mathscr{P}$ if necessary) it can be assumed without loss of generality that the numbers s_t increase at the rate of a geometric progression; namely, that there exist numbers $1 < \lambda_1 \leqslant \lambda_2 < \infty$, $\lambda_i = \lambda_i(\lambda_0)$, $i = 1, 2$, such that

$$\lambda_1 \leqslant s_{t+1}/s_t \leqslant \lambda_2, \qquad t = 0, 1, \ldots. \tag{23}$$

This assertion is also contained in [17] (Example 2 in §4). Note that condition (13) is not violated if the sequence is thinned out. Finally, under condition (23) we can give an equivalent description of the space B by using expansions (3) of a special form. Namely, for an $f \in L_p(R^n)$ and for the functions $\psi_t(\xi)$ in (15) let

$$\Lambda_0 f = \mathscr{F}^{-1} \psi_0 \mathscr{F} f, \qquad \Lambda_t f = \mathscr{F}^{-1}(\psi_t - \psi_{t-1}) \mathscr{F} f, \quad t = 1, 2, \ldots, \tag{24}$$

and consider the norm (quasinorm)

$$\|f\|_{\tilde{B}} = \left\{ \sum_{t=0}^{\infty} \left(s_t \|\Lambda_t f\|_{L_p(R^n)}\right)^\theta \right\}^{1/\theta} \qquad (0 < \theta < \infty),$$
$$\|f\|_{\tilde{B}} = \sup_{t \geqslant 0} \left[s_t \|\Lambda_t f\|_{L_p(R^n)}\right] \qquad (\theta = \infty). \tag{25}$$

Then, as shown in [17] (Theorem 11), the condition $f \in B$ is equivalent to the condition that $f \in L_p(R^n)$ and $\|f\|_{\tilde{B}} < \infty$, and $\|f\|_{\tilde{B}}$ is an equivalent norm in B. This description will be used in the proof of the theorem.

2. We must show that for $f \in B$ and $\varphi \in C^M(R^n)$

$$\|f\varphi\|_{\tilde{B}} \leqslant \left(\|\varphi\|_{C(R^n)} + B_1 \sum_{0 < |\alpha| \leqslant M} \|D^\alpha \varphi\|_{C(R^n)}\right) \|f\|_{\tilde{B}}, \tag{26}$$

where $M \geqslant 0$ is a sufficiently large integer.

Setting $\varphi_{-1}(x) \equiv 0$, we consider

$$\Lambda_t(\varphi f, x) = (2\pi)^{-n/2}\int_{R^n}[\varphi_t(y) - \varphi_{t-1}(y)]f(x-y)\varphi(x-y)\,dy$$

for $t = 0, 1, \ldots$ (we have passed from (24) to an equivalent expression as a convolution with the kernels $\varphi_t(x)$ in (16)). We apply to $\varphi(x-y)$ Taylor's formula with the remainder in the Lagrange form. Then

$$\Lambda_t(\varphi f, x) = \varphi(x)\Lambda_t(f, x) + \sum_{0<|\alpha|\leqslant M-1} d_\alpha D^\alpha\varphi(x)(\Lambda_t^\alpha f)(x)$$
$$+ (2\pi)^{-n/2}\sum_{|\alpha|=M} d_\alpha q_t^\alpha(f, \varphi, x),$$

where the d_α are the coefficients in Taylor's formula, and for $t = 0, 1, \ldots$

$$\Lambda_t^\alpha f(x) = (2\pi)^{-n/2}\int_{R^n} y^\alpha[\varphi_t(y) - \varphi_{t-1}(y)]f(x-y)\,dy, \tag{27}$$

$$q_t^\alpha(f, \varphi, x) = \int_{R^n} y^\alpha[\varphi_t(y) - \varphi_{t-1}(y)]D^\alpha\varphi(x-\beta y)f(x-y)\,dy, \tag{28}$$

where $\beta = \beta(y)$ satisfies $0 < \beta < 1$. The number M will be specified later.

By using the condition $\varphi(x) \in C^M(R^n)$ it is easy to show that (26) follows from the inequalities

$$\left\{\sum_{t=0}^{\infty}\left(s_t\|\Lambda_t^\alpha f\|_{L_p(R^n)}\right)^\theta\right\}^{1/\theta} \leqslant c_\alpha\|f\|_{\bar{B}}, \qquad |\alpha| \leqslant M-1, \tag{29}$$

$$\left\{\sum_{t=0}^{\infty}\left(s_t\|Q_t^\alpha(f)\|_{L_p(R^n)}\right)^\theta\right\}^{1/\theta} \leqslant c_\alpha\|f\|_{L_p(R^n)}, \qquad |\alpha| = M, \tag{30}$$

where

$$Q_t^\alpha(f, x) = \int_{R^n}|y^\alpha[\varphi_t(y) - \varphi_{t-1}(y)]|\cdot|f(x-y)|\,dy, \qquad t = 0, 1, \ldots. \tag{31}$$

The estimate (30) will be obtained by choosing $M = |\alpha| = M(\lambda, \lambda_0, L)$ sufficiently large, and (29) will be proved for all α with $|\alpha| \leqslant M-1$.

3. We first obtain (29). From (17) and the monotonicity of $\rho_{t,j}$, $t = 0, 1, \ldots,$ it follows that for $t = 0, 1, \ldots$

$$\|y^\alpha[\varphi_t(y) - \varphi_{t-1}(y)]\|_{L_1(R^n)} \leqslant c_\alpha\prod_{j=1}^n \rho_{t-1,j}^{-\alpha_j} \leqslant c_\alpha. \tag{32}$$

Here we set $\rho_{-1,j} = \rho_{0,j}$.

Moreover,

$$y^\alpha[\varphi_t(y) - \varphi_{t-1}(y)] = c(\alpha)\mathscr{F}^{-1}(D^\alpha[\psi_t - \psi_{t-1}]),$$

so that, by (13),

$$\operatorname{supp}\mathscr{F}(y^\alpha[\varphi_t(y) - \varphi_{t-1}(y)]) \subset P_t\setminus P_{t-1}^\lambda \subset P_t\setminus P_{t-L-1}. \tag{33}$$

We use the expansion $f(x) = \sum_0^\infty \lambda_\tau(f, x)$ in $L_p(R^n)$ and get that

$$\Lambda_t^\alpha f(x) = \sum_{\tau=t-L}^{t+L}\Lambda_t^\alpha(\Lambda_\tau f)(x), \qquad t = L, L+1, \ldots$$

(the remaining terms are equal to zero in view of (33)).

This, (27), and (32) tell us, by Minkowski's inequality for the convolution, that

$$\|\Lambda_t^\alpha f\|_{L_p(R^n)} \leqslant c_\alpha \sum_{\tau=t-L}^{t+L} \|\Lambda_\tau f\|_{L_p(R^n)}, \qquad t = L, L+1, \ldots,$$

and, similarly,

$$\|\Lambda_t^\alpha f\|_{L_p(R^n)} \leqslant c_\alpha \sum_{\tau=0}^{2L} \|\Lambda_\tau f\|_{L_p(R^n)}, \qquad t = 0, 1, \ldots, L-1.$$

Multiplying both sides by s_t and using the fact that $s_t \sim s_\tau$ for $t - L \leqslant \tau \leqslant t + L$ (see (23)), we get

$$\left\{ \sum_{t=0}^{\infty} \left(s_t \|\Lambda_t^\alpha f\|_{L_p(R^n)} \right)^\theta \right\}^{1/\theta} \leqslant \tilde{c}_\alpha \left\{ \sum_{\tau=0}^{\infty} \left(s_\tau \|\Lambda_\tau f\|_{L_p(R^n)} \right)^\theta \right\}^{1/\theta},$$

which coincides with (29). Here $\tilde{c}_\alpha$ depends on α, L, λ_1, λ_2, and θ.

4. We prove (30) for sufficiently large $M = |\alpha|$. By Minkowski's inequality for the convolution, (31) and (32) give us that

$$\|Q_t^\alpha(f)\|_{L_p(R^n)} \leqslant c_\alpha \prod_{j=1}^{n} \rho_{t-1,j}^{-\alpha_j} \|f\|_{L_p(R^n)}.$$

To derive (30) it remains to prove that for sufficiently large $M = |\alpha| = \alpha_1 + \cdots + \alpha_n$ the series

$$\sum_{t=0}^{\infty} \left(s_t \prod_{j=1}^{n} \rho_{t-1,j}^{-\alpha_j} \right)^\theta$$

converges. There exists a j_0 such that $\alpha_{j_0} \geqslant M/n$; hence

$$\prod_{j=1}^{n} \rho_{t-1,j}^{-\alpha_j} \leqslant c_\alpha \rho_{t-1,j_0}^{-M/n} \leqslant c_\alpha \rho_{0,j_0} \lambda^{(t-1)M/(nL)},$$

where the number $\lambda, 0 < \lambda < 1$, is taken from (13). At the same time, $s_t \leqslant s_0 \lambda_0'$ by (12). These inequalities with M sufficiently large imply the convergence of the series and the estimate (30). The theorem is proved.

3°. Denseness of the compactly supported functions in the space $B_{p\theta}^{(\mathscr{P},s)}(R^n)$ for $p, \theta < \infty$.

1. *The topological imbedding $S(R^n) \hookrightarrow B_{p\theta}^{(\mathscr{P},s)}(R^n)$ is valid under the conditions of Theorem* 15.

This assertion can be obtained from Theorem 5 of [17], and we confine ourselves here to a brief sketch of its derivation. Suppose that $\varphi(x) \in S(R^n)$ and let $\psi(\xi) = \mathscr{F}\varphi(\xi)$. It follows from (24) and (15) that for any $\alpha = (\alpha_1, \ldots, \alpha_n)$ with $\alpha_j \geqslant 0$

$$x^\alpha \Lambda_t(\varphi, x) = c(\alpha) \int_{P_t \setminus P_{t-1}^\lambda} e^{ix\xi} D^\alpha(\psi(\xi)[\psi_t(\xi) - \psi_{t-1}(\xi)])\, d\xi.$$

Since $\psi \in S(R^n)$ is rapidly decreasing along any of its derivatives, and $|D^\beta \psi_t(\xi)| \leqslant c_\beta$, this gives us the estimate

$$|x^\alpha \Lambda_t(\varphi, x)| \leqslant c_{\alpha,\gamma}(\varphi) \prod_{j=1}^{n} \rho_{t-1,j}^{-\gamma} \quad \forall \gamma \geqslant 1,$$

and from it, just as in the proof of Lemma 16, we get that

$$\|\Lambda_t(\varphi)\|_{L_p(R^n)} \leqslant c_\gamma(\varphi) \prod_{j=1}^{n} \rho_{t-1,j}^{-\gamma} \quad \forall \gamma \geqslant 1.$$

In view of (12) and (13) we conclude from this (just as in the fourth step in the proof of Theorem 15) that $\|\varphi\|_{\tilde{B}} < \infty$. This argument is not hard to mold into a proof that the operator imbedding $S(R^n)$ into B is bounded.

2. *The set of entire functions in $S(R^n)$ is dense in $B_{p\theta}^{(\mathscr{P},s)}(R^n)$ for $p, \theta < \infty$.*

This property was proved in [17] (Theorem 2).

3. *The set of functions in $C_0^\infty(R^n)$ is dense in $B_{p\theta}^{(\mathscr{P},s)}(R^n)$ for $p, \theta < \infty$.*

Indeed, $C_0^\infty(R^n)$ is dense in $S(R^n)$ in the $S(R^n)$ topology; in particular, $C_0^\infty(R^n)$ is dense in the set of entire functions in $S(R^n)$, hence also in B in the sense of the norm $\|\cdot\|_B$ for $p, \theta < \infty$.

REMARK. We see that the general anisotropic Nikol'skiĭ-Besov spaces have the properties of completeness, translation invariance of the norm (property 1(d)), boundedness of the operator of multiplication by a function in $C^M(R^n)$ (property 3(b), and even 3(c) if the equivalent norm $\|\cdot\|_{\tilde{B}}$ is used in B, see (26)), and also denseness of the compactly supported functions. This enables us to apply the results in Chapters I–III to the spaces B.

§5.2. General periodic Nikol'skiĭ-Besov spaces

1°. Equivalent definitions. The spaces $L_p(R^n)$ and $\mathfrak{M}_p(G, R^n)$ on which the definition of the space $B_{p\theta}^{(\mathscr{P},s)}(R^n)$ is based have natural periodic analogues: the periodic space $L_p^*(R^n, a)$ and the space $\mathfrak{M}_p^*(G, T_a^n)$ of trigonometric polynomials of the form

$$q(x) = \sum_{l \in L_G} q_l \exp(i2\pi lx/a), \qquad x = (x_1, \dots, x_n) \in T_a^n,$$

with norm $\|\cdot\|_{L_p^*(R^n,a)} = \|\cdot\|_{L_p(T_a)}$. Here $L_G = \{l \in \mathbf{Z}^n \colon 2\pi l/a \in G\}$. Therefore, the natural periodic analogue of Definition 7 has the following form.

DEFINITION 8. The space $B_a^* \equiv B_{p\theta}^{(\mathscr{P},s)*}(R^n, a)$ is the collection of all functions $f(x) \in L_p^*(R^n, a)$ which have an expansion

$$(q)\colon f \underset{L_p^*(R^n,a)}{=} \sum_{t=0}^{\infty} q_t(x), \qquad q_t \in \mathfrak{M}_p^*(P_t, T_a^n), \tag{34}$$

such that

$$|f|_{(q)}^* = \left\{ \sum_{t=0}^{\infty} \left(s_t \|q_t\|_{L_p(T_a^n)} \right)^\theta \right\}^{1/\theta} < \infty \qquad (0 < \theta < \infty), \tag{35}$$

or

$$|f|_{(q)}^* = \sup_{t \geqslant 0} \left[s_t \|q_t\|_{L_p(T_a^n)} \right] < \infty \qquad (\theta = \infty).$$

Let

$$\|f\|_{B_a^*} = \inf_{(q)} |f|_{(q)}^*. \tag{36}$$

However, we can also give another definition. For this, recall that the space $Z(R^n) = B_{p\theta}^{(\mathscr{P},s)}(R^n)$ has the properties 1(d) and 3(b), and consequently admits the definition of the periodic analogue $Z^*(R^n, a)$ introduced in §1.1. We denote this periodic analogue by $\tilde{B}_a^* \equiv \tilde{B}_{p\theta}^{(\mathscr{P},s)*}(R^n, a)$ and let its norm $\|\cdot\|_{\tilde{B}_a^*}$ be one of the equivalent norms considered in Theorem 4. If we prove that $B_a^* = \tilde{B}_a^*$, then by using the properties given in [17] and in §51. for the spaces B, along with the results in Chapters I–IV, we can at once obtain results for the spaces B_a^* analogous to those known for the spaces B.

THEOREM 16. *Suppose that the conditions of Theorem* 15 *hold, and* $a_0 > 0$. *For any* $a > 0$ *the spaces* B_a^* *and* $\tilde{B}_a^*$ *coincide, and their norms are equivalent, with constants independent of a for all* $a \geqslant a_0$.

PROOF. 1. We first carry out a proof for $\theta \geqslant 1$, when B is a normed space. Let $f(x) \in B_a^*$, $a \geqslant a_0$. We choose a function $\varphi(x) = \varphi_a(x) \in \Phi_a$ (see §4.3) and prove that $f\varphi \in B$ and

$$\|f\varphi\|_B \leqslant c_1\|f\|_{B_a^*}, \qquad c_1 = c_1(c_0, a_0, h, \lambda, L, \lambda_0), \tag{37}$$

with the constants taken from the conditions that $\varphi \in \Phi_a$, (12), and (13). This gives us the imbedding $B_a^* \hookrightarrow \tilde{B}_a^*$ and, by Theorem 4, the norm estimate

$$\|f\|_{\tilde{B}_a^*} \leqslant c_2\|f\|_{B_a^*}, \qquad c_2 = c_2(c_0, a_0, h, \lambda, L, \lambda_0). \tag{38}$$

To prove (37) we choose for $f \in B_a^*$ an expansion (34) such that $|f|_{(q)}^* \leqslant 2\|f\|_{B_a^*}$, and multiply both sides of it by $\varphi(x)$. Here a sequence converging in $L_p^*(R^n, a)$ passes into a sequence converging in $L_p(R^n)$ (see, for example, §4.4). Therefore, we have the expansion

$$(\tilde{q})\colon f\varphi \underset{L_p(R^n)}{=} \sum_{t=0}^{\infty} q_t\varphi,$$

and

$$\|q_t\varphi\|_{L_p(R^n)} \leqslant \|\varphi\|_{C^0(R^n,a)}\|q_t\|_{L_p(T_a)} \leqslant c_0\|q_t\|_{L_p(T_a)}.$$

Moreover, for $\mathscr{F}(q_t\varphi) = \mathscr{F}q_t * \mathscr{F}\varphi$ (the convolution is understood in the generalized sense, for example, in $S'(R^n)$) we have that

$$\operatorname{supp}\mathscr{F}(q_t\varphi) \subset \operatorname{supp}\mathscr{F}q_t + \operatorname{supp}\mathscr{F}\varphi \subset P_{t+r} \qquad (t = 0, 1, \dots),$$

where $r = r(\varepsilon, a_0, \lambda, L)$ is chosen to be sufficiently large (take into account the condition $\operatorname{supp}\mathscr{F}\varphi \subset K_{\varepsilon/a}^n$ (see §4.3), and also (13)).

Thus,

$$\tilde{q}_{t+r} \equiv q_t\varphi \in \mathfrak{M}_p(P_{t+r}, R^n), \qquad t = 0, 1, \dots,$$

and

$$|f\varphi|_{(\tilde{q})} = \left\{\sum_{t=0}^{\infty}\left(s_{t+r}\|\tilde{q}_{t+r}\|_{L_p(R^n)}\right)^\theta\right\}^{1/\theta} \leqslant c_0\left\{\sum_{t=0}^{\infty}\left(s_{t+r}\|q_t\|_{L_p(T_a)}\right)^\theta\right\}^{1/\theta} \leqslant c_0\lambda_0^r|f|_{(q)}^*.$$

At the end we have taken condition (12) into account. The estimate (37) follows from this and the choice of the expansion (q).

2. Before proceeding to the proof of the reverse imbedding, we consider the function $\varphi(x) \in \Phi_a$, $a \geqslant a_0$, used in the first step of the proof, and choose for it a function $0 \leqslant \varphi_1(x) \in C_0^\infty(R^n)$ with $\operatorname{supp}\varphi_1 \subset T_{2a}$ such that for $M \geqslant 1$

$$\|\varphi_1\|_{C^M(R^n)} \leqslant d(M, n, a_0, h), \qquad Q_a(\varphi_1\varphi) \equiv 1. \tag{39}$$

This can be done, for example, as follows. Suppose that $0 \leqslant \psi(x) \in C_0^\infty(R^n)$, $\operatorname{supp}\psi \subset T_{2a}$, $\|\psi\|_{C^M(R^n)} \leqslant d_1(M, n, a_0)$ and $(Q_a\psi)(x) \equiv 1$ (the existence of such functions follows from Lemma 2). Then let $\varphi_1(x) = \psi(x)/\varphi(x)$. It is not hard to see that the properties (39) hold for it.

3. Suppose now that $f \in \tilde{B}_a^*$, $a \geqslant a_0$. We prove that $f \in B_a^*$ and

$$\|f\|_{B_a^*} \leqslant c_3\|f\|_{\tilde{B}_a^*}, \qquad c_3 = c_3(M, n, a_0, h, \lambda_0, \lambda, L, c_0). \tag{40}$$

This gives us the imbedding $\tilde{B}_a^* \hookrightarrow B_a^*$.

It follows from the condition $f \in \tilde{B}_a^*$ that $\varphi_1 f \in B$, and

$$\|\varphi_1 f\|_B \leqslant d_2(M, n, a_0, h)\|f\|_{\tilde{B}_a^*} \tag{41}$$

in view of (39), Remark 4, and Theorem 4 (if we take into account that 3(b) holds for B according to Theorem 15).

We consider an expansion

$$(q_1)\colon \varphi_1 f \underset{L_p(R^n)}{=} \sum_{t=0}^{\infty} q_t, \qquad q_t \in \mathfrak{M}_p(P_t, R^n), \tag{42}$$

for which

$$|\varphi_1 f|_{(q_1)} = \left\{\sum_{t=0}^{\infty}\left(s_t\|q_t\|_{L_p(R^n)}\right)^\theta\right\}^{1/\theta} \leqslant 2\|\varphi_1 f\|_B. \tag{43}$$

Note that for any function $g \in L_p(R^n)$ and $\varphi \in \Phi_a$

$$\|Q_a(g\varphi)\|_{L_p^*(R^n,a)} \leqslant \|\varphi\|_{C^0(R^n,a)}\|g\|_{L_p(R^n)}. \tag{44}$$

Therefore, multiplying both sides of (42) by $\varphi(x)$ and summing the translations by al over $l \in \mathbf{Z}^n$, we get the expansion

$$(q)\colon f \underset{L_p^*(R^n,\,a)}{=} \sum_{t=0}^{\infty} q_t^*, \qquad q_t^*(x) = Q_a(q_t\varphi)(x). \tag{45}$$

We have also taken into account the equality $Q_a(\varphi_1\varphi f) \equiv f$, which follows from (39) and the periodicity of $f(x)$.

Note that $(q_t\varphi)(x)$ decreases rapidly as $x \to \infty$ together with φ, and that $\mathcal{F}(q_t\varphi) \in C(R^n)$ and $\operatorname{supp}\mathcal{F}(q_t\varphi) \subset P_{t+r}$ (see the argument in the first step of the proof). Therefore, the Poisson summation formula (see [2], Theorem 2.4, Corollary 6) is applicable; it says that the Fourier coefficients $(q_t^*)_l$ in the trigonometric system $\{\exp(i2\pi lx/a)\}_{l\in\mathbf{Z}^n}$ coincide with $\mathcal{F}(q_t\varphi)(2\pi l/a)$, i.e., are nonzero only for $2\pi l/a \in P_{t+r}$. Hence, $q_t^* \in \mathfrak{M}_p^*(P_{t+r}, T_a^n)$ and, by (44),

$$\|q_t^*\|_{L_p(T_a^n)} \leqslant c_0\|q_t\|_{L_p(R^n)}, \qquad t = 0, 1, \ldots.$$

But then from (45), (43), and the condition $s_{t+1} \leqslant \lambda_0 s_t$ we get that

$$|f|_{(q)} = \left\{\sum_{t=0}^{\infty}\left(s_{t+r}\|q_t^*\|_{L_p(T_a^n)}\right)^\theta\right\}^{1/\theta} \leqslant 2c_0\lambda_0^r\|\varphi_1 f\|_B.$$

The estimate (40) follows from this and (41).

4. The estimates (37) and (40) prove the theorem at this point for $1 \leqslant \theta \leqslant \infty$. But if $0 < \theta < 1$, then we should use another definition of the norm in $C^M(R^n, a)$ for $\varphi(x) \in \Phi_a$ according to 2° in §4.1, adapting it to the case of quasinormed spaces. The remaining arguments do not change.

REMARK. We have considered the general anisotropic spaces $B_{p\theta}^{(\mathscr{P},s)}(R^n)$ for $1 \leqslant p \leqslant \infty$. Triebel, Peetre, and others have also considered the isotropic spaces $B_{p\theta}^r(R^n)$, $r \in R^1$, $0 < p, \theta \leqslant \infty$ (see Triebel's book [27]). Their definition is equivalent to Definition 7, in which it is necessary to take $P_t = \{\xi \in R^n\colon |\xi_j| \leqslant 2^t\}$, $s_t = 2^{t/r}$, $t = 0, 1, \ldots$, and assume that $q_t \in \mathfrak{M}_p(P_{t+1} \setminus P_{t-1}, R^n)$ in (3) (see [27], §2.2.2). We get natural periodic analogues $B_a^* = B_{p\theta}^{r*}(R^n, a)$ of these spaces if we modify Definition 8 in the same way. An analogue of Theorem 15 is known for the spaces $B_{p\theta}^r(R^n)$ (see [27], §2.6.1) and it is known that the norm is translation invariant. This enables us to apply all the results in Chapters I–III (which, as we noted in 2° of §4.1, are valid for quasinormed spaces) to these spaces for $0 < p \leqslant \infty$. The periodic spaces thus obtained ($\tilde{B}_a^*$ in our notation) coincide with the spaces B_a^*. This follows from Theorem 16, whose proof works also when $0 < p < 1$ (see the fourth step of the proof). All the results in Chapters I–III on the interconnection of periodic and nonperiodic spaces are thereby applicable for the spaces $B_{p\theta}^r(R^n)$ and $B_{p\theta}^{r*}(R^n, a)$, $0 < p, \theta < \infty$, $r \in R^1$.

Triebel [18] worked out another approach to the study of 2π-periodic spaces, which involves regarding them as subspaces of weighted Besov spaces with rapidly decreasing weights and enables us to carry imbedding theorems for weighted spaces over to the 2π-periodic case.

2°. Examples of application of the results in Chapters I–III to general Nikol'skiĭ-Besov spaces. We give some examples of how it is possible to use the theorems in Chapters I–III and known properties of nonperiodic Nikol'skiĭ-Besov spaces to get corresponding properties for periodic spaces.

1. *Imbeddings with different metrics.* Suppose that $1 \leqslant p \leqslant \infty$, $0 < \theta \leqslant \infty$, $\mathscr{P} = \{P_t\}_0^\infty$ is a covering of R^n, $s = \{s_t\}_0^\infty$ is a system of positive numbers, and (12) and (13) hold. Further, suppose that $p' \geqslant p$, $0 < \theta' \leqslant \infty$, and the system $\{s'_t\}_0^\infty = s'$ of positive numbers also satisfies (12). Let $a_t = s_t^{-1} s'_t |P_t|^{1/p - 1/p'}$. Then under the conditions

$$\sup_{t \geqslant 0} a_t < \infty \quad (\theta \leqslant \theta'), \qquad \sum_{t=0}^{\infty} a_t^q < \infty \quad \left(\theta > \theta', q = \frac{\theta' \theta}{\theta - \theta'}\right)$$

we have the imbedding

$$B_{p\theta}^{(\mathscr{P},s)*}(R^n, a) \hookrightarrow B_{p'\theta'}^{(\mathscr{P},s')*}(R^n, a),$$

and the norm of the imbedding operator is uniformly bounded for all $a \geqslant a_0 > 0$.

This result follows from Theorem 6 in [17], where a corresponding assertion is given for the nonperiodic case, and from Theorem 7 with account taken of the properties of B proved in §§5.1 and 5.2.

2. *Description of traces.* Suppose that $1 \leqslant m < n$ and $x = (u, v) \in R^n$, $u \in R^m$. Under the assumptions (12) and (13) we write $P'_t = P_t|_{R^m}$, $s'_t = s_t \Pi_{j=m+1}^n \rho_{t,j}^{-1/p}$ (see (1)), $\mathscr{P}' = \{P'_t\}_0^\infty$ and $s' = \{s'_t\}_0^\infty$. For every function

$$f(u, v) \in B_{p\theta}^{(\mathscr{P},s)*}(R^n, a)$$

to have a trace on R^m it is necessary and sufficient that

$$\sup_{t \geqslant 0} (s'_t)^{-1} < \infty \qquad (0 < \theta \leqslant 1),$$

$$\sum_{t=0}^{\infty} (s'_t)^{-\theta'} < \infty \qquad (1 < \theta \leqslant \infty, 1/\theta + 1/\theta' = 1). \tag{46}$$

Under this condition we have the imbedding

$$B_{p\theta}^{(\mathcal{P},s)*}(R^n, a) \rightleftarrows B_{p\theta}^{(\mathcal{P}',s')*}(R^m, a). \tag{47}$$

(a) If $1 < \theta \leqslant \infty$ or $0 < \theta \leqslant 1$ and $\lim_{t\to\infty} s'_t = \infty$, then the extension operator in (47) is linear and continuous.

(b) If $0 < \theta \leqslant 1$ and $\underline{\lim}_{t\to\infty} s'_t < \infty$, then

$$B_{p\theta}^{(\mathcal{P}',s')*}(R^m, a) = L_p^*(R^m, a)$$

and the bounded extension operator in (47) is nonlinear; moreover, there does not exist a bounded linear extension operator.

Indeed, the sufficiency of the conditions (46) and the description of traces in (47), as well as part (a) of this assertion follow at once from Theorem 12 in [17] and the theorems in Chapter III. The coincidence $B^* = L_p$ under the conditions of part (b) also follows from the corresponding nonperiodic result (see, for example, [17], §6.2) and Theorem 7 (or 8).

The necessity of the conditions (46) requires some additional arguments, and we do not dwell on them here. Finally, the impossibility of a linear extension under the conditions of part (b) is simpler to prove in the periodic case and then carry over to nonperiodic spaces by using Theorem 11. The situation here is analogous to that described in the Introduction. But these assertions are proved in a very complicated way and require treatment in a separate paper.

3. *Description of a periodic space in terms of differences.* In §5.1 we mentioned that under the condition (7)

$$B_{p\theta}^{(\mathcal{P},s)}(R^n) = B_{p\theta}^{r}(R^n). \tag{48}$$

We define $B_{p\theta}^{r*}(R^n, a)$ as in §4.4, with the norms expressed in terms of finite differences. Then under condition (7)

$$B_{p\theta}^{(\mathcal{P},s)*}(R^n, a) = B_{p\theta}^{r*}(R^n, a). \tag{49}$$

Indeed, by Theorem 7, it follows from (48) that the periodic analogues (in the sense of Chapter I) coincide for the indicated spaces. But, by Theorem 16, $B_{p\theta}^{(\mathcal{P},s)*}(R^n, a)$ also is a periodic analogue for $B_{p\theta}^{(\mathcal{P},s)}(R^n)$, and, according to §4.4, the periodic analogue of $B_{p\theta}^{r}(R^n)$ coincides with $B_{p\theta}^{r*}(R^n, a)$. We have obtained (49) without resorting to the theory of approximation of functions by trigonometric polynomials.

Bibliography

1. S. M. Nikol'skiĭ, *Approximation of functions of several variables and embedding theorems*, 2nd rev. aug. ed., "Nauka", Moscow, 1977; English transl. of lst ed., Springer-Verlag, 1974.

2. Elias M. Stein and Guido Weiss, *Introduction to Fourier analysis on Euclidean spaces*, Princeton Univ. Press, Princeton, N. J., 1971.

3. V. I. Burenkov and M. L. Gol'dman, *On extension of functions in* L_p, Trudy Inst. Mat. Steklov. **150** (1979), 31–51; English transl. in Proc. Steklov Inst. Math. **1981**, no. 4(150).

4. G. G. Magaril-Il'yaev, *Imbedding of generalized Sobolev classes and inequalities for derivatives*, Author's Summary of Candidate's Dissertation, Computing Center Sci. Res. Inst. Complex Automation, Moscow, 1980. (Russian)

5. M. Z. Berkolaĭko and V. I. Ovchinnikov, *Inequalities for entire functions of exponential type in symmetric spaces*, Trudy Inst. Mat. Steklov. **161** (1983), 3–18; English transl. in this issue.

6. S. M. Nikol'skiĭ, *On a property of the classes* H_p, Ann. Univ. Sci. Budapest. Eötvös Sect. Math. **3 / 4** (1960/61), 205–216. (Russian)

7. G. N. Yakovlev, *Boundary properties of functions in the class* $W_p^{(l)}$ *on domains with corner points*, Dokl. Akad. Nauk SSSR **140** (1961), 73–76; English transl. in Soviet Math. Dokl. **2** (1961).

8. V. I. Burenkov, *On additivity of the classes* $W_p^r(\Omega)$, Trudy Mat. Inst. Steklov. **89** (1967), 31–55; English transl. in Proc. Steklov Inst. Math. **89** (1967).

9. ______, *On additivity of the spaces* W_p^r *and* B_p^r *and on imbedding theorems for domains of general form*, Trudy Mat. Inst. Steklov **105** (1969), 30–45; English transl. in Proc. Steklov Inst. Math. **105** (1969).

10. ______, *On the property of additivity of the space* $W_p^{(r)}(\Omega)$, Imbedding Theorems and Their Applications (Proc. Sympos., Baku, 1966), "Nauka", Moscow, 1970, pp. 47–52. (Russian)

11. Yu. V. Kuznetsov, *Some inequalities for fractional seminorms*, Trudy Mat. Inst. Steklov. **131** (1974), 147–157; English transl. in Proc. Steklov Inst. Math. **131** (1974).

12. ______, *On pasting together functions in the spaces* $W_{p,\theta}^r$, Trudy Mat. Inst. Steklov. **140** (1976), 191–200; English transl. in Proc. Steklov Inst. Math. **1979**, no. 1(140).

13. ______, *Some properties of the spaces* $W_{p,\theta}^r$ *and* $B_{p,\theta}^r$, Theory of Cubature Formulas and Applications of Functional Analysis to Some Problems in Mathematical Physics (Materials School-Conf., Tashkent, 1974; S. L. Sobolev, editor), Inst. Mat. Sibirsk. Otdel. Akad. Nauk SSSR, Novosibirsk, 1975, pp. 89–93. (Russian)

14. ______, *Some properties of spaces of functions with fractional order of differentiation*, Author's Summary of Candidate's Dissertation, Moscow Regional Pedagogical Institute, Moscow, 1975. (Russian)

15. M. L. Gol'dman, *A description of traces of anisotropic generalized Liouville classes*, Dokl. Akad. Nauk SSSR **233** (1977), 17–20; English transl. in Soviet Math. Dokl. **18** (1977).

16. G. A. Kalyabin, *Imbedding theorems for generalized Besov and Liouville spaces*, Dokl. Akad. Nauk SSSR **232** (1977), 1245–1248; English transl. in Soviet Math. Dokl. **18** (1977).

17. M. L. Gol'dman, *The method of coverings for describing general spaces of Besov type*, Trudy Mat. Inst. Steklov. **156** (1980), 47–81; English transl. in Proc. Steklov Inst. Math. **1983**, no. 2(156).

18. Hans Triebel, *Periodic spaces of Besov-Hardy-Sobolev type and related maximal inequalities for trigonometric polynomials*, Forschungsergebnisse N/80/11, Friedrich Schiller Univ., Jena, 1980.

19. S. G. Kreĭn, Yu. I. Petunin and E. M. Semenov, *Interpolation of linear operators*, "Nauka", Moscow, 1978; English transl., Amer. Math. Soc., Providence, R. I., 1982.

20. Jöran Bergh and Jörgen Löfström, *Interpolation spaces: an introduction*, Stringer-Verlag, 1976.

21. N. S. Nikol'skaya, *Approximation of differentiable functions of several variables by Fourier sums in the* L_p*-metric*, Sibirsk. Mat. Zh. **15** (1974), 395–412; English transl. in Siberian Math. J. **15** (1974).

22. È. M. Galeev, *Approximation of classes of periodic functions with several bounded derivatives*, Author's Summary of Candidate's Dissertation, Moscow State Univ., Moscow, 1978. (Russian)

23. V. N. Temlyakov, *Approximation of periodic functions of several variables with a bounded mixed derivative*, Dokl. Akad. Nauk SSSR **248** (1979), 527–531; English transl. in Soviet Math. Dokl. **20** (1979).

24. Din' Zung and G. G. Magaril-Il'yaev, *Problems of Bernstein and Favard type and the mean* ε*-dimension of certain classes of functions*, Dokl. Akad. Nauk SSSR **249** (1979), 783–786; English transl. in Soviet Math. Dokl. **20** (1979).

25. M. L. Gol'dman, *Description of the space of traces for functions in a generalized Hölder class*, Dokl. Akad. Nauk SSSR **231** (1976), 525–528; English transl. in Soviet Math. Dokl. **17** (1976).

26. ______, *Description of traces for certain function spaces*, Trudy Mat. Inst. Steklov. **150** (1979), 99–127; English transl. in Proc. Steklov Inst. Math. 1981, no. 4(150).

27. Hans Triebel, *Spaces of Besov-Hardy-Sobolev type*, Teubner, Leipzig, 1978.

28. V. I. Burenkov, *On the extension of functions with preservation of seminorms*, Dokl. Akad. Nauk SSSR **228** (1976), 779–782; English transl. in Soviet Math. Dokl. **17** (1976).

29. O. V. Besov, V. P. Il'in, and S. M. Nikol'skiĭ, *Integral representations of functions and imbedding theorems*, "Nauka", Moscow, 1975; English transl., Vols. 1, 2, Wiley, 1979.

30. V. I. Burenkov and M. L. Gol'dman, *On estimates of the norms of operators in spaces of periodic and nonperiodic functions*, Dokl. Akad. Nauk SSSR **267** (1982), 1289–1293; English transl. in Soviet Math. Dokl. **26** (1982).

Translated by H. H. MCFADEN

Труды Матем. Инст. Стеклов.
161(1983)

Proceedings of the Steklov
Institute of Mathematics
1984, Issue 3

ON THE DENSENESS OF THE COMPACTLY SUPPORTED FUNCTIONS IN WEIGHTED SPACES

UDC 517.518.235

L. N. DOMYSHEVA

ABSTRACT. The denseness of the infinitely differentiable compactly supported functions in the seminormed Sobolev spaces $L^r_{p,\varphi}(Q^n)$ is proved in the case of bounded or unbounded n-dimensional parallelepipeds Q^n and under broad assumptions on the weight function φ.

Bibliography: 8 titles.

For a function in a weighted seminormed space $L^{r_n}_{p,\varphi}(J^n)$ we construct a sequence of compactly supported infinitely differentiable functions that converges in the seminorm to the given function, and we determine the rate of convergence of this sequence. The question of denseness of the compactly supported functions in weighted spaces has been studied in [1]–[6] and [8].

Let R be the set of real numbers, $\overset{+}{R}$ the subset of nonnegative numbers, $1 \leqslant p < +\infty$, R^n the Euclidean space of points $x = (x^{(n-1)}, x_n)$, $x^{(n-1)} = (x_1, \ldots, x_{n-1})$, $0 < \alpha \leqslant +\infty$, $0 < \beta \leqslant +\infty$, and $J^n_{\alpha,\beta} = \{x\colon |x_i| < \alpha,\ i = 1,\ldots,n-1;\ 0 < x_n < \beta\}$. Denote by J^n some bounded or unbounded "parallelepiped" $J^n = J^n_{a,b}$ fixed throughout the paper, and assume that the parameters α and β of the other parallelepipeds $J^n_{\alpha,\beta}$ considered always satisfy $0 < \alpha \leqslant a \leqslant +\infty$ and $0 < \beta \leqslant b \leqslant +\infty$. Further, $J^{(n-1)} = \{x\colon |x_i| < a,\ i = 1,\ldots,n-1\}$, $\varphi\colon J^n \to \overset{+}{R}$, r_n is a nonnegative integer, and μ is Lebesgue measure.

Let $L^{r_n}_{p,\varphi}(J^n_{\alpha,\beta})$ denote the set of measurable functions $f\colon J^n_{\alpha,\beta} \to R$ having on the open set $J^n_{\alpha,\beta}$ a generalized derivative $D^{r_n}f = D^{r_n}_{x_n^{r_n}}f$ of order r_n with respect to the variable x_n ([7], §4.1) and with finite seminorm

$$\left|f\colon L^{r_n}_{p,\varphi}\left(J^n_{\alpha,\beta}\right)\right| \overset{\text{def}}{=} \left|\varphi D^{r_n} f;\ L_p\left(J^n_{\alpha,\beta}\right)\right|$$

where, as usual, $|f;\ L_p(E)| = (\int_E |f(x)|^p\,dx)^{1/p}$ for any measurable function $f\colon E \to R$ with $E \subset R^n$. If $r_n > 0$, then $|f;\ L^{r_n}_{p,\varphi}(J^n_{\alpha,\beta})|$ is obviously a seminorm. The space $L^r_{p,1}(J^n_{\alpha,\beta}) = L^{r_n}_p(J^n_{\alpha,\beta})$ is a Sobolev space. The function φ is called a *weight* or a *weight function*.

A function $g\colon J^n \to R$ is said to be *compactly supported relative to the hyperplane* $R^{n-1} = \{x\colon x_n = 0\}$ if there is a δ, $0 < \delta < b$, such that $g(x) = 0$ when $x \in J^n_{a,\delta}$.

1980 *Mathematics Subject Classification*. Primary 41A25, 46E35.

0081-5438/84 \$1.00 + \$.25 per page

LEMMA 1. *Suppose that $f \in L^{r_n}_{p,\varphi}(J^n)$ and $0 < \lambda < b$. Then there exists a function $f^* \in L^{r_n}_{p,\varphi}(J^n)$ compactly supported relative to the hyperplane R^{n-1} such that*

$$f^*(x) = 0 \quad \text{on } J^n_{a,\lambda} \tag{1}$$

and

$$|f - f^*;\, L^{r_n}_{p,\varphi}(J^n)| = |f;\, L^{r_n}_{p,\varphi}(J^n_{a,\lambda})|. \tag{2}$$

Let $0 < \lambda < \lambda_0 < b$. By the existence of the generalized derivative $D^{r_n}f$ (on J^n) it can be assume that all the functions $f, D_{x_n}f, \dots, D^{r_n-1}_{x^{r_n-1}_n}f$ are absolutely continuous with respect to x_n on the set $\overline{J^n_{a,\lambda_0} \setminus J^n_{a,\lambda}}$ (see [7], §4.1), where the overbar denotes closure. Therefore, the traces

$$f(x^{(n-1)}, \lambda),\, D_{x_n}f(x^{(n-1)}, \lambda), \dots, D^{r_n-1}_{x^{r_n-1}_n}f(x^{(n-1)}, \lambda)$$

are defined (in the sense of convergence almost everywhere). These traces can be represented as limits of sequences of measurable fuctions that converge almost everywhere (in the sense of the $(n-1)$-dimensional measure), so they are measurable with respect to the variable $x^{(n-1)}$.

Let

$$f_1(x) = f(x) - \sum_{k=0}^{r_n-1} \frac{1}{k!} D^k_{x^k_n} f(x^{(n-1)}, \lambda)(x_n - \lambda)^k. \tag{3}$$

The function f_1 differs from f by a measurable function which is a polynomial of degree less than r_n in the variable x_n. Therefore,'f_1 is measurable, and the generalized derivative $D^{r_n}_{x^{r_n}_n} f_1(x) = D^{r_n}_{x^{r_n}_n} f(x)$, $x \in J^n$, exists. Consequently,

$$|f - f_1,\, L^{r_n}_{p,\varphi}(J^n)| = 0. \tag{4}$$

Moreover, by (3),

$$D^k_{x^k_n} f_1(x^{(n-1)}, \lambda) = 0, \qquad k = 0, 1, \dots, r_n - 1,\ x^{(n-1)} \in J^{n-1}.$$

Next let

$$f^*(x) \overset{\text{def}}{=} \begin{cases} f_1(x), & x \in J^n \setminus J^n_{a,\lambda}, \\ 0, & x \in J^n_{a,\lambda}. \end{cases} \tag{5}$$

Clearly the function f^* is compactly supported relative to the hyperplane R^{n-1}, since condition (1) holds because of the very definition (5) of f^*. Since

$$|f^*;\, L^{r_n}_{p,\varphi}(J^n)| = |f_1;\, L^{r_n}_{p,\varphi}(J^n \setminus J^n_{a,\lambda})| = |f;\, L^{r_n}_{p,\varphi}(J^n \setminus J^n_{a,\lambda})| < +\infty,$$

it follows that $f^* \in L^{r_n}_{p,\varphi}(J^n)$. Finally,

$$\begin{aligned} |f - f^*;\, L^{r_n}_{p,\varphi}(J^n)|^p &= |f - f^*;\, L^{r_n}_{p,\varphi}(J^n_{a,\lambda})|^p + |f - f_1;\, L^{r_n}_{p,\varphi}(J^n \setminus J^n_{a,\lambda})|^p \\ &\underset{(4),(5)}{=} |f;\, L^{r_n}_{p,\varphi}(J^n_{a,\lambda})|^p. \end{aligned}$$

We introduce more notation. Suppose that a function $f\colon J^n \to R$ is compactly supported relative to the hyperplane R^{n-1}, with $f(x) = 0$ for all $x \in J^n_{a,\lambda}$ $0 < \lambda < b$, and that $f \in L^{r_n}_{p,\varphi}(J^n)$. For every δ, $0 < \delta < \lambda$, let

$$\begin{aligned} &\Omega_{p,\varphi,x_n}(\delta;\, D^{r_n}f;\, J^n) \\ &\quad = \sup_{0<t<1} \left[\int_{J^n} \varphi^p(x) \left| D^{r_n}f(x^{(n-1)}, x_n) - D^{r_n}f(x^{(n-1)}, x_n - t\delta) \right|^p dx \right]^{1/p}, \end{aligned} \tag{6}$$

where it is assumed that $f = 0$ outside J^n.

The function $\Omega_{p,\varphi,x_n}(\delta; D^{r_n}f; J^n)$ is the weighted modulus of continuity with respect to the variable x_n of the derivative $D^{r_n}f$ in the space $L^{r_n}_{p,\varphi}(J^n)$.

Two restrictions are imposed on the weight function:

1) For any λ with $0 < \lambda < b$ there exists a constant $c_\lambda > 0$ such that

$$\varphi\left(x^{(n-1)}, x_n\right) \leqslant c_\lambda \varphi\left(x^{(n-1)}, x_n - h\right) \tag{7}$$

for all $x = (x^{(n-1)}, x_n) \in J^n \setminus J^n_{a,\lambda}$ and all h with $0 < h < \lambda/2$.

2) For any compact set $A \subset J^n$

$$\varphi \in L_p(A). \tag{8}$$

We remark that in the case where $b < +\infty$ condition (7) is satisfied by all weight functions for which the "degeneracy" is concentrated near the hyperplane R^{n-1} in the sense that for any λ with $0 < \lambda < b$ there exist $c'_\lambda > 0$ and $c''_\lambda > 0$ such that $c'_\lambda < \varphi(x) < c''_\lambda$, $x \in J^n \setminus J^n_{a,\lambda}$. For $0 < b \leqslant +\infty$ condition (7) is satisfied, for example, by power functions, powers of logarithms, and the exponential function.

LEMMA 2. *Suppose that $f \in L^{r_n}_{p,\varphi}(J^n)$, f is compactly supported relative to the hyperplane R^{n-1}, i.e., there exists a λ_0 such that*

$$f(x) = 0, \qquad 0 < x_n < \lambda_0, \tag{9}$$

and the weight function φ satisfies conditions (7) *and* (8). *Then*

$$\lim_{\delta \to 0} \Omega_{p,\varphi,x_n}(\delta; D^{r_n}f; J^n) = 0. \tag{10}$$

Let $\varepsilon > 0$. By the countable additivity of the integral, $f \in L^{r_n}_{p,\varphi}(J^n)$ implies that there exist ξ and η with $0 < \xi < a$ and $0 < \eta < b$ such that for $B_1 = J^n \setminus J^n_{\xi,\eta}$ we have

$$\left\{\int_{B_1} \varphi^p(x)|D^{r_n}f(x)|^p\, dx\right\}^{1/p} < \varepsilon. \tag{11}$$

The set $J^n_{\xi,\eta}$ is bounded; therefore, by the absolute continuity of the integral, there exists a $\delta > 0$ such that for any measurable set $E \subset J^n_{\xi,\eta}$ with $\mu E < \delta$

$$\left\{\int_{E} \varphi^p(x)|D^{r_n}f(x)|^p\, dx\right\}^{1/p} < \varepsilon. \tag{12}$$

Note that if $\lambda < \lambda_1 \overset{\text{def}}{=} \min\{\delta/3(2\xi)^{n-1}, \eta\}$, then the set $B_2 \overset{\text{def}}{=} J^n_{\xi,\eta} \setminus J^n_{\xi,\eta-\lambda}$ satisfies

$$\mu B_2 = \mu J^n_{\xi,\lambda} = \lambda(2\xi)^{n-1} < \delta/3.$$

Suppose in what follows that the number λ is fixed and such that

$$\lambda < \min\{\lambda_1, \lambda_0/2\}. \tag{13}$$

According to the C-property of measurable functions, for the number δ indicated above there exists a compact set

$$A \subset J^n_{\xi,\eta} \setminus \left(J^n_{\xi,\eta} \cup B_2\right)$$

such that the function $D^{r_n}f$ is continuous on it, and if

$$B_3 \overset{\text{def}}{=} J^n_{\xi,\eta} \setminus \left(J^n_{\xi,\eta} \cup B_2 \cup A\right),$$

then $\mu B_3 < \delta/3$. Setting $B = B_2 \cup B_3 \cup J^n_{\xi,\eta}$, we have

$$\mu B < \delta \tag{14}$$

and

$$J^n = A \cup B_1 \cup B. \tag{15}$$

Suppose that $0 < t < 1$ and $0 < h < \lambda$. We set

$$A_t = \left\{x \colon x \in A, \left(x^{(n-1)}, x_n - th\right) \in A\right\},$$
$$A_t^* = \left\{x \colon x \in A, \left(x^{(n-1)}, x_n - th\right) \notin A\right\};$$

then

$$A = A_t \cup A_t^*. \tag{16}$$

If $x = (x^{(n-1)}, x_n) \in A_t^*$, then $(x^{(n-1)}, x_n - th) \in B$, and thus

$$\mu A_t^* \leqslant \mu B < \delta. \tag{17}$$

If $A_t^{**} \overset{\text{def}}{=} \{x \colon x \notin A, (x^{(n-1)}, x_n - th) \in A\}$, then $A_t^{**} \subset B$ and, consequently,

$$\mu A_t^{**} \leqslant \mu B < \delta. \tag{18}$$

Let $\omega_{x_n}(\delta; D^{r_n}f; A)$ denote the usual uniform modulus of continuity of the function $D^{r_n}f$ with respect to the variable x_n on the compact set A, and let (see (8))

$$c_0 \overset{\text{def}}{=} \left(\int_A \varphi^p(x)\,dx\right)^{1/p}. \tag{19}$$

Now suppose $0 < h < \lambda/2$; then, by virtue of (9) and (13), for $x \in J^n_{a,\lambda}$

$$f\left(x^{(n-1)}, x_n\right) = f\left(x^{(n-1)}, x_n - th\right) = 0, \tag{20}$$

while for $x \in J^n \setminus J^n_{a,\lambda}$ we have, by (7),

$$\varphi\left(x^{(n-1)}, x_n\right) \leqslant c_\lambda \varphi\left(x^{(n-1)}, x_n - th\right), \qquad 0 < t < 1. \tag{21}$$

Therefore, for all h and t with $0 < h < \lambda/2$ and $0 < t < 1$

$$\left\{\int_{J^n} \varphi^p(x)\left|D^{r_n}f\left(x^{(n-1)}, x_n\right) - D^{r_n}f\left(x^{(n-1)}, x_n - th\right)\right|^p dx\right\}^{1/p}$$

$$\underset{(20)}{=} \left\{\int_{J^n\setminus J^n_{a,\lambda}}\right\}^{1/p} \underset{(15)}{\leqslant} \left\{\int_A\right\}^{1/p} + \left\{\int_{B_1\setminus J^n_{a,\lambda}}\right\}^{1/p} + \left\{\int_{B\setminus J^n_{a,\lambda}}\right\}^{1/p}$$

$$\underset{(16)}{\leqslant} \left\{\int_{A_t} \varphi^p(x)\left|D^{r_n}f\left(x^{(n-1)}, x_n\right) - D^{r_n}f\left(x^{(n-1)}, x_n - th\right)\right|^p dx\right\}^{1/p}$$

$$+ \left\{\int_{A_t^*} \varphi^p(x)\left|D^{r_n}f(x)\right|^p dx\right\}^{1/p} + \left\{\int_{A_t^*} \varphi^p(x)\left|D^{r_n}f\left(x^{(n-1)}, x_n - th\right)\right|^p dx\right\}^{1/p}$$

$$+ \left\{\int_{B_1} \varphi^p(x)\left|D^{r_n}f(x)\right|^p dx\right\}^{1/p} + \left\{\int_{B_1\setminus J^n_{a,\lambda}} \varphi^p(x)\left|D^{r_n}f\left(x^{(n-1)}, x_n - th\right)\right|^p dx\right\}^{1/p}$$

$$+ \left\{\int_{B} \varphi^p(x)\left|D^{r_n}f(x)\right|^p dx\right\}^{1/p} + \left\{\int_{B\setminus J^n_{a,\lambda}} \varphi^p(x)\left|D^{r_n}f\left(x^{(n-1)}, x_n - th\right)\right|^p dx\right\}^{1/p}$$

$$
\begin{aligned}
&\underset{(21),(17),(12)}{\leqslant} \omega_{x_n}(h;\, D^{r_n}f;\, A)\left\{\int_A \varphi^p(x)\,dx\right\}^{1/p} \\
&\qquad + \varepsilon + c_\lambda\left\{\int_B \varphi^p(x)\left|D^{r_n}f(x)\right|^p dx\right\}^{1/p} \\
&\qquad + \varepsilon + c_\lambda\left\{\int_{B_1\cup B} \varphi^p(x)\left|D^{r_n}f(x)\right|^p dx\right\} \\
&\qquad + \varepsilon + c_\lambda\left\{\int_{B_1\cup B\cup A_l^{**}} \varphi^p(x)\left|D^{r_n}f(x)\right|^p dx\right\}^{1/p} \\
&\underset{\substack{(19),(11),(12)\\(14),(18)}}{\leqslant} c_0\omega_{x_n}(h;\, D^{r_n}f;\, A) + 3\varepsilon + c_\lambda\varepsilon + 2c_\lambda\varepsilon + 3c_\lambda\varepsilon \\
&= c_0\omega_{x_n}(h;\, D^{r_n}f;\, A) + 3(2c_\lambda + 1)\varepsilon.
\end{aligned}
$$

Since $\lim_{h\to 0}\omega_{x_n}(h;\, D^{r_n}f;\, A) = 0$, (10) follows from the inequality just obtained.

THEOREM. *Suppose that $f \in L^{r_n}_{p,\varphi}(J^n)$, $r_n \geqslant 1$, $0 < \delta < b$, and conditions* (7) *and* (8) *hold. Then there exists a function $f^*_\delta \in L^{r_n}_{p,\varphi}(J^n)$ infinitely differentiable with respect to the variable x_n and compactly supported relative to the hyperplane R^{n-1} such that $f^*_\delta(x) = 0$ on $J^n_{a,\delta}$ and*

$$\left|f - f^*_\delta;\, L^{r_n}_{p,\varphi}(J^n)\right| \leqslant \Omega_{p,\varphi,x_n}(\delta;\, D^{r_n}f^*;\, J^n) + \left|f;\, L^{r_n}_{p,\varphi}(J^n_{a,2\delta})\right|,$$

where f^* is the function defined by (5) *for $\lambda = 2\delta$.*

COROLLARY. *The set of functions infinitely differentiable with respect to the variable x_n and compactly supported relative to the hyperplane R^{n-1} is dense in the seminormed space $L^{r_n}_{p,\varphi}(J^n)$.*

Poulsen [1] proved the denseness of the compactly supported functions in the corresponding normed weighted space under different assumptions on the weight function with the help of general function methods and without constructing concrete compactly supported functions approximating a given function.

Suppose that $\omega\colon R \to R$ is a function infinitely differentiable with respect to the variable x_n and such that $\operatorname{supp}\omega(t) \subset [-1, 0]$ and

$$c \overset{\text{def}}{=} \int_{-1}^{0} \omega(t)\,dt > 0. \tag{22}$$

Let $f \in L^{r_n}_{p,\varphi}(J^n)$, and let f^* be the function defined by(5) with $\lambda = 2\delta$. Consider the Sobolev average f^*_δ of f^* with the radius δ with respect to x_n on J^n (f^* isassumed to be zero outside J^n):

$$
\begin{aligned}
f^*_\delta(x) &\overset{\text{def}}{=} \frac{1}{c}\int_{x_n-\delta}^{x_n} \omega\left(\frac{y_n - x_n}{\delta}\right) f^*\left(x^{(n-1)}, y_n\right) dy_n \\
&= \frac{1}{c}\int_{-1}^{0} \omega(t) f^*\left(x^{(n-1)}, x_n - t\delta\right) dt, \qquad x \in J^n.
\end{aligned}
$$

For $r_n \geqslant 1$ the function f^* can be assumed to be continuous in x_n; therefore, f^*_δ is infinitely differentiable with respect to x_n, and since f^* has a generalized derivative $D^{r_n}f$, it follows that

$$D^{r_n}(f^*_\delta) = (D^{r_n}f^*)_\delta. \tag{23}$$

Using (22), (23) and the generalized Minkowski inequality, we get

$$\begin{aligned}
&|f^* - f^*_\delta;\ L^{r_n}_{p,\varphi}(J^n)| \\
&= \frac{1}{c}\left\{\int_{J^n} \varphi^p(x)\,dx \left|\int_{-1}^{0} \omega(t)\left[D^{r_n}f^*(x^{(n-1)}, x_n - t\delta) - D^{r_n}f^*(x)\right] dt\right|^p\right\}^{1/p} \\
&\leqslant \frac{1}{c}\int_{-1}^{0} \omega(t)\,dt\left\{\int_{J^n} \varphi^p(x)\left|D^{r_n}f^*(x^{(n-1)}, x_n - t\delta) - D^{r_n}f^*(x)\right|^p dx\right\}^{1/p} \\
&\leqslant \Omega_{p,\varphi,x_n}(\delta;\ D^{r_n}f^*;\ J^n).
\end{aligned}$$

By (2), this gives us

$$\begin{aligned}
|f - f^*_\delta;\ L^{r_n}_{p,\varphi}(J^n)| &\leqslant |f - f^*;\ L^{r_n}_{p,\varphi}(J^n)| + |f^* - f^*_\delta;\ L^{r_n}_{p,\varphi}(J^n)| \\
&\leqslant \Omega_{p,\varphi,x_n}(\delta;\ D^{r_n}f^*;\ J^n) + |f;\ L^{r_n}_{p,\varphi}(J^n_{2\delta})|.
\end{aligned}$$

According to (10), this inequality leads at once to the assertion of the corollary, because $f \in L^{r_n}_{p,\varphi}(J^n)$ implies that

$$\lim_{\delta\to 0} |f;\ L^{r_n}_{p,\varphi}(J^n_{2\delta})| = 0.$$

REMARK. The proof given above for denseness of the functions which are infinitely differentiable with respect to x_n and compactly supported relative to the hyperplane R^{n-1} remains valid also for $r_n = 0$, i.e., for the space $L_{p,\varphi}(J^n) = L^0_{p,\varphi}(J^n)$, provided it is required in addition that for almost all $x^{(n-1)} \in J^{n-1}$ the function $f \in L_{p,\varphi}(J^n)$ is locally integrable with respect to x_n. For example, this certainly holds if for any λ with $0 < \lambda < b$

$$1/\varphi \in L_q(J^n \setminus J^n_{a,\lambda}), \qquad 1/p + 1/q = 1,$$

since in this case

$$\int_{J^n\setminus J^n_{a,\lambda}} |f(x)|\,dx \leqslant |f;\ L_{p,\varphi}(J^n)|\,|1/\varphi;\ L_q(J^n \setminus J^n_{a,\lambda})| < +\infty.$$

BIBLIOGRAPHY

1. Ebbe Thue Poulsen, *Boundary values in function spaces*, Math. Scand. **10** (1962), 45–52.

2. V. R. Portnov, *A theorem on denseness of the compactly supported functions in weighted spaces*, Dokl. Akad. Nauk SSSR **160** (1965), 545–548; English transl. in Soviet Math. Dokl. **6** (1965).

3. G. N. Yakolev, *On the denseness of the compactly supported functions in weighted spaces*, Dokl. Akad. Nauk SSSR **170** (1966), 797–798; English transl. in Soviet Math. Dokl. **7** (1966).

4. O. V. Besov, Ya. Kadlets [J. Kadlec] and A. Kufner, *On some properties of weighted spaces*, Dokl. Akad. Nauk SSSR **171** (1966), 514–516; English transl. in Soviet Math. Dokl. **7** (1966).

5. L. D. Kudryavtsev, *On the denseness of compactly supported functions in weighted spaces*, Dokl. Akad. Nauk SSSR **239** (1978), 46–49; English transl. in Soviet Math. Dokl. **19** (1978).

6. P. I. Lizorkin, *The closure of the set of compactly supported functions in the weighted spaces* $W^l_{p,\Phi}$, Dokl. Akad. Nauk SSSR **239** (1978), 789–792; English transl. in Soviet Math. Dokl. **19** (1978).

7. S. M. Nikol′skiĭ, *Approximation of functions of several variables and embedding theorems*, 2nd rev. aug. ed., "Nauka", Moscow, 1977; English transl. of 1st ed., Springer-Verlag, 1974.

8. Alois Kufner, *Weighted Sobolev spaces*, Teubner, Leipzig, 1980.

Translated by H. H. McFADEN

Труды Матем. Инст. Стеклов.
161(1983)

Proceedings of the Steklov
Institute of Mathematics
1984, Issue 3

ESTIMATES OF THE CAPACITY OF SETS WITH RESPECT TO GENERALIZED LIZORKIN-TRIEBEL CLASSES AND WEIGHTED SOBOLEV CLASSES

UDC 517.51 + 517.946

G. A. KALYABIN

ABSTRACT. Estimates are established for the capacity of sets $E \subset R^n$: a lower estimate in terms of a certain function of $(\text{meas}\, E)^{1/n}$, and an upper estimate in terms of the same function of diam E; and the estimating function is effectively expressed in terms of the defining parameters of the corresponding function spaces. Analogous results were known previously for the ordinary Besov spaces $B_p^l(R^n)$ and Liouville spaces $L_p^l(R^n)$.

Bibliography: 12 titles.

Estimates are established for the capacity of measurable sets $E \subset R^n$ with respect to Lizorkin-Triebel classes $L_{p,\theta}^{(\beta,N)}(R^n)$: a lower estimate in terms of a certain function of $h_E \doteq (\text{meas}\, E)^{1/n}$,(1) and an upper estimate in terms of the same function of $\rho_E \doteq \text{diam}\, E$. In the case where $\rho_E \leqslant Ah_E$ these estimates become two-sided.

We remark that similar inequalities are well known for capacities with respect to the ordinary Liouville spaces $L_p^l(R^n)$ and Besov spaces $B_p^l(R^n)$ (see, for example, [1], where there is also a detailed bibliography of references dealing with capacity problems).

It is characteristic that the estimating function found here does not depend at all on the parameter $\theta \in [1, \infty]$.

From the results obtained we derive estimates of the capacity of a "nonhomogeneous anisotropic capacitor" whose properties are described by general weight functions depending only on the "transversal" variable $y \in R_+^1$ and which has the set $E \subset R^n$ as a "lower plate" and a hyperplane in R_+^{n+1} parallel to R^n as an "upper plate".

1. Definitions and statement of the main results

1°. Everywhere below, p and θ denote parameters with the limits $1 < p < \infty$ and $1 \leqslant \theta \leqslant \infty$; p' is the exponent conjugate to p (i.e., $p' \doteq p/(p-1)$); n, r, and s are positive integers; j and k are indices running through the set $\{1,\dots,n\}$ and the set $\mathbf{N}$ of all positive integers, respectively; ε, δ, c, c_0, $c_1,\dots$ are positive constants; x is a vector $(x_1,\dots,x_n) \in R^n$.

1980 *Mathematics Subject Classification*. Primary 31C15, 46E35.

(1) Here and in what follows, the expression $P \doteq Q$ means that either P is defined by the expression Q or is an abbreviation for it.

For $N > 0$ and $g(x) \in L_p \doteq L_p(R^n)$ let $S_N g(x)$ be the Nth-order Dirichlet sum of a function g with respect to all the variables x_j (see [2], §8.10), and for definiteness assume that $S_0 g(x) \doteq 0$.

Let two sequences $\{\beta_k\}$ and $\{N_k\}$ of positive numbers be given such that for some A_0, A_1, A_2 ($A_0 > 1, A_2 \geqslant A_1 > 1$) and all k

$$N_{k+1} \geqslant A_0 N_k, \qquad A_1 \beta_k \leqslant \beta_{k+1} \leqslant A_2 \beta_k. \tag{1.1}$$

Denote by $\Lambda_{p,\theta} \doteq L^{(\beta,N)}_{p,\theta}(R^n)$ the set of all functions with finite norm

$$\|g\|_{\Lambda_{p,\theta}} \doteq \left\| \left\| \left\{ \beta_k \left(S_{N_k} g(x) - S_{N_{k-1}} g(x) \right) \right\} \right\|_{l_\theta} \right\|_p, \tag{1.2}$$

where $\| \cdot \|_p$ is the norm in $L_p(R^n)$, and $N_0 \doteq 0$.

In the case where $N_k \doteq A_1^{k/l}$ and $\beta_k = A_1^k$ the class $\Lambda_{p,\theta}$ is the usual Lizorkin-Triebel space $L^l_{p,\theta}(R^n)$, which coincides when $\theta = 2$ with the Liouville class $L^l_p(R^n)$, and when $\theta = p$ with the Besov class $B^l_p(R^n)$ (see [8], and Chapter 8 of [2]). In the general case the classes $L^{(\beta,N)}_{p,\theta}(R^n)$ were introduced and studied by the author in [3]–[5].

The n-dimensional Lebesgue measure of a measurable set $E \subset R^n$ is denoted by meas E, and for brevity we set

$$h_E \doteq (\operatorname{meas} E)^{1/n}, \qquad \rho_E \doteq \operatorname{diam} E \doteq \sup\{\|x - \tilde{x}\| : x \in E, \tilde{x} \in E\}. \tag{1.3}$$

The capacity of a set $E \subset R^n$ with respect to the class $\Lambda_{p,\theta}$ is defined to be the quantity

$$\operatorname{Cap}(E, \Lambda_{p,\theta}) \doteq \inf\left\{ \|g\|^p_{\Lambda_{p,\theta}} : g(x) \geqslant 1 \text{ on } E \right\}. \tag{1.4}$$

Our aim in this article is to prove the following assertion.

THEOREM 1. *Define a function of the positive parameter t by*

$$\gamma_1(t) \doteq \left(\sum_{k=1}^{\infty} \left(\beta_k \left(N_k^{-1} + t \right)^{n/p} \right)^{-p'} \right)^{-p/p'}. \tag{1.5}$$

There exists a constant $c_1 > 0$, depending only on n, p, and the constants A_0, A_1, A_2 in (1.1), *such that*

$$c_1^{-1} \gamma_1(h_E) \leqslant \operatorname{Cap}\left(E, L^{(\beta,N)}_{p,\theta}(R^n)\right) \leqslant c_1 \gamma_1(\rho_E). \tag{1.6}$$

The lower estimate is proved in 2.1° and the upper estimate in 2.2.°.

2°. As shown in [6], the classes $L^{(\beta,N)}_{p,\theta}(R^n)$ are closely connected with the space of traces on R^n of the functions in the weighted Sobolev classes $\mathscr{W}_p \doteq W^{r,s}_{p,\xi,\varphi}(G^{n+1}_a)_0$.

By definition, a function $f(x, y)$ defined in the strip

$$G^{n+1}_a \doteq \{(x, y) : x \in R^n, 0 < y < a\}, \tag{1.7}$$

belongs to the class $\mathscr{W}_p$ if it has generalized derivatives $\partial^r f / \partial y^r$ and $\partial^s f / \partial x_j^s$ in G^{n+1}_a and satisfies the "upper" boundary conditions

$$\lim_{y \to a-0} \partial^l f(x, y) / \partial y^l \equiv 0 \qquad (l \in \{0, 1, \dots, r-1\}), \tag{1.8}$$

and the integrals

$$I_r(f) \doteq \int_{G^{n+1}_a} \xi(y) |\partial^r f(x, y) / \partial y^r|^p \, dx \, dy, \tag{1.9}$$

$$J_s(f) \doteq \int_{G^{n+1}_a} \varphi(y) \left(\sum_{j=1}^{n} |\partial^s f(x, y) / \partial x_j^s| \right)^p dx \, dy$$

are finite. The norm of f in $\mathscr{W}_p$ is defined by

$$\|f\|_{\mathscr{W}_p} \doteq (I_r(f) + J_s(f))^{1/p}. \tag{1.10}$$

We assume that the weight functions (weights, for brevity) $\xi(y)$ and $\varphi(y)$ are positive and continuous on $(0, a]$; they may have singularities near $y = 0$, but such that the functions

$$\zeta(y) \doteq \int_0^y \xi^{1/1-p}(t) t^{(r-1)p'}\, dt, \qquad \psi(y) \doteq \int_0^y \varphi(t)\, dt \tag{1.11}$$

are finite.

The first of the conditions (1.11) is necessary and at the same time sufficient for any function $f(x, y)$ satisfying (1.8) and such that $I_r(f) < \infty$ to have a limit (after possible modification on a set of $(n + 1)$-dimensional measure 0) as $y \to +0$ in the sense of convergence in $L_p(R^n)$ and for almost all $x \in R^n$ [7].

As for the finiteness of $\psi(y)$, if this condition fails, then the limit of any function $f \in \mathscr{W}_p$ as $y \to +0$ is equal to 0 for almost all $x \in R^n$.

We define the capacity of $E \subset R^n$ with respect to $\mathscr{W}_p$ by the formula

$$\operatorname{Cap}(E, \mathscr{W}_p) \doteq \inf\left\{ \|f\|^p_{\mathscr{W}_p} : \lim_{y \to +0} f(x, y) \geqslant 1 \text{ on } E \right\}. \tag{1.12}$$

In §3 the following corollary is derived from Theorem 1.

THEOREM 2. *Let $r = 1$, and for $t > 0$ let (see (1.11))*

$$\gamma_2(t) \doteq \left\{ \int_0^a \left[\left(\zeta^{1/p'}(y) \psi^{1/p}(y)\right)^{1/s} + t \right]^{-np'/p} d\zeta(y) \right\}^{-p/p'}. \tag{1.13}$$

There exists a constant $c_2 > 0$ independent of E such that

$$c_2^{-1} \gamma_2(h_E) \leqslant \operatorname{Cap}\left(E, W^{1,s}_{p,\xi,\varphi}(G_a^{n+1})_0\right) \leqslant c_2 \gamma_2(\rho_E). \tag{1.14}$$

3°. For the case of a higher derivative with respect to the transversal variable (i.e., for $r > 1$) the author has not succeeded in getting analogous estimates for the capacity under minimal restrictions on the weight (expressed by the finiteness of the functions (1.11)), but under additional conditions of monotonicity type on the weight functions the desired result is established in the following theorem, which is also a corollary of Theorem 1 (see 3.4°).

THEOREM 3. *Assume that for some constants $\varepsilon > 0$ and κ the functions*

$$\varphi(y) y^{1-\varepsilon}, \qquad \xi^{-1}(y) y^{pr-1-\varepsilon}, \qquad \xi(y) y^{\kappa} \tag{1.15}$$

are increasing on $(0, a)$, and for $t > 0$ define

$$\gamma_3(t) \doteq \left[\int_0^a \xi^{1/1-p}(y) y^{p'(r-1)} \left(y^{r/s} \left(\xi^{-1}(y) \varphi(y)\right)^{1/ps} + t \right)^{-np/p} dy \right]^{-p/p'}. \tag{1.16}$$

There exists a constant $c_3 > 0$, not depending on E, such that

$$c_3^{-1} \gamma_3(h_E) \leqslant \operatorname{Cap}\left(E, W^{r,s}_{p,\xi,\varphi}(G_a^{n+1})_0\right) \leqslant c_3 \gamma_3(\rho_E). \tag{1.17}$$

We remark that a weak one-sided growth restriction is imposed on the weight function $\varphi(y)$ in Theorem 3; in particular, $\varphi(y)$ can be an arbitrary increasing function. In essence, the first weight function $\xi(y)$ is assumed to be close to a power.

4°. Before proceeding to the proofs we make some remarks and refinements concerning the assertions formulated.

1. It is clear from the defining formulas (1.5), (1.13), and (1.16) that the estimating functions $\gamma_l(t)$, $l \in \{1,2,3\}$, are positive and increasing for $t > 0$, but the functions $\gamma_l(t)t^{-n}$ are decreasing, and the limits

$$\lim_{t \to +\infty} \gamma_l(t)t^{-n} = \delta_l > 0 \tag{1.18}$$

exist.

2. As shown in [9] and [10], the condition $\gamma_1(+0) > 0$, i.e., convergence of $\sum \beta_k^{-p'} N_k^{np'/p}$, is necessary and sufficient for all the functions in $L_{p,\theta}^{(\beta,N)}(R^n)$ to be continuous (after modifications on a set of measure 0 in R^n). Arguments in §3 will tell us that the conditions $\gamma_l(+0) > 0$, $l \in \{2,3\}$, i.e., the relations

$$\int_0^a \left(\zeta^{1/p'}(y)\psi^{1/p}(y)\right)^{-np'/ps} d\zeta|(y) < \infty,$$

$$\int_0^a y^{\lambda_1}(\xi(y))^{\lambda_2}(\varphi(y))^{\lambda_3}\, dy < \infty \tag{1.19}$$

$$\left(\lambda_1 \doteq p'(r-1) - \frac{rnp'}{ps}, \lambda_2 \doteq \frac{p'}{p}\left(\frac{n}{ps} - 1\right), \lambda_3 \doteq -\frac{np'}{p^2 s}\right),$$

under the respective conditions of Theorems 2 and 3 are criteria for continuity on R^n of the trace of any function in $\mathscr{W}_p$.

3. It will be shown also that if $\theta > 1$ under the conditions of Theorem 1, then for any set $E \subset R^n$ it is possible to construct an entire function $g(x)$ of exponential type ρ_E^{-1} in all the x_j (see [2], Chapter 3) that is nonnegative on R^n and such that

$$g(x) \geqslant 1, \quad x \in E, \qquad \|g\|_{\Lambda_{p,\theta}}^p \leqslant \tilde{c}_1\gamma_1(\rho_E), \tag{1.20}$$

where the constant $\tilde{c}_1$ can now depend on θ.

4. Similarly, under the conditions of Theorems 2 and 3 a function $f \in \mathscr{W}_p$ such that

$$f(x,0) \geqslant 1, \quad x \in E, \qquad \|f\|_{\mathscr{W}_p}^p \leqslant c_l\gamma_l(\rho_E), \quad l \in \{2,3\} \tag{1.21}$$

can always be chosen to be *nonnegative* and infinitely differentiable on $R^n \times [0, a]$.

2. Proof of Theorem 1

1°. PROPOSITION 1. *Suppose that* $\{\beta_k\}$, $\{N_k\}$, $\gamma_1(t)$ *and* h_E *have the same meanings as in* 1.1°, *and* $g_k(x)$ *is a sequence of entire functions of exponential type* N_k *in all the variables* x_j. *Then for any set* $E \subset R^n$ *of positive measure*

$$h_E^{-n}\int_E \sum_{k=1}^{\infty} |g_k(x)|\, dx \leqslant c\gamma_1^{-1/p}(h_E)\left\| \sup_k |\beta_k g_k(x)| \right\|_p, \tag{2.1}$$

where the constant c depends only on n, p, A_0, A_1 and A_2.

This proposition easily gives the left-hand inequality in (1.6). Indeed, let $g(x) \in \Lambda_{p,\theta}$ be such that $g(x) \geqslant 1$ on E, and let $g_k(x) \doteq S_{N_k}g(x) - S_{N_{k-1}}g(x)$ in Proposition 1. Then by (2.1), (1.3), and (1.2),

$$\begin{aligned} 1 &\leqslant (\text{meas } E)^{-1}\int_E |g(x)|\, dx \leqslant (\text{meas } E)^{-1}\int_E \sum_{k=1}^{\infty} |g_k(x)|\, dx \\ &\leqslant c\gamma_1^{-1/p}(h_E)\left\| \sup_k |\beta_k g_k(x)| \right\|_p = c\gamma_1^{-1/p}(h_E)\|g\|_{\Lambda_{p,\infty}} \\ &\leqslant c\gamma_1^{-1/p}(h_E)\|g\|_{\Lambda_{p,\theta}}. \end{aligned} \tag{2.2}$$

Consequently, if $g(x) \geqslant 1$ on E, then $\|g\|^p_{\Lambda_{p,\theta}} \geqslant c^{-1}\gamma_1(h_E)$ for all $\theta \in [1, \infty]$, and this is the lower estimate in (1.6).

PROOF OF PROPOSITION 1. Denote by m a positive integer such that

$$N_{m-1} < h_E^{-1} \leqslant N_m \qquad (N_0 \doteq 0), \tag{2.3}$$

and by $G(x)$ the supremum of $|\beta_k g_k(x)|$ over k. By (1.1), for $k \geqslant m$

$$|g_k(x)| \leqslant \beta_k^{-1}G(x) \leqslant \beta_m^{-1}A_1^{m-k}G(x), \tag{2.4}$$

and from this, applying Hölder's inequality, we get that

$$\begin{aligned} h_E^{-n}\int_E \sum_{k=m}^{\infty} |g_k(x)|\, dx &\leqslant h_E^{-n}\beta_m^{-1}\frac{A_1}{A_1 - 1}\int_E G(x)\, dx \\ &\leqslant \frac{A_1}{A_1 - 1}\beta_m^{-1}h_E^{-n/p}\|G\|_p. \end{aligned} \tag{2.5}$$

But $h_E^{-n/p} \leqslant 2^{n/p}(N_m^{-1} + h_E)^{-n/p}$, by (2.3), and it follows fromthe defining formula (1.5) that

$$\beta_m^{-1}\left(N_m^{-1} + h_E\right)^{-n/p} \leqslant \left\|\left\{\beta_k^{-1}\left(N_k^{-1} + h_E\right)^{-n/p}\right\}\right\|_{l'_p} = \gamma_1^{-1/p}(h_E), \tag{2.6}$$

and, thus, for the left-hand part of (2.5)

$$h_E^{-n}\int_E \sum_{k=m}^{\infty} |g_k(x)|\, dx \leqslant \frac{2^{n/p}}{A_1 - 1}\|G\|_p\gamma_1^{-1/p}(h_E). \tag{2.7}$$

To estimate the first $m - 1$ terms in (2.1) we need

LEMMA 1 (see [4], Lemma 4 in §2). *Let $F(x)$ be an entire function of exponential type $N_j > 0$ in the variables x_j. Then there exists a constant c_n depending only on n such that for all $x, \tilde{x} \in R^n$*

$$|F(\tilde{x})| \leqslant c_n\left\{\prod_{j=1}^{n}\left(1 + N_j|\tilde{x}_j - x_j|\right)\right\}MF(x), \tag{2.8}$$

where M denotes the Hardy-Littlewood maximal operator:

$$MF(x) \doteq \sup_{\tau_j > 0}\left\{(\operatorname{meas}\Delta_\tau)^{-1}\int_{\Delta_\tau} |F(x + z)|dz\colon \Delta_\tau \doteq \left\{z \in R^n\colon |z_j| < \tau_j\right\}\right\}. \tag{2.9}$$

Now fix a point $\tilde{x} \in R^n$ and consider the sequence of cubes

$$Q_k(\tilde{x}) \doteq \left\{x \in R^n\colon |x_j - \tilde{x}_j| < N_k^{-1}\right\} \tag{2.10}$$

and the "corridors" between them:

$$\Gamma_k(\tilde{x}) \doteq Q_k(\tilde{x}) \setminus Q_{k+1}(\tilde{x}). \tag{2.11}$$

By Lemma 1, $|g_k(\tilde{x})| \leqslant c_n Mg_k(x)$, $x \in \Gamma_k(\tilde{x})$, and, consequently,

$$|g_k(\tilde{x})|^p \leqslant c_n^p\left(\int_{\Gamma_k(\tilde{x})} (Mg_k(x))^p\, dx\right)(\operatorname{meas}\Gamma_k(\tilde{x}))^{-1}. \tag{2.12}$$

Taking into account now that $\Gamma_k(\tilde{x}) \geqslant A_0^{-n}(2N_k^{-1})^n$ (by (2.10), (2.11), and (1.1)) and that $\beta_k Mg_k(x) \leqslant MG(x)$ for all k (by the definition of $G(x)$), we get from (2.12) that

$$N_k^{-n}\beta_k^p|g_k(\tilde{x})|^p \leqslant c\int_{\Gamma_k(\tilde{x})} (MG(x))^p\, dx. \tag{2.13}$$

Using Hölder's inequality and summing (2.13) over $k \in \{1,\dots,m-1\}$, we arrive at the estimate

$$\left(\sum_{k=1}^{m-1} |g_k(\tilde{x})|\right)^p \leqslant \left(\sum_{k=1}^{m-1} \left(\beta_k N_k^{-n/p}\right)^{-p'}\right)^{p/p'} \sum_{k=1}^{m-1} N_k^{-n}\beta_k^p |g_k(\tilde{x})|^p$$
$$\leqslant c\left(\sum_{k=1}^{m-1} \left(\beta_k N_k^{-n/p}\right)^{-p'}\right)^{p/p'} \int_{Q_1(\tilde{x})} (MG(x))^p\, dx. \quad (2.14)$$

According to the classical Hardy-Littlewood theorem ([12], Chapter 1), $\|MG(x)\|_p \leqslant c_{n,p}\|G(x)\|_p$; on the other hand, $(h_E + N_k^{-1})^{n/p} \leqslant 2^{n/p}N_k^{-n/p}$, for $k < m$ by (2.3), and it therefore follows from (2.4) that for any $\tilde{x} \in R^n$ (see also (1.5))

$$\sum_{k=1}^{m-1} |g_k(\tilde{x})| \leqslant c\gamma_1^{-1/p}(h_E)\|G(x)\|_p \quad (2.15)$$

and thus

$$h_E^{-n}\int_E \sum_{k=1}^{m-1} |g_k(x)|dx \leqslant \sup_{\tilde{x}\in R^n} \sum_{k=1}^{m-1} |g_k(\tilde{x})| \leqslant c\gamma_1^{-1/p}(h_E)\|G\|_p. \quad (2.16)$$

Combining this inequality with (2.7), we get

$$h_E^{-n}\int_E \sum_{k=1}^{\infty} |g_k(x)|\, dx \leqslant c\gamma_1^{-1/p}(h_E)\|G(x)\|_p, \quad (2.17)$$

which coincides with (2.1), and this concludes the proof of Proposition 1 and hence of the lower estimate in (1.6).

2°. To prove the upper estimate in (1.6) it suffices to construct for any $\rho > 0$ a function $g(x)$ such that for some $c_1 > 0$, depending only on n, p, A_0, and A_1,

$$\|g\|^p_{\Lambda_{p,1}} \leqslant c_1\gamma_1(\rho), \qquad g(x) \geqslant 1, \quad x \in Q_\rho \doteq \left\{x \in R^n\colon |x_j| < \rho\right\}, \quad (2.18)$$

because every set of diameter ρ is a subset of the cube Q_ρ after translation by some vector.

First choose a positive integer m such that

$$N_{m-1} < \tfrac{1}{2}\rho^{-1} \leqslant N_m \qquad (N_0 \doteq 0). \quad (2.19)$$

We introduce the numbers

$$\tilde{N}_k \doteq N_k, \quad k \neq m, \qquad \tilde{N}_m \doteq \min\left(N_m, \rho^{-1}\right), \quad (2.20)$$

which, by (1.1), are such that

$$\tilde{N}_{k+1} \geqslant \tilde{A}_0\tilde{N}_k, \qquad \tilde{A}_0 \doteq \min(A_0, 2) > 1 \quad (2.21)$$

for all k. Further, let

$$\chi(t) \in C_0^\infty(R^1_+); \quad \chi(t) \geqslant 0; \quad \chi(t) = 0, \quad t \notin \left(\tilde{A}_0^{-1}, 1\right); \quad \chi(t) \not\equiv 0. \quad (2.22)$$

For $\tau \in R^1$ we set

$$\omega(\tau) = b\int_0^\infty (\cos \tau t)\chi(t)\, dt, \quad (2.23)$$

where the factor $b > 0$ will be chosen later. The following properties of $\omega(\tau)$ are significant for our purposes:

1) $\omega(\tau)$ is continuous on R^1 and

$$|\omega(\tau)| \leqslant c(1 + |\tau|)^{-1}. \quad (2.24)$$

2) The Fourier transform of $\omega(\tau)$ differs by only a factor from $\chi(t)+\chi(-t)$ and, consequently, is equal to 0 outside the set

$$\Delta \doteq \left\{t \in R^1 : \tilde{A}_0^{-1} < |t| < 1\right\}.$$

3) $\omega(\tau) > 0$ for $|\tau| \leqslant \pi/2$, since for these τ the integrand in (2.23) is nonnegative and is strictly positive in some intervals. By suitably choosing the factor b we can ensure that

$$\min\left\{\omega(\tau) : |\tau| \leqslant \pi/2\right\} \geqslant 1. \tag{2.25}$$

We let

$$\Omega(x) \doteq \prod_{j=1}^{n} \omega(x_j) \qquad (x \doteq (x_1, \ldots, x_n) \in R^n), \tag{2.26}$$

define numbers a_k by the formula

$$a_k \doteq \delta_m^{-1} \beta_k^{-p'} \tilde{N}_k^{np'/p}, \quad k \in \{1, \ldots, m\}, \qquad \delta_m \doteq \sum_{k=1}^{m} \beta_k^{-p'} \tilde{N}_k^{np'/p}, \tag{2.27}$$

and prove that all the conditions (2.18) hold for the function

$$g(x) \doteq \sum_{k=1}^{m} a_k \Omega(\tilde{N}_k x). \tag{2.28}$$

According to (2.20) we have $|\tilde{N}_k x_j| \leqslant 1$ for $k \leqslant m$ and $x \in Q_p$, and, using (2.26)–(2.28), we get that

$$g(x) \geqslant \sum_{k=1}^{m} a_k = 1, \qquad x \in Q_p. \tag{2.29}$$

Further, it follows from property 2) of $\omega(\tau)$, the condition (2.21), and the defining formula (2.26) that the Fourier transform of $\Omega(\tilde{N}_k x)$ is concentrated in the set $\tilde{\Delta}_k \doteq \{\nu \in R^n : \tilde{N}_{k-1} < |\nu_j| < \tilde{N}_k\}$, which for $k < m$ coincides with $\Delta_k \doteq \{\nu \in R^n : N_{k-1} < |\nu_j| < N_k\}$ and for $k = m$ is contained in it. We can thus assert that for the function (2.28)

$$S_{N_k} g(x) - S_{N_{k-1}} g(x) = \begin{cases} a_k \Omega(\tilde{N}_k x), & k \leqslant m, \\ 0, & k > m. \end{cases} \tag{2.30}$$

The following lemma (established in [4], Lemma 1 in §2) is used to obtain the desired norm estimate of $g(x)$ in the class $L_{p,1}^{(\beta,N)}(R^n)$.

LEMMA 2. *Suppose that the numbers $\tilde{N}_k$ satisfy condition* (2.21), *where $\tilde{A}_0 > 1$. Then there exists a constant c depending only on n, p, and $\tilde{A}_0$ such that for any numerical sequence $\{u_k\}$*

$$\left\| \sum_{k=1}^{\infty} u_k \left(\prod_{j=1}^{n} \left(1 + \tilde{N}_k |x_j|\right) \right)^{-1} \right\|_p \leqslant c \left\| \left\{ u_k \tilde{N}_k^{-n/p} \right\} \right\|_{l_p}. \tag{2.31}$$

If in Lemma 2 we set $u_k \doteq a_k \beta_k$, $k \leqslant m$, and $u_k \doteq 0$, $k > m$, and use the inequality $|\Omega(x)| \leqslant c \prod_1^n (1 + |x_j|)^{-1}$, which follows from (2.24) and (2.26), then from (1.2), (2.30), and (2.31) we get

$$\|g\|_{L_{p,1}^{(\beta,N)}(R^n)} \leqslant c \sum_{k=1}^{m} a_k^p \beta_k^p \tilde{N}_k^{-n}. \tag{2.32}$$

But we find directly from the definition (2.27) of a_k by simple computations that

$$\sum_{k=1}^{m} a_k^p \beta_k^p \tilde{N}_k^{-n} = \delta_m^{1-p} = \left(\sum_{k=1}^{m} \beta_k^{-p'} \tilde{N}_k^{np'/p} \right)^{1-p}. \tag{2.33}$$

Let us now establish the inequality (see (1.5) and (2.27))

$$\gamma_1^{1/(1-p)}(\rho) \left(= \sum_{k=1}^{\infty} \beta_k^{-p'} \left(N_k^{-1} + \rho \right)^{-np'/p} \right) \leqslant c\delta_m. \tag{2.34}$$

For $k \leqslant m$ it follows from (2.19) and (2.20) that

$$\left(N_k^{-1} + \rho \right)^{-np'/p} \leqslant \left(\max\left(N_k^{-1}, \rho \right) \right)^{-np'/p} = \tilde{N}_k^{np'/p},$$

and so

$$\sum_{k=1}^{m} \beta_k^{-p'} \left(N_k^{-1} + \rho \right)^{-np'/p} \leqslant \sum_{k=1}^{m} \beta_k^{-p'} N_k^{np'/p} = \delta_m. \tag{2.35}$$

By (2.19), $\rho^{-1} \leqslant 2N_m$; hence

$$\rho^{-1} \leqslant 2 \min\left(\rho^{-1}, N_m \right) = 2\tilde{N}_m$$

for all $k \in \mathbf{N}$.

But, in view of (1.11), $(N_k^{-1} + \rho)^{-np'/p} \leqslant \rho^{-np'/p} \leqslant (2\tilde{N}_m)^{np'/p}$ for $k > m$, and thus

$$\sum_{k=m+1}^{\infty} \beta_k^{-p'} \left(N_k^{-1} + \rho \right)^{-np'/p} \leqslant (2\tilde{N}_m)^{np'/p} \beta_m^{-p'} (A_1 - 1)^{-1}$$

$$\leqslant c(n, p, A_1)\delta_m. \tag{2.36}$$

Combining (2.35) and (2.36), we get (2.34), which, combined with (2.33) and (2.32), leads us to the desired inequality $\|g\|_{\Lambda_{p,1}}^p \leqslant c_1\gamma_1(\rho)$ (see (2.18)), and this concludes the proof of Theorem 1.

3°. We now show (see Remark 3 in 1.4°) that for $\theta > 1$ and for any set E with diameter $\rho > 0$ it is possible to construct an entire function $g(x)$ of exponential type ρ^{-1} such that

$$g(x) \geqslant 0, \quad x \in R^n, \qquad g(x) \geqslant 1, \quad x \in E, \qquad \|g\|_{\Lambda_{p,\theta}} \leqslant \tilde{c}_1\gamma_1(\rho), \tag{2.37}$$

where $\tilde{c}_1$ may now depend on θ.

As in the preceding subsection, it suffices to consider the case when $E \doteq Q_\rho$ (see (2.18)).

The construction of the required function $g(x)$ will be based on the following assertion, which was proved in §2 of [5] (see also [3]).

PROPOSITION 2. *If $1 < \theta < \infty$, then the norm (1.2) of a function $g(x)$ in the class $\Lambda_{p,\theta}$ admits the estimate*

$$\|g\|_{\Lambda_{p,\theta}} \leqslant c_0 \inf\left\{ \left\| \|\{\beta_k g_k(x)\}\|_{l_\theta} \right\|_p : g_k(x) \text{ is an entire function of exponential type } N_k \text{ in } x_j, \text{ and } \left\| g(x) - \sum_{k=1}^{l} g_k(x) \right\|_p \underset{l\to\infty}{\to} 0 \right\}, \tag{2.38}$$

where c_0 depends only on θ, p, n, A_0, A_1, and A_2.

Of course, the infimum on the right-hand side of (2.38) is in fact a quantity equivalent to $\|g\|_{\Lambda_{p,\theta}}$ for $\theta \in (1, \infty)$, because $S_{N_k} g(x) - S_{N_{k-1}} g(x)$ are entire functions of exponential type N_k in x_j.

According to the properties (2.37) we again define the function $g(x)$ by (2.26)–(2.28), but this time instead of the $\omega(\tau)$ in (2.23) we take some entire function of exponential type 1 which is nonnegative on R^1 and satisfies (2.24) and (2.25). It is easy to check that all these conditions are satisfied by the function

$$\omega(\tau) \doteq \frac{\pi^2}{2}\left(\frac{\sin(\tau/2)}{\tau}\right)^2. \tag{2.39}$$

Here, of course, we choose the functions $a_k \Omega(\tilde{N}_k x)$ as the $g_k(x)$ in (2.38) (see (2.28)) and, arguing as before, get the inequalities

$$g(x) \geqslant 1, \quad x \in Q, \qquad \left\| \|\{\beta_k g_k(x)\}\|_{l_1} \right\|_p^p \leqslant c_1 \gamma_1(\rho). \tag{2.40}$$

The quantity $\|\{u_k\}\|_{l_\theta}$ cannot increase as θ increases, so, by (2.38), the $g(x)$ we have constructed satisfies the chain of inequalities ($1 < \theta \leqslant \infty$, $\tilde{\theta} \doteq \min(\theta, 2)$)

$$\begin{aligned} \|g\|^p_{\Lambda_{p,\theta}} \leqslant \|g\|^p_{\Lambda_{p,\tilde{\theta}}} &\leqslant c_0 \left\| \|\{\beta_k g_k(x)\}\|_{l_{\tilde{\theta}}} \right\|_p^p \\ &\leqslant c_0 \left\| \|\{\beta_k g_k(x)\}\|_{l_1} \right\|_p^p \leqslant \tilde{c}_1 \gamma_1(\rho). \end{aligned} \tag{2.41}$$

The nonnegativity of $g(x)$ on the whole space R^n is clear from the nonnegativity of the function (2.39) and of the numbers (2.27).

§3. Spaces of traces for weighted Sobolev classes

1°. The proof of Theorems 2 and 3 are based on the descriptions obtained in [6] for the spaces $\mathscr{T}r\mathscr{W}_p$ of traces on R^n of the functions in the classes $W^{r,s}_{p,\xi,\varphi}(G^{n+1}_a)_0$.

PROPOSITION 3 ([6], Theorem 1). *In the notation of* 1.2° *let*

$$N(y) \doteq \left(\zeta^{1/p'}(y)\psi^{1/p}(y)\right)^{-1/s}, \qquad y \in (0, a], \tag{3.1}$$

and suppose that a decreasing numerical sequence $\{t_m\}$, $t_m \in (0, a)$, $m \in \{0, 1, \ldots, \}$, *is chosen such that for some* λ, μ ($0 < \lambda \leqslant \mu < 1$) *and all* $m \geqslant 1$

$$\lambda \leqslant \max\left(\zeta(t_m)\zeta^{-1}(t_{m-1}), \psi(t_m)\psi^{-1}(t_{m-1})\right) \leqslant \mu. \tag{3.2}$$

If $m = 1$ *or* $\zeta(t_m) \geqslant \lambda\zeta(t_{m-1})$, *then the index* m *is put in the set* $\mathfrak{A}_1$; *the set of remaining positive integers* m *is denoted by* $\mathfrak{A}_2$ (*by* (3.2), $\psi(t_m) \geqslant \lambda\psi(t_{m-1})$ *for* $m \in \mathfrak{A}_2$); *let* $\hat{N}_m \doteq N(t_m)$, $m \geqslant 0$ *and* $\hat{N}_0 \doteq 0$.

A function $g(x) \in L_p(R^n)$ *is an element of* $\mathscr{T}r\mathscr{W}_p$ *for* $r = 1$, *i.e., the trace on* R^n *of some function* $f(x, y) \in W^{1,s}_{p,\xi,\varphi}(G^{n+1}_a)_0$, *if and only if the quantities*

$$\begin{aligned} \sigma_1(g) &\doteq \sum_{m \in \mathfrak{A}_1} \zeta^{1-p}(t_m) \left\| S_{\hat{N}_m} g(x) - S_{\hat{N}_{m-1}} g(x) \right\|_p^p, \\ \sigma_2(g) &\doteq \sum_{m \in \mathfrak{A}_2} \psi(t_m) \sum_{j=1}^n \left\| \frac{\partial^s}{\partial x_j^2}\left(S_{\hat{N}_m} g(x) - S_{\hat{N}_{m-1}} g(x)\right) \right\|_p^p \end{aligned} \tag{3.3}$$

are finite and there exists a constant c independent of $g(x)$ *such that*

$$\begin{aligned} c^{-1}(\sigma_1(g) + \sigma_2(g)) &\lesssim \inf\left\{ \|f\|^p_{\mathscr{W}_p} : \lim_{y \to 0} \|f(x, y) - g(x)\|_p = 0 \right\} \\ &\leqslant c(\sigma_1(g) + \sigma_2(g)). \end{aligned} \tag{3.4}$$

We note that a sequence $\{t_m\}$ with the properties (3.2) always exists: for example, we can choose $t_0 \in (0, a)$ arbitrarily and then make the recursive definition

$$t_{m+1} \doteq \max\{y: \zeta(y) \leqslant \zeta(t_m)/2 \,\&\, \psi(y) \leqslant \psi(t_m)/2\}. \tag{3.5}$$

The conditions (3.2) are satisfied with the constants $\lambda = \mu = 1/2$ when $\{t_m\}$ is so constructed.

The connection of the spaces $\mathscr{T}rW^{1,s}_{p,\xi,\varphi}(G^{n+1}_a)_0$ with the classes $\Lambda_{p,\theta}$ is cleared up in the following assertion.

Proposition 4. *Under the conditions of Theorem 2 there exists a sequence* $\{y_k\} \downarrow 0$ *such that the numbers*

$$\beta_k \doteq \zeta^{-1/p'}(y_k)$$

satisfy (1.1), *and the imbeddings*

$$L^{(\beta,N)}_{p,\theta_-}(R^n) \subset \mathscr{T}rW^{1,s}_{p,\xi,\varphi}\left(G^{n+1}_a\right)_0 \subset L^{(\beta,N)}_{p,\theta_+}(R^n) \tag{3.7}$$

are valid, where $\theta_- = \min(2, p)$, $\theta_+ \doteq \max(2, p)$ *and* $N_k \doteq N(y_k)$ (*see* (3.1)).

Proof of Proposition 4. We start the construction of $\{y_k\}$ from the numbers $\{t_m\}$ in (3.5).

Define the positive integers l_m ($m \in \mathbf{N}$) by the formula

$$l_m \doteq \begin{cases} 1, & m = 1, \\ \min\left\{l \in \mathbf{N}: \zeta(t_m)\left(\frac{3}{2}\right)^l \geqslant \frac{3}{4}\zeta(t_{m-1})\right\}, & m > 1. \end{cases} \tag{3.8}$$

By definition (see Proposition 3), $\zeta(t_m) = \zeta(t_{m-1})/2$ if $m \in \mathfrak{A}_1$ and $m > 1$, and $\zeta(t_m) < \zeta(t_{m-1})/2$ for $m \in \mathfrak{A}_2$; hence it is clear from (3.8) that $l_m = 1$ for $m \in \mathfrak{A}_1$ and $l_m > 1$ for $m \in \mathfrak{A}_2$. Let

$$q_m \doteq l_1 + \cdots + l_m, \qquad m \in \mathbf{N}. \tag{3.9}$$

Furthermore, $q_1 = 1$ by (3.8). For a positive integer $k = q_m - d$, $0 \leqslant d < l_m$, we now define the number y_k by the condition

$$\zeta(y_k) \doteq \zeta_k \doteq \zeta(t_m)\left(\tfrac{3}{2}\right)^d. \tag{3.10}$$

In particular, if we set $d = 0$, then we find that $y_{q_m} = t_m$.

By construction, if $m \in \mathfrak{A}_1$ and $k = q_m$, then $k - 1 = q_{m-1}$; consequently, $\zeta_{k-1} = 2\zeta_k$. Let $m \in \mathfrak{A}_2$, and let $k = q_m - l_m + 1 = q_{m-1} + 1$. Then the defining formulas (3.8) and (3.10) imply that

$$\tfrac{3}{2}\zeta_k = \zeta_{q_m}\left(\tfrac{3}{2}\right)^{l_m} \geqslant \tfrac{3}{4}\zeta_{q_{m-1}} \geqslant \zeta_{q_m}\left(\tfrac{3}{2}\right)^{l_m - 1} = \zeta_k \tag{3.11}$$

and thus $\frac{1}{2}\zeta_{k-1} \leqslant \zeta_k < \frac{3}{4}\zeta_{k-1}$ for the indicated k.

Thus, taking (3.10) into account, we conclude that for all $k \in N$

$$\tfrac{1}{2}\zeta_{k-1} \leqslant \zeta_k \leqslant \tfrac{3}{4}\zeta_{k-1}, \tag{3.12}$$

and this is equivalent to the conditions (1.1) for the numbers (3.6).

Further, as already noted, if $m \in \mathfrak{A}_1$, then $N_{q_m} = \hat{N}_m$ and $N_{q_{m-1}} = \hat{N}_{m-1}$, and comparison of (3.3) and (3.6) gives us

$$\sigma_1(g) = \sum_{\substack{k = q_m \\ m \in \mathfrak{A}_1}} \left(\beta_k \left\|S_{N_k}g(x) - S_{N_{k-1}}g(x)\right\|_p\right)^p. \tag{3.13}$$

Let us now consider the indices $m \in \mathfrak{A}_2$, i.e., the indices such that (see Proposition 3 and (3.5)) $m > 1$, $\zeta_{q_m} < \zeta_{q_{m-1}}/2$, and $\psi_{q_{m-1}} \doteq \psi(y_{q_{m-1}}) = 2\psi_{q_m}$; here the number l_m (see (3.8)) is greater than 1, while the numbers $N_k \doteq N(y_k)$ (see (3.1) and (3.10)) satisfy the two-sided lacunarity conditions

$$\left(\tfrac{3}{2}\right)^{1/p's} N_k \leqslant N_{k-1} \leqslant \left(\tfrac{3}{2}\right)^{1/p's} 2^{1/ps} N_k \qquad (k \in \{q_{m-1}+1, \ldots, q_m\}). \tag{3.14}$$

Using norm estimates (based on the generalized Paley-Littlewood theorem [11]) of the derivatives in $L_p(R^n)$ (see [12], Chapter 5), and taking account of (3.14), (3.1), and (3.6), we arrive at the inequalities

$$\begin{aligned} \psi_{q_m} \sum_{j=1}^{n} & \left\| \frac{\partial^s}{\partial x_j^s} \left(S_{\hat{N}_m} g(x) - S_{\hat{N}_{m-1}} g(x) \right) \right\|_p^p \\ & \leqslant c \psi_{q_m} \left\| \left(\sum_{k=q_{m-1}+1}^{q_m} N_k^{2s} \left| S_{N_k} g(x) - S_{N_{k-1}} g(x) \right|^2 \right)^{1/2} \right\|_p^p \\ & \leqslant c \left\| \left(\sum_{k=q_{m-1}+1}^{q_m} \left| \beta_k \left(S_{N_k} g(x) - S_{N_{k-1}} g(x) \right) \right|^2 \right)^{1/2} \right\|_p^p, \end{aligned} \tag{3.15}$$

$$\begin{aligned} & \left\| \left(\sum_{k=q_{m-1}+1}^{q_m} \left| \beta_k \left(S_{N_k} g(x) - S_{N_{k-1}} g(x) \right) \right|^2 \right)^{1/2} \right\|_p^p \\ & \leqslant c \psi_{q_{m-1}} \left\| \left(\sum_{k=q_{m-1}+1}^{q_m} N_{k-1}^{2s} \left| S_{N_k} g(x) - S_{N_{k-1}} g(x) \right|^2 \right)^{1/2} \right\|_p^p \\ & \leqslant c \psi_{q_m} \sum_{j=1}^{n} \left\| \frac{\partial^s}{\partial x_j^s} \left(S_{\hat{N}_m} g(x) - S_{\hat{N}_{m-1}} g(x) \right) \right\|_p^p. \end{aligned} \tag{3.16}$$

Comparison of (3.15), (3.16), and (3.3) shows that

$$\begin{aligned} c^{-1} \sigma_2(g) \leqslant \sum_{m \in \mathfrak{A}_2} & \left\| \left(\sum_{k=q_{m-1}+1}^{q_m} \left| \beta_k \left(S_{N_k} g(x) - S_{N_{k-1}} g(x) \right) \right|^2 \right)^{1/2} \right\|_p^p \\ & \leqslant c \sigma_2(g), \end{aligned} \tag{3.17}$$

where c does not depend on $g(x)$.

Using further the fact that $q_{m-1} + 1 = q_m$ for $m \in \mathfrak{A}_1$ and combining (3.17) and (3.13), we conclude that there exists a c_0 independent of $g(x)$ such that

$$\begin{aligned} c_0^{-1} \left(\sigma_1(g) + \sigma_2(g) \right)^{1/p} & \leqslant \left\| \left\| \left\{ \beta_k \left(S_{N_k} g(x) - S_{N_{k-1}} g(x) \right) \right\} \right\|_{\mathfrak{b}} \right\|_p \\ & \leqslant c_0 \left(\sigma_1(g) + \sigma_2(g) \right)^{1/p}, \end{aligned} \tag{3.18}$$

where $\mathfrak{b}$ denotes the space of sequences $\{u_k\}$ with finite norm

$$\|\{u_k\}\|_{\mathfrak{b}} \doteq \left(\sum_{m=1}^{\infty} \left(\sum_{k=q_{m-1}+1}^{q_m} |u_k|^2 \right)^{p/2} \right)^{1/p} \qquad (q_0 \doteq 0). \tag{3.19}$$

But it is immediately clear from the definition of $\|\cdot\|_{\mathfrak{b}}$ that

$$\begin{aligned} \|\{u_k\}\|_{l_p} \leqslant \|\{u_k\}\|_{\mathfrak{b}} \leqslant \|\{u_k\}\|_{l_2} \qquad (p \geqslant 2), \\ \|\{u_k\}\|_{l_2} \leqslant \|\{u_k\}\|_{\mathfrak{b}} \leqslant \|\{u_k\}\|_{l_p} \qquad (p \leqslant 2), \end{aligned} \tag{3.20}$$

and thus the imbeddings (3.7) follow from (3.18) and Proposition 3 (see (3.4)).

This concludes the proof of Proposition 4.

2°. To pass from the sum appearing in the definition (1.5) of the function $\gamma_1(t)$ to the integrals in (1.13) and (1.16) we need

LEMMA 3. *Let $F(y)$ and $G(y)$ be positive functions defined on $(0, a]$, with $F(y)$ increasing and tending to 0 as $y \to 0$, and with $G(y)$ decreasing on $(0, a]$. Suppose that the numbers $y_k \in (0, a]$ form a decreasing sequence such that for all k*

$$\lambda F(y_k) \leqslant F(y_{k+1}) \leqslant \mu F(y) \qquad (0 < \lambda \leqslant \mu < 1). \tag{3.21}$$

Then

$$c_1 \sum_{k=1}^{\infty} G(y_k)F(y_k) \leqslant \int_b^a G(y)\,dF(y) \leqslant c_2 \sum_{k=1}^{\infty} G(y_k)F(y_k), \tag{3.22}$$

where $c_1 \doteq (1-\mu)$ and $c_2 \doteq \max((\lambda^{-1}-1), F(a)/F(y_1))$ are positive constants independent of $G(y)$.

PROOF. Since $G(y)$ is positive and decreasing and $F(y)$ is increasing, the numbers

$$I_k \doteq \int_{y_{k+1}}^{y_k} G(y)\,dF(y) \tag{3.23}$$

are positive and satisfy the inequalities

$$G(y_k)(F(y_k) - F(y_{k+1})) \leqslant I_k \leqslant G(y_{k+1})(F(y_k) - F(y_{k+1})). \tag{3.24}$$

But, according to (3.21),

$$\begin{aligned} F(y_k) \leqslant (1-\mu)^{-1}(F(y_k) - F(y_{k+1})), \\ F(y_k) - F(y_{k+1}) \leqslant (\lambda^{-1}-1)F(y_{k+1}). \end{aligned} \tag{3.25}$$

From the left-hand relations in (3.24) and (3.25) we get by summing over k that

$$\sum_{k=1}^{\infty} G(y_k)F(y_k) \leqslant (1-\mu)^{-1} \sum_{k=1}^{\infty} I_k \leqslant (1-\mu)^{-1} \int_0^a G(y)\,dF(y), \tag{3.26}$$

which is equivalent to the left-hand estimate in (3.22).

Similarly, from the right-hand inequalities in (3.24) and (3.25),

$$\int_0^{y_1} G(y)\,dF(y) = \sum_{k=1}^{\infty} I_k \leqslant (\lambda^{-1}-1) \sum_{k=2}^{\infty} G(y_k)F(y_k), \tag{3.27}$$

and to get the right-hand estimate in (3.22) it remains to observe that

$$\int_{y_1}^a G(y)\,dF(y) \leqslant G(y_1)F(a) = (G(y_1)F(y_1))(F(a)/F(y_1)),$$

which concludes the proof of Lemma 3.

3°. If we now set

$$F(y) \doteq \zeta(y), \qquad G(y) \doteq \left((\zeta^{1/p'}(y)\psi^{1/p}(y))^{1/s} + t\right)^{-np'/p} \tag{3.28}$$

in Lemma 3 and take Proposition 4 into account along with the defining formulas (1.5) and (1.13), then we conclude from (3.22) that if the sequences $\{\beta_k\}$ and $\{N_k\}$ are chosen according to (3.6) and (3.1), then

$$c^{-1}\gamma_1(t) \leqslant \gamma_2(t) \leqslant c\gamma_1(t), \tag{3.29}$$

where c does not depend on t.

Suppose now that $E \subset R^n$, $h_E \doteq (\text{meas } E)^{1/n} > 0$, $f(x, y) \in \mathscr{W}_p$, and $\lim_{y\to 0} f(x, y) \geqslant 1$, $x \in E$. Then the left-hand inequality in (3.29), the left-hand estimate in (1.6), and the right-hand imbedding in (3.7) give us that

$$\gamma_2(h_E) \leqslant c\gamma_1(h_E) \leqslant c\|f(x,0)\|^p_{L^{(\beta,N)}_{p,\theta_+}(R^n)} \leqslant c\|f(x, y)\|^p_{\mathscr{W}_p}. \tag{3.30}$$

Conversely, suppose that $E \subset R^n$ is a set of diameter ρ; then by 2.3° there exist a positive integer m, numbers $a_k \geqslant 0$, and an entire function $\Omega(x) \geqslant 0$ of exponential type 1 such that (see (2.28), (2.19), and (2.39))

$$g(x) \doteq \sum_{k=1}^{m} a_k\Omega(\tilde{N}_k x) \geqslant 1, \quad x \in E, \qquad \|g\|^p_{L^{(\beta,N)}_{p,\theta_-}(R^n)} \leqslant \tilde{c}_1\gamma_1(\rho). \tag{3.31}$$

Observe now that a function $f_0(x, y) \in \mathscr{W}_p$ such that (see (3.7))

$$\|f_0(x, y)\|_{\mathscr{W}_p} \leqslant c\|g\|_{\Lambda_{p,\theta_-}}, \qquad f_0(x, y) = g(x), \tag{3.32}$$

can be defined by the formula (see the arguments in §2 of [6])

$$f_0(x, y) \doteq \sum_{k=1}^{m} a_k\Omega(\tilde{N}_k x)\eta_k(y), \tag{3.33}$$

where the $\eta_k(y)$ are smooth functions equal to 0 for $y \geqslant a/2$, and to 1 in a neighborhood of $y = 0$; nothing prevents us from choosing the $\eta_k(y)$ to be nonnegative and infinitely differentiable, and, doing this, we obtain a function $f_0(x, y) \in C^\infty(G_a^{n+1})$ that is nonnegative and such that (see (3.32), (3.31), and the right-hand inequality in (3.29))

$$\begin{gathered} f_0(x,0) \geqslant 1, \quad x \in E, \\ \|f_0(x, y)\|^p_{\mathscr{W}_p} \leqslant c\|g\|_{\Lambda_{p,\theta_-}} \leqslant c\gamma_1(\rho) \leqslant c\gamma_2(\rho), \end{gathered} \tag{3.34}$$

where c does not depend on ρ.

Recalling the definition (1.12) of the capacity $\operatorname{Cap}(E, \mathscr{W}_p)$, we observe that the inequalities (3.30) and (3.34) prove Theorem 2.

4°. The proof of Theorem 3 can be carried out by the same scheme and is even simpler, since the description of the traces on R^n of functions in $\mathscr{W}_p$ is simpler in that case.

PROPOSITION 5. *Under the conditions of Theorem* 3 *let*

$$y_k = 2^{-k}a, \quad \beta_k = y_k^{(1/p)-r}\xi^{1/p}(y_k), \quad N_k = \left(y_k^{-r}\left(\xi(y_k)\varphi^{-1}(y_k)\right)^{1/p}\right)^{1/s}. \tag{3.35}$$

Then the conditions (1.1) *hold, and*

$$\mathscr{T}r W^{r,s}_{p,\xi,\varphi}(G_a^{n+1})_0 = L^{(\beta,N)}_{pp}(R^n). \tag{3.36}$$

PROOF. Let (see (1.11) and (1.16))

$$F(y) \doteq \zeta(y) \doteq \int_0^y (\xi(u))^{1/(1-p)} u^{(r-1)p'}\,du,$$
$$G(y) = \left(y^{r/s}(\xi^{-1}(y)\varphi(y))^{1/ps} + t\right)^{-np'/p}. \tag{3.37}$$

The function $F(y)$ is obviously increasing, and since

$$T(y) \doteq y^{r/s}(\xi^{-1}(y)\varphi(y))^{1/ps} = \left[(\varphi(y)y^{1-\varepsilon})(\xi^{-1}(y)y^{pr-1-\varepsilon})y^{2\varepsilon}\right]^{1/ps}, \tag{3.38}$$

it follows from the fact that the first two functions in (1.15) are increasing that $G(y)$ is decreasing and that (see (3.35))

$$N_{k+1} = T^{-1}(y_{k+1}) \geqslant 2^{2\varepsilon np'/p^2 s} T^{-1}(y_k) \geqslant A_0 N_k \qquad (A_0 > 1), \tag{3.39}$$

i.e., (1.1) holds for the $\{N_k\}$ constructed in Proposition 5.

Further, the fact that the functions $\xi^{-1}(y)y^{pr-1-\varepsilon}$ and $\xi(y)y^{\kappa}$ are increasing (see (1.15)) implies the relations

$$c^{-1}\zeta(y/2) \leqslant \zeta(y) \leqslant c\zeta(y/2), \quad c^{-1}\tilde{\zeta}(y/2) \leqslant \tilde{\zeta}(y) \leqslant c\tilde{\zeta}(y/2)$$
$$\left(\tilde{\zeta}(y) = (\xi(y))^{1/(1-p)} y^{1+(r-1)p'}\right), \tag{3.40}$$
$$c^{-1}\tilde{\zeta}(y) \leqslant \zeta(y) \leqslant c\tilde{\zeta}(y),$$

where the constant $c > 1$ depends only on p, r, ε, and κ.

This establishes that the numbers $\beta_k \doteq (\zeta(y_k))^{-1/p'}$ (see (3.40), (3.37), and (3.35)) also satisfy (1.1), and (3.21) holds for the sequence $F(y_k)$.

Using Lemma 3 and the last inequality in (3.40), we conclude that the functions $\gamma_3(t)$ in (1.16) and $\gamma_1(t)$ in (1.5) constructed from the numbers (3.35) satisfy the estimates

$$c^{-1}\gamma_1(t) \leqslant \gamma_3(t) \leqslant c\gamma_1(t), \tag{3.41}$$

where c does not depend on t.

On the basis of (3.36) (which is proved in §3 of [6]) and the arguments used in 3.3° we get the estimates (1.17) from (3.41), and this concludes the proof of Theorem 3.

In conclusion we remark that the validity of Remark 2 in 1.4° about conditions for continuity of the traces on R^n of functions in $\mathscr{W}_p$ follows immediately from (3.7) and (3.36) and a criterion for the imbedding $L_{p,\theta}^{(\beta,N)}(R^n) \hookrightarrow C(R^n)$, $1 \leqslant \theta \leqslant \infty$ (see [10]).

BIBLIOGRAPHY

1. V. G. Maz'ya and V. P. Khavin, *Nonlinear potential theory*, Uspekhi Mat. Nauk **27** (1972), no. 6 (163), 67–138; English transl. in Russian Math. Surveys **27** (1972).

2. S. M. Nikol'skiĭ, *Approximation of functions of several variables and embedding theorems*, 2nd rev. aug. ed., "Nauka", Moscow, 1977; English transl. of 1st ed., Springer-Verlag, 1974.

3. G. A. Kalyabin, *A characterization of functions in spaces of Besov-Lizorkin-Triebel type*, Dokl. Akad. Nauk SSSR **236** (1977), 1056–1059; English transl. in Soviet Math. Dokl. **18** (1977).

4. ______, *A description of traces for anisotropic spaces of Triebel-Lizorkin type*, Trudy Mat. Inst. Steklov. **150** (1979), 160–173; English transl. in Proc. Steklov Inst. Math. **1981**, no. 4 (150).

5. ______, *A description of functions in classes of Besov-Lizorkin-Triebel type*, Trudy Mat. Inst. Steklov. **156** (1980), 82–109; English transl. in Proc. Steklov Inst. Math. 1983, no. 2(156).

6. ______, *The trace problem for weighted anisotropic spaces of Liouville type*, Izv. Akad. Nauk SSSR Ser. Mat. **41** (1977), 1138–1160; English transl. in Math. USSR Izv. **11** (1977).

7. L. D. Kudryavtsev, *Properties of the boundary values of functions in weighted spaces and their applications to boundary value problems*, Continuum Mechanics and Related Problems in Analysis (N. I. Muskhelishvili 80th Birthday Collection), "Nauka", Moscow, 1972, pp. 259–265. (Russian)

8. P. I. Lizorkin, *Properties of functions in the spaces* $\Lambda^r_{p,\theta}$, Trudy Inst. Mat. Steklov. **131** (1974), 158–181; English transl. in Proc. Steklov Inst. Math. **131** (1974).

9. G. A. Kalyabin, *Multiplicativity conditions for function spaces of Besov and Lizorkin-Triebel type*, Dokl. Akad. Nauk SSSR **251** (1980), 25–26; English transl. in Soviet Math. Dokl. 25–26; English transl. in Soviet Math. Dokl. **21** (1980).

10. ______, *A criterion for multiplicativity and imbedding in C for spaces of Besov-Lizorkin-Triebel type*, Mat. Zametki **30** (1981), 517–526; English transl. in Math. Notes **30** (1981).

11. Paul Krée, *Sur les multiplicateurs dans* $\mathscr{F}L^p$, Ann. Inst. Fourier (Grenoble) **16** (1966), fasc. 2, 31–89.

12. Elias M. Stein, *Singular integrals and differentiability properties of functions*, Princeton Univ. Press, Princeton, N. J., 1970.

Translated by H. H. McFADEN

Труды Матем. Инст. Стеклов.
161(1983)

Proceedings of the Steklov
Institute of Mathematics
1984, Issue 3

ON WEIGHTED LIZORKIN-TRIEBEL SPACES. SINGULAR INTEGRALS, MULTIPLIERS, IMBEDDING THEOREMS

UDC 517.518.22

V. M. KOKILASHVILI

ABSTRACT. Weighted Lizorkin-Triebel spaces are studied in this article. It is proved that the maximal singular operator generated by a multiple singular integral is bounded in a weighted L^p-space with $1 < p < \infty$. A complete description is given for the class of weight functions such that singular integral operators are bounded in the L^p-spaces with the corresponding weight. Multiplier theorems are proved, and imbedding theorems with different metrics and different dimensions are established for weighted Lizorkin-Triebel spaces by the multiplier method.

Bibliography: 25 titles.

In this article we study weighted function spaces $\mathscr{L}^{p,\theta}_{w,m}$ of Lizorkin-Triebel type which generalize the Nikol'skiĭ classes $H_p^{(r)}$, the Besov classes $B^r_{p,\theta}$, and the Liouville classes L^r_p. The Lizorkin-Triebel spaces (see [1]–[4]) and generalizatins of them (see [5] and [6]) arose from the Littlewood-Paley theory on "dissection" of Fourier transforms of functions in the L^p-spaces ($1 < p < \infty$) (see, for example, [7], Chapter IV). To construct these spaces we use a method of partitioning a Euclidean space and the corresponding "dissection" of the given function.

We prove theorems on multipliers of Fourier transforms, and use the multiplier method for the spaces $\mathscr{P}^{p,\theta}_{w,m}$ to construct an imbedding theory with different metrics and different dimensions. Estimates in weighted function spaces for various integral operators such as operators generated by multiple Hilbert transformations and integrals of potential type are important in the method of investigation.

§1. On weight functions

Some classes of weight functions used below are defined in this section. Henceforth, $\mathbf{J}^n$ denotes the collection of all n-dimensional parallelepipeds with faces parallel to the coordinate axes, and $\mathbf{I}^n$ denotes the collection of all n-dimensional cubes with faces parallel to the coordinate axes.

DEFINITION 1. A nonnegative measurable function $w\colon R^r \to R^1$ is said to be a *function of class* $\mathscr{A}^{q,p}(\mathbf{J}^n, c)$, $1 < p \leqslant q < \infty$, if for any $J \in \mathbf{J}^n$

$$\int_J w^q(x)\,dx < \infty, \qquad \int_J w^{-p'}(x)\,dx < \infty, \quad p' = \frac{p}{p-1}, \tag{1}$$

1980 *Mathematics Subject Classification*. Primary 42B15, 42B20, 46E35; Secondary 42B25, 44A15, 46E30.

0081-5438/84 \$1.00 + \$.25 per page

and

$$\sup\left(\frac{1}{|J|}\int_J w^q(x)\,dx\right)^{1/q}\left(\frac{1}{|J|}\int_J w^{-p'}(x)\,dx\right)^{1/p'}=c<\infty, \tag{2}$$

where the supremum is over all $J\in\mathbf{J}^n$.

Let $\mathscr{A}^{q,p}(\mathbf{J}^n,c)=\mathscr{A}^p(\mathbf{J}^n,c)$ for $p=q$. Also, let

$$\mathscr{A}^{p,q}(\mathbf{J}^n)=\bigcup_{c>0}\mathscr{A}^{q,p}(\mathbf{J}^n,c)\quad\text{and}\quad\mathscr{A}^p(\mathbf{J}^n)=\mathscr{A}^{p,p}(\mathbf{J}^n).$$

Muckenhoupt [8] introduced the class of fucntions satisfying the conditions (1) and (2), with the integration and supremum taken over the n-dimensional cubes with faces parallel to the coordinate axes, in connection with a description of the class of weight functions such that the Hardy-Littlewood-Wiener maximal operator is bounded in the Lebesgue spaces with the corresponding weight. This class will be denoted by $\mathscr{A}^{q,p}(\mathbf{I}^n)$. Analogous problems for singular integral operators and generalized Hardy-Littlewood-Wiener maximal operators were considered in [9]–[12].

The class $\mathscr{A}^{q,p}(\mathbf{J}^n)$ is smaller than the class $\mathscr{A}^{q,p}(\mathbf{I}^n)$. For example, the function

$$w(x)=\left(\sum_{j=1}^{k}|x_j|\right)^{\alpha}\qquad(1\leqslant k\leqslant n) \tag{3}$$

belongs to $\mathscr{A}^{q,p}(\mathbf{J}^n)$ if and only if $-1/p<\alpha<1/p'$, while such a w is in $\mathscr{A}^{q,p}(\mathbf{I}^n)$ if and only if $-n/p<\alpha<n/p'$.

It will be shown below that the class $\mathscr{A}^p(\mathbf{J}^n)$ completely describes the class of weight functions such that operators generated by multiple singular integrals are bounded on the L^p-spaces ($1<p<\infty$) with the corresponding weight.

We now describe some properties of weight functions in $\mathscr{A}^{q,p}(\mathbf{J}^n)$ which will be used in the framework of the present article.

PROPOSITION 1 [14]. *A nonnegative measurable function $w\colon R^n\to R^1$ is in $\mathscr{A}^{q,p}(\mathbf{J}^n)$ if and only if there exists a constant $c>0$ such that*

$$\left(\frac{1}{|I|}\int_I w^q(x_1,\dots,x_i,\dots,x_n)\,dx_i\right)^{1/q}\left(\frac{1}{|I|}\int_I w^{-p'}(x_1,\dots,x_i,\dots,x_n)\,dx_i\right)^{1/p'}<c$$

for any interval $I\in\mathbf{I}^1$ and for arbitrary $\tilde{x}^i=(x_i,\dots,x_{i-1},x_{i+1},\dots,x_n)\in R^{n-1}$ $(i=1,\dots,n)$.

Note that the sufficiency part of this assertion is a corollary to Lebesgue's theorem on differentiation of an indefinite integral.

DEFINITION 2. We write $\rho\in\mathscr{A}_p(\mathbf{J}^n)$ for a nonnegative measurable function ρ if $\rho^{1/p}\in\mathscr{A}^p(\mathbf{J}^n)$.

PROPOSITION 2 ([8], Lemma 5). *If $\rho\in\mathscr{A}_p(\mathbf{I}^n)$ $(1<p<\infty)$, then $\rho\in\mathscr{A}_s(\mathbf{I}^n)$ for any $s>p$; if $\rho\in\mathscr{A}_p(\mathbf{I}^n)$ $(1<p<\infty)$, then there exists a $p_0<p$ such that $\rho\in\mathscr{A}_{p_0}(\mathbf{I}^n)$.*

PROPOSITION 3 ([9], Lemma 3). *If $1<p<\infty$ and $\rho\in\mathscr{A}_p(\mathbf{I}^1,c)$, then there exists a constant $c_1>0$ depending only on c and p such that for any interval $I\in\mathbf{I}^1$*

$$|I|^{p-1}\int_{x\notin I}\frac{\rho(x)}{|x-a_1|^p}\,dx\leqslant c_1|I|^{-1}\int_I\rho(x)\,dx, \tag{4}$$

where a_I is the center of I.

From this it follows, in particular, that

$$\int_{R^1} \frac{\rho(x)}{(1+|x|)^p}\,dx < \infty. \tag{5}$$

PROPOSITION 4. *If $\rho \in \mathscr{A}_p(\mathbf{J}^n, c)$, then there exist a p_0 with $1 < p_0 < p$ and a $c_0 > 0$ depending only on p and c such that $\rho \in \mathscr{A}_{p_0}(\mathbf{J}^n, c_0)$.*

Proposition 4 follows immediately from Propositions 1 and 3.

PROPOSITION 5. *If $\rho \in \mathscr{A}_p(\mathbf{J}^n)$, then*

$$\int_{R^n} \frac{\rho(x)}{\prod_1^n(1+|x_j|)^p}\,dx < \infty. \tag{6}$$

This proposition follows from Propositions 3 and 1.

Let $w\colon R^n \to R^1$ be a nonnegative measurable function. By $L^p_w(R^n)$ we denote the Banach space of measurable functions f on R^n with the norm

$$\|f\|_{L^p_w(R^n)} = \left(\int_{R^n} |f(x)|^p w^p(x)\,dx\right)^{1/p}.$$

This norm will sometimes be denoted by $\|f\|_{p,w}$. Henceforth, p' denotes the number $p/(p-1)$.

Let (Y, Σ, ν) be a space with a σ-finite measure, and $L^\theta(Y, \Sigma, \nu)$ the Banach space of measurable (Σ) functions $F\colon Y \to R^1$ with the norm

$$\|F\|_{\theta,\nu} = \left(\int_Y |F(x)|^\theta\,d\nu\right)^{1/\theta} < \infty.$$

Suppose now that $\mathbf{T}$ is the algebra of Lebesgue-measurable sets in R^n, m is Lebesgue measure, and $\lambda = m \times \nu$ is the product measure defined on the σ-algebra $\mathbf{T} \times \Sigma$. Let $X = L^\theta(Y, \Sigma, \nu)$, $1 \leqslant \theta < \infty$. By $L^p_w[R^n, X]$ we denote the set of λ-measurable functions $f\colon R^n \times Y \to R^1$ such that

$$|f, L^p_w[R^n, X]| = \left(\int_{R^n} \|f(t,\cdot)\|_X^p w^p(t)\,dt\right)^{1/p} < \infty.$$

PROPOSITION 6. *If $f \in L^p_w[R^n, X]$, $1 < p < \infty$, $w \in \mathscr{A}^p(\mathbf{I}^n)$, and $X = L^\theta(Y, \Sigma, \nu)$, then there exists a $p_p < p$ such that $f \in L^{p_0}_{\text{loc}}[R^n, X]$.*

PROOF. Let $w^p(x) = \rho(x)$. By Proposition 2, there exists a $p_0 < p$ such that $\rho \in \mathscr{A}_{p/p_0}$. Hölder's inequality gives us that

$$\int_I \left(\int_Y |f(x,y)|^\theta\,d\nu(y)\right)^{p_0/\theta} dx \leqslant \left(\int_I \left(\int_Y |f(x,y)|^\theta\,d\nu(y)\right)^{p/\theta} \rho(x)\,dx\right)^{p_0/p}$$
$$\times \left(\int_I (\rho(x))^{-p_0/(p-p_0)}\,dx\right)^{(p-p_0)/p_0}.$$

Furthermore, the second factor on the right-hand side is finite, because $\rho \in \mathscr{A}_{p/p_0}(\mathbf{I}^n)$. Consequently, $f \in L^{p_0}_{\text{loc}}[R^n, X]$.

PROPOSITION 7. *If* $f \in L^p_w[R^n, X]$, $X = L^\theta(Y, \Sigma, \nu)$, $1 < p$, $\theta < \infty$, *and* $w \in \mathscr{A}^p(\mathbf{J}^n)$, *then there exists a* $p_0 < p$ *such that* $f(\cdot, y)$ *is in* $L^{p_0}_{\text{loc}}(R^n)$ *and*

$$\int_{R^n} |f(x, y)| \prod_{k=1}^{n} \left(1 + |x_k|\right)^{-1} dx < \infty$$

for almost all $y \in Y$.

PROOF. Let $D \subset Y$ be a set of finite ν-measure, and let I be an arbitrary n-dimensional cube. For $p_0 < \theta$

$$\int_I \left(\int_D |f(x, y)|^{p_0} d\nu \right) dx \leqslant \int_I \left(\int_D |f(x, y)|^{\theta} d\nu \right)^{p_0/\theta} dx,$$

where p_0 is the number indicated in Proposition 6. For $p_0 \geqslant \theta$

$$\int_I \left(\int_D |f(x, y)|^{\theta} d\nu \right) dx \leqslant \int_I \left(\int_D |f(x, y)|^{p_0} d\nu \right)^{\theta/p_0} dx \cdot (\nu D)^{(p_0 - \theta)/p_0}.$$

By Hölder's inequality

$$\int_D \int_{R^n} |f(x, y)| \prod_{k=1}^{n} \left(1 + |x_k|\right)^{-1} dx\, dy$$
$$\leqslant \|f, L^p_w[R^n, X]\| \left(\int_{R^n} \frac{w^{-p'}(x)}{\prod_{k=1}^{n} \left(1 + |x_k|\right)^{p'}} dx \right)^{1/p} (\nu D)^{1/p}.$$

According to Proposition 5, the second factor on the right-hand side is finite. Consequently, the left-hand side is also finite, and

$$\int_R |f(x, y)| \prod_{k=1}^{n} \left(1 + |x_k|\right)^{-1} dx < \infty$$

ν-almost everywhere on D.

Finally, we give a proposition which is not connected with weight functions but which will be useful in what follows.

PROPOSITION 8. *Let* $\varphi\colon R^1_+ \to R^1$ *be a nonnegative locally integrable function. For any* $x > 0$

$$x \int_x^\infty \frac{\varphi(s)}{s^2} ds \leqslant 2 \tilde{M}(\varphi)(\xi),$$

where

$$\tilde{M}(\varphi)(y) = \sup_I \frac{1}{|I|} \int_I \varphi(s)\, ds$$

is the supremum over all intervals $I \subset R^1_+$ *containing* y, *and* ξ *is an arbitrary point in* $(0, x]$.

This proposition is known. For the convenience of the reader we give a proof. Clearly, for $u > x$

$$\frac{1}{u} \int_0^u \varphi(s)\, ds \leqslant \tilde{M}(\varphi)(\xi),$$

where ξ is any point in $(0, x]$. Hence,

$$\int_x^u \varphi(s)\,ds \leqslant \tilde{M}(\varphi)(\xi)u.$$

We multiply both sides by u^{-3}, integrate from x to ∞, and interchange the order of integration. This gives us that

$$\int_x^\infty \frac{\varphi(s)}{s^2}\,ds \leqslant 2x^{-1}\tilde{M}(\varphi)(\xi).$$

§2. Weighted inequalities for maximal singular integral operators

In this section we study the question of whether integral operators generated by multiple singular integrals are bounded on weighted Lebesgue spaces. Multiple singular integrals more general than multiple Hilbert transforms will be considered. The kernel of a multiple singular integral is a direct product of Calderón kernels.

THEOREM 1. *Suppose that the measurable functions a_i: $R^1 \to R^1$ $(i = 1,\dots,n)$ satisfy a Lipschitz condition*

$$|a_i(s) - a_i(t)| \leqslant c|s - t|, \qquad s, t \in R^1. \tag{7}$$

Let

$$C^*(f)(x) = \sup_{\varepsilon>0}\left|\int_{E_x^\varepsilon} f(y)\prod_{i=1}^n \frac{a_i(x_i) - a_i(y_i)}{(x_i - y_i)^2}\,dy\right|, \qquad x \in R^n,$$

where $E_x^\varepsilon = \{y \in R^n, |x_i - y_i| > \varepsilon_i, i = 1,\dots,n\}$ and $\varepsilon = = (\varepsilon_1,\dots,\varepsilon_n)$, $\varepsilon_i > 0$ $(i = 1,\dots,n)$. If $w \in \mathscr{A}^p(\mathbf{J}^n)$, then for any $f \in L_w^p(R^n)$

$$\int_{R^n}(C^*(f)(x))^p w^p(x)\,dx \leqslant b\int_{R^n}|f(x)|^p w^p(x)\,dx,$$

where the constant $b > 0$ does not depend on f.

The proof of this theorem will be based on a number of theorems and lemmas to be stated and proved (in part) below.

Let (Y, Σ, ν) be a space with a τ-finite measure, $\mathbf{T}$ the σ-algebra of Lebesgue-measurable sets in R^n, and $\lambda = m \times \nu$ the product measure defined on the σ-algebra $\mathbf{T} \times \Sigma$. Let f: $R^n \times Y \to R^1$ be a λ-measurable function. We set

$$M(f)(x, \cdot) = \sup \frac{1}{|I|}\int_I |f(t, \cdot)|\,dt,$$

where the supremum is over all n-dimensional cubes whose faces are parallel to the coordinate axes and which contain the point x, i.e., $I \in \mathbf{I}^n$.

THEOREM A (see [13] and [14]). *Suppose that $1 < p, \theta < \infty$. The inequality*

$$\int_{R^n}\left(\int_Y (M(f)(x, y))^\theta\,d\nu(y)\right)^{p/\theta} w^p(x)\,dx$$
$$\leqslant b\int_{R^n}\left(\int_Y |f(x, y)|^\theta\,d\nu(y)\right)^{p/\theta} w^p(x)\,dx \tag{8}$$

holds for some $b > 0$ and any $(T \times \Sigma)$-measurable function $f\colon R^n \times Y \to R^1$ making the right-hand side of (8) *finite if and only if*

$$\sup\left(\frac{1}{|I|}\int_I w^p(x)\,dx\right)^{1/p}\left(\frac{1}{|I|}\int_I w^{-p'}(x)\,dx\right)^{1/p'} < \infty,$$

where the supremum is over all cubes $I \in \mathbf{I}^n$.

This theorem generalizes a result of Fefferman and Stein [15], who proved (8) in the case where $w(x) \equiv 1$ and ν is a measure concentrated at points of the set of positive integers.

Theorem A is more general than the theorem of Stein and Fefferman even in the case without a weight.

THEOREM B [16]. *Let $w \in A^p(\mathbf{I}^1)$, $1 < p < \infty$. Then for any $f \in L^p_w[R^n, X]$*

$$\left(\int_{R^1}\left(\int_Y\left(\sup_{\varepsilon>0}\left|\int_{|x-t|>\varepsilon}\frac{a(x)-a(t)}{(x-t)^2}f(t,y)\,dt\right|\right)^\theta d\nu(y)\right)^{p/\theta} w^p(x)\,dx\right)^{1/p}$$
$$\leqslant v\|f, L^p_w[R^1, X]\| \tag{9}$$

with a constant $b > 0$ independent of f.

First we prove several lemmas.

LEMMA 1. *Suppose that the measurable functions $k_i\colon R^2 \to R^1$ $(i = 1, 2)$ satisfy the conditions*

1) $|k_i(s_1, s_2)| \leqslant c|s_1 - s_2|^{-1}$ $(s_1, s_2 \in R^1)$,

2) $T_1^*(\varphi)(s) = \sup_{\eta>0}|\int_{|t-s|>\eta} k_1(s,t)\varphi(t)\,dt|$ *is bounded in $L^p_\rho(R^1)$ for $1 < p < \infty$ for any $\rho \in \mathscr{A}^p(\mathbf{I}^1)$*

Then for any $w \in \mathscr{A}^p(\mathbf{J}^2)$, $f \in L^p_w(R^2)$, and almost every $(s_1, s_2) \in R^2$

$$\int_{R^1} k_1(x_1, t_1)\left(\int_{|s_2-t_2|>\varepsilon_2} k_2(s_2, t_2) f(t_1, t_2)\,dt_2\right)dt_1$$
$$= \int_{|s_2-t_2|>\varepsilon_2} k_2(s_2, t_2)\left(\int_{R^1} k_1(s_1, t_1) f(t_1, t_2)\,dt_1\right)dt_2. \tag{10}$$

PROOF. From the condition $f \in L^p_w(\mathbf{J}^2)$ in the lemma, Hölder's inequality, and Proposition 5 it follows that

$$\int_{R^2}\frac{|f(s_1, s_2)|}{(1+|s_1|)(1+|s_2|)}\,ds_1\,ds_2 < \infty.$$

By Fubini's theorem,

$$\int_{|s_1-t_1|>\varepsilon_1} k_1(s_1, t_1)\left(\int_{|s_2-t_2|>\varepsilon_2} k_2(s_2, t_2) f(t_1, t_2)\,dt_2\right)dt_1$$
$$= \int_{|s_2-t_2|>\varepsilon_2} k_2(s_2, t_2)\left(\int_{|s_1-t_1|>\varepsilon_1} k_1(s_1, t_1) f(t_1, t_2)\,dt_1\right)dt_2. \tag{11}$$

We now show that

$$\lim_{\varepsilon_1\to 0}\int_{|s_2-t_2|>\varepsilon_2} k_2(s_2,t_2)\left(\int_{|s_1-t_1|>\varepsilon_1} k_1(s_1,t_1)f(t_1,t_2)\,dt_1\right)dt_2$$
$$=\int_{|s_2-t_2|>\varepsilon_2} k_2(s_2,t_2)\left(\int_{R'} k_1(s_1,t_1)f(t_1,t_2)\,dt_1\right)dt_2 \qquad (12)$$

for almost all $(s_1, s_2)\in R^2$. Obviously,

$$\left|\int_{|s_1-t_1|>\varepsilon_1} k_1(s_1,t_1)f(t_1,t_2)\,dt_1\right| \leqslant T_1^*(f)(s_1,t_2).$$

By a condition of the lemma,

$$\int_{R^2}(T_1^*(f)(s_1,t_2))^p w^p(x_1,t_2)\,ds_1\,dt_2 \leqslant b\int_{R^2}|f(x_1,t_2)|^p w^p(s_1,t_2)\,ds_1\,dt_2.$$

Therefore,

$$\int_{R^1}(T_1^*(f)(s_1,t_2))^p w^p(s_1,t_2)\,dt_2<\infty$$

for almost all $s_1\in R^1$. Consequently, by Proposition 5,

$$\int_{R^1}\frac{T_1^*(f)(s_1,t_2)}{1+|t_2|}\,dt_2<\infty$$

for almost all $s_1\in R^1$. We conclude from this that the function

$$k_2(s_2,t_2)\int_{|s_1-t_1|>\varepsilon_1} k_1(s_1,t_1)f(t_1,t_2)\,dt_1$$

has for almost all $(s_1, s_2)\in R^2$ an integrable majorant on the set $\{t_2\colon |s_2-t_2|>\varepsilon_2\}$. Therefore, (12) holds by the dominated convergence theorem. Passage to the limit in (11) as $\varepsilon_1\to 0$ leads to the lemma, by (12).

We introduce some notation:

$$k_i(s,t)=\frac{a_i(s)-a_i(t)}{(s-t)^2},\qquad C_i^*(\varphi)(s)=\sup\left|\int_{|s-t|>\varepsilon} k_i(s,t)\varphi(t)\,dt\right|,$$
$$C_i(\varphi)(s)=S\int_{-\infty}^{+\infty} k_i(s,t)\varphi(t)\,dt,$$
$$C_{i,\varepsilon}(\varphi)(s)=\int_{|s-t|>\varepsilon} k_i(s,t)\varphi(t)\,dt \qquad (i=1,2).$$

For given $x=(x_1,x_2)$ and $\varepsilon=(\varepsilon_1,\varepsilon_2)$ let

$I_i=I_{x_i}=(x_i-\varepsilon_i, x_i+\varepsilon_i)$, $\bar I_i=\bar I_{x_i}=(x_i-\varepsilon/4, x_i+\varepsilon/4)$, $\bar J=\bar I_1\times\bar I_2$.

LEMMA 2. *Let $f\in L_w^p(R^2)$, $1<p<\infty$, $w\in\mathscr{A}^p(\mathbf{J}^2)$. For any $\varepsilon\in R_+^2$ and almost all $x=(x_1,x_2)\in R^2$ and $\bar x=(\bar x_1,\bar x_2)\in\bar J$*

$$\begin{aligned}C_{1,\varepsilon_1}(C_{2,\varepsilon_2}(f))(x_1,x_2)\leqslant b\big(&|C_2(C_1(f))(\bar x_1,\bar x_2)|\\&+|C_2(C_1(\chi_{I_1}f))(\bar x_1,\bar x_2)|+|C_2(\chi_{I_2}C_1(f))(\bar x_1,\bar x_2)|\\&+M_2(C_1(f))(\bar x_1,x_2)+|C_2(\chi_{I_2}(C_1(\chi_{I_1}f)))(\bar x_1,\bar x_2)|\\&+M_2(C_1(\chi_{I_2}f))(\bar x_1,x_2)+M_1(C_2^*(f))(x_1,x_2)\big)\end{aligned}\qquad (13)$$

where b is an absolute constant, and $M_i(f)$ is the Hardy-Littlewood function with respect to the variable x_i.

PROOF. It is not hard to show that the kernels k_i $(i = 1, 2)$ satisfy the condition

$$|k(x, y) - k(x, y_0)| \leqslant c|y - y_0|\,|x - y_0|^{-2}$$

for $|x - y_0| > 2|y - y_0|$, where c does not depend on x, y, nor y_0 in R^1. If we use this fact, Lemma 1, and Proposition 8, we have that

$$\begin{aligned}
&|C_{1,\varepsilon_1}C_{2,\varepsilon_2}(f)(x_1, x_2) + C_2(\chi_{I_2}C_1(f))(\bar{x}_1, \bar{x}_2)\\
&\quad + C_2(C_1(\chi_{I_2}f))(\bar{x}_1, \bar{x}_2)\\
&\quad - C_2(C_1(f))(\bar{x}_1, \bar{x}_2) - C_2(C_1(\chi_{I_2}\chi_{I_2}f))(\bar{x}_1, \bar{x}_2)|\\
&\quad = \Bigg| \int_{|x_1 - t_1| > \varepsilon_1} k_1(x_1, t) C_{2,\varepsilon_2}(f)(t, x_2)\, dt\\
&\quad + \int_{|x_2 - s| > \varepsilon_2} k_2(\bar{x}_2, s)\left(\int_{-\infty}^{\infty} k_1(\bar{x}_1, t) f(t, s)\, dt\right) ds\\
&\quad + \int_{|x_2 - s| > \varepsilon_2} k_2(\bar{x}_2, s) C_1(\chi_{I_1}f)(\bar{x}_1, s)\, ds\\
&\quad - \int_{|x_2 - s| > \varepsilon_2} k_2(x_2, s) C_1(\chi_{I_1}f)(\bar{x}_1, s)\, ds\\
&\quad + \int_{|x_2 - s| > \varepsilon_2} k_2(x_2, s) C_1(\chi_{I_1}f)(\bar{x}_1, s)\, ds\\
&\quad - \int_{-\infty}^{\infty} k_2(\bar{x}_2, s) C_1(f)(\bar{x}_1, s)\, ds\\
&\quad - \int_{|x_2 - s| > \varepsilon_2} k_2(x_2, s) C_1(f)(\bar{x}_1, s)\, ds\\
&\quad + \int_{|x_2 - s| > \varepsilon_2} k_2(x_2, s) C_1(f)(\bar{x}_1, s)\, ds \Bigg|\\
&\quad = \Bigg| \int_{|x_2 - s| > \varepsilon_2} (k_2(x_2, s) - k_2(\bar{x}_2, s)) C_1(f)(\bar{x}_1, s)\, ds\\
&\quad + \int_{|x_2 - s| > \varepsilon_2} (k_2(\bar{x}_2, s) - k_2(x_2, s)) C_1(\chi_{I_1}f)(\bar{x}_1, s)\, ds\\
&\quad + \int_{|x_1 - t| > \varepsilon_2} (k_1(s_1, t) - k_1(\bar{x}_1, t)) C_{2,\varepsilon_2}(f)(t, x_2)\, dt \Bigg|\\
&\quad \leqslant b\Bigg(|x_2 - \bar{x}_2| \int_{|x_2 - s| > \varepsilon_2} \frac{C_1(f)(\bar{x}_1, s)}{|x_2 - s|^2}\, ds\\
&\quad + |x_2 - \bar{x}_2| \int_{|x_2 - s| > \varepsilon_2} \frac{C_1(f\chi_{I_1})(x_1, s)}{|x_2 - s|^2}\, ds\\
&\quad + |x_1 - \bar{x}_1| \int_{|x_1 - t| > \varepsilon_2} \left| \int_{|x_2 - s| > \varepsilon_2} k_2(x_2, s) f(t, s)\, ds \right| \frac{dt}{|x_1 - t|^2} \Bigg)\\
&\quad \leqslant b\big(M_2(C_1(f))(\bar{x}_1, x_2)\\
&\quad + M_2(C_1(f\chi_{I_1}))(\bar{x}_1, x_2) + M_1(C_2^*(f))(x_1, x_2)\big).
\end{aligned} \tag{14}$$

Consequently,

$$|C_{1,\varepsilon_1}C_{2,\varepsilon_2}(f)(x_1, x_2)| \leqslant bV,$$

where bV denotes the right-hand side of (13).

PROOF OF THEOREM 1. To avoid awkward computations we prove the theorem for $n = 2$. Let F be the set of functions $f\colon R^2 \to R^1$ representable in the form

$$f(x) = \sum_{i=1}^{m} a_i \Phi_i(x_1)\Psi_i(x_2),$$

where Φ_i and Ψ_i are infinitely differentiable functions with compact support. As follows from Theorem A, for any $f \in F$

$$C^*(f)(x_1, x_2) < \infty$$

almost everywhere in R^2. For each point $x \in R^2$ there exists an $\varepsilon = (\varepsilon_1, \varepsilon_2) \in R^2_+$ such that

$$C^*(f)(x_1, x_2) < 2C_{1,\varepsilon_1}(C_{2,\varepsilon_2}(f))(x_1, x_2). \tag{15}$$

Consequently, by (13),

$$C^*(f)(x) < bV,$$

where the ε_1 and ε_2 in the expression V are chosen in the corresponding way. Let us integrate the last inequality with respect to the variable $(\bar{x}_1, \bar{x}_2)$ on the set $\bar{J} = \bar{I}_1 \times \bar{I}_2$. We get

$$C^*(f)(x_1, x_2) < b\sum_{i=1}^{7} V_i(x_1, x_2), \tag{16}$$

where the constant b does not depend on (x_1, x_2),

$$V_1(x) = \frac{1}{|\bar{J}|}\int_{\bar{I}_1}\int_{\bar{I}_2} |C_2(C_1(f))(\bar{x}_1, \bar{x}_2)|\, d\bar{x}_1\, d\bar{x}_2,$$

$$V_2(x) = \frac{1}{|\bar{J}|}\int_{\bar{I}_1}\int_{\bar{I}_2} |C_2(C_1(\chi_{I_2}f))(\bar{x}_1, \bar{x}_2)|\, d\bar{x}_1\, d\bar{x}_2,$$

$$V_3(x) = \frac{1}{|\bar{I}_1|}\int_{\bar{I}_1} M_2(C_1(f))(\bar{x}_1, x_2)\, d\bar{x}_1,$$

$$V_4(x) = \frac{1}{|\bar{J}|}\int_{\bar{I}_1}\int_{\bar{I}_2} |C_2(C_1(\chi_{I_2}f))(\bar{x}_1, \bar{x}_2)|\, d\bar{x}_1\, d\bar{x}_2,$$

$$V_5(x) = \frac{1}{|\bar{J}|}\int_{\bar{I}_1}\int_{\bar{I}_2} |C_2(C_1(\chi_{I_1}X_{I_2}f))(\bar{x}_1, \bar{x}_2)|\, d\bar{x}_1\, d\bar{x}_2,$$

$$V_6(x) = \frac{1}{|\bar{I}_1|}\int_{\bar{I}_1} M_2(C_1(\chi_{I_2}f))(\bar{x}_1, x_2)\, d\bar{x}_1,$$

and

$$V_7(x) = M_1(C_2^*(f))(x_1, x_2).$$

Obviously,

$$V_1(x) < M_1(M_2(C_2(C_1 f)))(x_1, x_2).$$

Since $w \in \mathscr{A}^p(\mathbf{J}^2)$, Theorems A and B give us that

$$\|V_1\|_{p,w} \leqslant v\|f\|_{p,w}. \tag{17}$$

Suppose that $p_0 < p$. We use Hölder's inequality. Then

$$V_2(x) \leqslant \frac{1}{|\bar{I}_1|}\int_{\bar{I}_1}\left(\frac{1}{|\bar{I}_2|}\int_{\bar{I}_2}\left|C_2\left(C_1(\chi_{I_2}f)\right)(\bar{x}_1,\bar{x}_2)\right|^{p_0}d\bar{x}_2\right)^{1/p_0}d\bar{x}_1.$$

Estimation of the inner integral on the basis of Theorem A gives us

$$V_2(x) \leqslant b|\bar{I}_1|^{-1}\int_{\bar{I}_1}\left(|\bar{I}_2|^{-1}\int_{\bar{I}_2}|C_1(f)(\bar{x}_1,\bar{x}_2)|^{p_0}d\bar{x}_2\right)^{1/p_0}d\bar{x}_1.$$

Therefore,

$$V_2(x) \leqslant bM_1(M_2(C_1^{p_0}(f)))^{1/p_0}(x_1,x_2). \tag{18}$$

Choose the number p_0 so that $w^{p_0} \in \mathscr{A}^{p/p_0}(J_2)$. This is possible, because the condition $w \in \mathscr{A}^p$ is equivalent to the condition $w^p \in \mathscr{A}_p$. On the other hand, there exists a $p_0 < p$ such that $w^p \in \mathscr{A}_{p_0}$, i.e., $w^{p_0} \in \mathscr{A}^{p/p_0}$. Let us raise (18) to the pth power and integrate with respect to the measure $w^p(x)\,dx$.

As a result of applying Theorems A and B we conclude that

$$\int_{R^2}|V_2 j(x)|^p w^p(x)\,dx \leqslant b\int_{R^2}|f(x)|^p(x)\,dx. \tag{19}$$

Further,

$$V_3(x) \leqslant M_1(M_2(C_1(f)))(x_1,x_2)$$

and, by Proposition 1 and a theorem of Muckenhoupt [8],

$$\|V_3\|_{p,w} \leqslant b\|f\|_{p,w}. \tag{20}$$

We proceed to estimate $\|V_4\|_{p,w}$. Obviously,

$$V_4(x) \leqslant |\bar{I}_1|^{-1}\int_{\bar{I}_1}M_2\left(C_2\left(C_1(\chi_{I_1}f)\right)\right)(\bar{x}_1,x_2)\,d\bar{x}_1.$$

By Hölder's inequality

$$V_4(x) \leqslant |\bar{I}_1|^{-1/p_0}\left(\int_{\bar{I}_1}\left(M_2\left(C_2\left(C_1(\chi_{I_1}f)\right)\right)(\bar{x}_1,x_2)\right)^{p_0}d\bar{x}_1\right)^{1/p_0}.$$

Hence,

$$\begin{aligned}&\int_{R^2}|V_4(x)|^p w^p(x)\,dx\\ &\leqslant b\int_{R^1}|\bar{I}_1|^{-p/p_0}\left(\int_{R^1}\left(\int_{\bar{I}_1}\left(M_2\left(C_2\left(C_1(\chi_{I_1}f)\right)\right)\right)^{p_0}(\bar{x}_1,x_2)\,d\bar{x}_1\right)^{p/p_0}\right.\\ &\quad\times w^p(x_1,x_2)\,dx_1\,dx_2.\end{aligned}$$

Applying Theorem A, we have that

$$\int_{R^2} |V_4(x)|^p w^p(x)\,dx$$
$$\leqslant b\int_{R^1} |\bar{I}_1|^{-p/p_0}\left(\int_{R^1}\left(\int_{\bar{I}_1} |C_2(C_1(\chi_{I_1}f))(\bar{x}_1, x_2)|^{p_0}\,d\bar{x}_1\right)^{p/p_0} w^p(x_1,x_2)\,dx_2\right)dx_1.$$

By Theorem B,

$$\int_{R^2} |V_4(x)|^p w^p(x)\,dx$$
$$\leqslant b\int_{R^1} |\bar{I}_1|^{-p/p_0}\left(\int_{R^1}\left(\int_{\bar{I}_1} |C_1(\chi_{I_1}f)(\bar{x}_1, x_2)|^{p_0}\,d\bar{x}_1\right)^{p/p_0} w^p(x_1,x_2)\,dx_2\right)dx_1$$
$$\leqslant b\int_{R^1} |\bar{I}_1|^{-p/p_0}\int_{R^1}\left(\int_{I_1} |f(\bar{x}_1, x_2)|^{p_0}\,d\bar{x}_1\right)^{p/p_0} w^p(x_1,x_2)\,dx_1\,dx_2.$$

Therefore,

$$\int_{R^2}|V_4(x)|^p w^p(x)\,dx \leqslant b\int_{R^2}\left(M_1(|f|^{p_0})(x_1,x_2)\right)^{p/p_0} w^p(x_1,x_2)\,dx_1\,dx_2.$$

The boundedness of M_1 on the weighted L^p-space and the properties of a function $w \in \mathscr{A}^p(\mathbf{J}^2)$ now give us

$$\int_{R^2} |V_4(x)|^p w^p(x)\,dx \leqslant b\int_{R^2} |f(x)|^p w^p(x)\,dx. \tag{21}$$

Further,

$$V_5(x) \leqslant |\bar{I}_1|\int_{\bar{I}_1}\left(|\bar{I}_2|\int_{\bar{I}_2} |C_2(C_1(\chi_{I_1}\chi_{I_2}f))(\bar{x}_1,\bar{x}_2)|^{p_0}\,d\bar{x}_2\right)^{1/p_0} d\bar{x}_1.$$

By the boundedness of C_i on $L^{p_0}(R^1)$ and the last inequality,

$$V_5(x) \leqslant b|\bar{I}_1|^{-1}\int_{\bar{I}_1}\left(|I_2|^{-1}\int_{\bar{I}_2} |C_1(\chi_{I_1}f)(\bar{x}_1,\bar{x}_2)|^{p_0}\,d\bar{x}_2\right)^{1/p_0} d\bar{x}_1$$
$$\leqslant b|\bar{I}_2|^{-1}\int_{\bar{I}_2}\left(|\bar{I}_1|^{-1}\int_{\bar{I}_1} |f(\bar{x}_1,\bar{x}_2)|^{p_0}\,d\bar{x}_2\right)^{1/p_0} d\bar{x}_1$$
$$\leqslant b\left(M_2\left(M_1\left(|f|^{p_0}\right)(x_1,x_2)\right)\right)^{1/p}.$$

Arguing exactly as in estimating $\|V_2\|_{p,w}$, we see that

$$\|V_5\|_{p,w} \leqslant b\|f\|_{p,w}. \tag{22}$$

Hölder's inequality and Theorem A allow us to write the following chain of inequalities:

$$\int_{R^2} V_6^p(x)w^p(x)\,dx$$

$$\leqslant \int_{R^1} |\bar{I}_1|^{-p/p_0}\left(\int_{R^1}\left(\int_{\bar{I}_1}\left(M_2(C_1(\chi_{I_1}f))(\bar{x}_1,x_2)\right)^{p_0}d\bar{x}_1\right)^{p/p_0} w^p(x_1,x_2)\,dx_2\right)dx_1$$

$$\leqslant b\int_{R^1} |\bar{I}_1|^{-p/p_0}\left(\int_{R^1}\left(\int_{R'}|C_1(\chi_{I_1}f)(\bar{x}_1,x_2)|^{p_0}d\bar{x}_1\right)^{p/p_0} w^p(x_1,x_2)\,dx_2\right)dx_1.$$

The operator C_1 is bounded on $L^{p_0}(R^1)$. Therefore,

$$\int_{R^2} V_6^p(x)w^p(x)\,dx$$

$$\leqslant b\int_{R^1} |\bar{I}_1|^{-p/p_0}\left(\int_{R^1}\left(\int_{I_1}|f(\bar{x}_1,x_2)|^{p_0}d\bar{x}_1\right)^{p/p_0} w^p(x_1,x_2)\,dx_2\right)dx_1.$$

Hence,

$$\int_{R^2} V_6^p(x)w^p(x)\,dx \leqslant v\int_{R^2}\left(M_1(|f|^{p_0})(x_1,x_2)\right)^{p/p_0} w^p(x_1,x_2)\,dx_1\,dx_2.$$

Arguing just as above, we arrive at

$$\|V_6\|_{p,w} \leqslant b\|f\|_{p,w}. \tag{23}$$

Finally, since M_1 and C_2^* are bounded on $L_w^p(R^1)$,

$$\|V_7\|_{p,w} \leqslant b\|M_1(C_2^*(f))\|_{p,w} \leqslant b\|f\|_{p,w}. \tag{24}$$

We thus conclude from (17) and (19)–(24) that

$$\|C^*(f)\|_{p,w} \leqslant b\|f\|_{p,w}, \qquad f \in F, \tag{25}$$

with a constant b independent of f.

Now assume that f is an arbitrary function in $L_w^p(R^2)$.

Suppose that $f_n \in F$ and $\lim_{n\to\infty}\|f_n - f\|_{p,w} = 0$. By Propositions 1 and 3 and Hölder's inequality, for almost all $x \in R^2$

$$|C_\varepsilon(f_n)(x) - c_\varepsilon(f)(x)| \leqslant b\|f_n - f\|_{p,w}\left(\int_{x_1-\varepsilon_1}^{x_1+\varepsilon_1}\int_{x_2-\varepsilon_2}^{x_2+\varepsilon_2} w^{-p'}(y)\,dy_1\,dy_2\right)^{1/p'}.$$

This implies that

$$\lim_{n\to\infty} C_\varepsilon(f_n)(x) = C_\varepsilon(f)(x)$$

everywhere except a set of measure zero, and the exceptional set does not depend on ε.

By (25), $C^*(f_n)$ is a Cauchy sequence in L_w^p. Assume that the sequence $(f_n)_1^\infty$ is chosen so that $(C^*(f_n(x))_1^\infty)$ also converges almost everywhere. Then

$$|C_\varepsilon(f)(x)| = \lim_{n\to\infty}|C_\varepsilon(f_n)(x)| \leqslant \lim_{n\to\infty} C^*(f_n)(x) \tag{26}$$

almost everywhere in R^2, and the set where this estimate does not hold does not depend on ε. By (26),

$$C^*(f)(x) \leqslant \lim_{n\to\infty} C^*(f_n)(x)$$

almost everywhere on R^2. Hence,

$$\|C^*(f)(x)\|_{p,w} \leqslant \|\lim C^*(f_n)(x)\|_{p,w} \leqslant \lim_{n\to\infty} \|C^*(f_n)(x)\|_{p,w} \leqslant b\|f\|_{p,w}.$$

The theorem is proved.

Theorem 1 is new even in the case without a weight. For $n = 1$ and $w(x) \equiv 1$ it was established by Calderón [17]. There are also further generalizations of the Calderón result in the case without a weight (see, for example, [18], [19], etc.).

THEOREM 2. *Let $1 < p < \infty$ and $w \in \mathscr{A}^p(\mathbf{J}^2)$. For an arbitrary $f \in L^p_w(R^n)$ the limit*

$$C(f)(x) = \lim_{\varepsilon\to 0} \int_{E^\varepsilon_x} f(t) \prod_{i=1}^n \frac{a_i(x_i) - a_i(t_i)}{(x_i - t_i)^2}\, dt$$

exists, and $C(f)(x)$ can be represented as an iterated integral almost everywhere in R^n.

PROOF. The first part of the theorem can be derived from the preceding theorem in a well-known way (see [7], Chapter II, §4.6.3). Now let $f \in L^p_w(R^2)$, $w \in \mathscr{A}^p(\mathbf{J}^2)$. We show that almost everywhere in R^2

$$\begin{aligned} C(f)(x) &= \int_{R^1} \frac{a_1(x_1) - a_1(y_1)}{(x_1 - y_1)^2} \left(\int_{R^1} \frac{a_2(x_2) - a_2(y_2)}{(x_2 - y_2)^2} f(y_1, y_2)\, dy_2 \right) dy_1 \\ &= \int_{R^1} \frac{a_2(x_2) - a_2(y_2)}{(x_2 - y_2)^2} \left(\int_{R^1} \frac{a_1(x_1) - a_1(y_1)}{(x_1 - y_1)^2} f(y_1, y_2)\, dy_1 \right) dy_2. \end{aligned} \tag{27}$$

By the existence of a limit in the one-dimensional case and Lemma 1,

$$\begin{aligned} &\lim_{\varepsilon_2\to 0} \lim_{\varepsilon_1\to 0} C_{1,\varepsilon_1}(C_{2,\varepsilon_2}(f))(x) \lim_{\varepsilon_2\to 0} C_1(C_{2,\varepsilon_2}(f))(x) \\ &\quad = \lim_{\varepsilon_2\to 0} C_{2,\varepsilon_2}(C_1(f))(x) = C_2(C_1(f))(x) \end{aligned}$$

almost everywhere in R^2.

Finally, from the existence of the limit $\lim_{\varepsilon\to 0} C_\varepsilon(f)(x)$ and of one of the iterated limits almost everywhere in R^2 we conclude that (27) holds. Theorems 1 and 2 were proved in another way for multiple Hilbert transformations ($a_i(x) \equiv x$) in the case without a weight in [20]–[22].

COROLLARY 1. *If $f \in L^p_{\text{loc}}(R^2)$ $(1 < p < \infty)$ and*

$$\int_{R^n} |f(x)| \prod_{j=1}^2 \left(1 + |x_j|\right)^{-1} dx < \infty, \tag{28}$$

then $C(f)(x)$ exists and (27) holds almost everywhere in R^2.

COROLLARY 2. *Suppose that $f \in L^p_w[R^n, X]$, $X = L^\theta(Y, \Sigma, \nu)$, $1 < \theta < \infty$. Then for almost all $(x, y) \in R^n \times Y$ in the sense of the measure $\lambda = m \times \nu$ the integral*

$$C(f)(x, y) = \int_{R^n} f(t, y) \prod_{i=1}^n \frac{a_i(x_i) - a_i(t_i)}{(x_i - t_i)^2}\, dt$$

exists, and $C(f)$ can be represented as in iterated integral.

Indeed, by Proposition 7, there exists a set $Y_0 \subset Y$ with $\nu Y_0 = 0$ such that $f(\cdot, y) \in L^{p_0}_{\text{loc}}(R^n)$ for $y \in Y \setminus Y_0$. Therefore, for each $y \in Y \setminus Y_0$ there exists a set $E_0 \subset R^n$ such that $mE_0 = 0$ and $C(f)(x, y)$ exists almost everywhere in R^n. Fubini's theorem gives us that $C(f)$ exists and can be represented as an iterated integral for λ-almost all $(x, y) \in R^n \times Y$.

§3. Weighted inequalities for multiple singular integrals of vector-valued functions

In this section we give a description of the class of weight functions w ensuring that the operator

$$H^{(\alpha)}(f)(x) = \int_{R^n} f(t, \cdot) \prod_{k=1}^{n} \frac{\operatorname{sgn}(x_k - t_k)}{|x_k - t_k|^{\alpha}}, \quad \alpha = 1 - \frac{1}{p} + \frac{1}{q}, 1 < p \leqslant q < \infty,$$

is bounded from $L^p_w[R^n, X]$ to $L^q_w[R^n, X]$ where $X = L^{\theta}(Y, \Sigma, \nu)$, $1 < \theta < \infty$.

In what follows, $f(t)$ will be written instead of $f(t, \cdot)$, and $\|f(t, \cdot)\|_X = \|f(t)\|$.

THEOREM 3. *The inequality*

$$\left(\int_{R^n} \|H^{\alpha}(f)(t)\|^q w^q(t)\, dt\right)^{1/q} \leqslant b\left(\int_{R^n} \|f(t)\|^p w^p(t)\, dt\right)^{1/p} \tag{29}$$

holds for any $f \in L^p_w[R^n, X]$ with some constant $b > 0$ independent of f if and only if

$$\sup\left(\frac{1}{|J|}\int_J w^q(x)\, dx\right)^{1/q}\left(\frac{1}{|J|}\int_J w^{-p'}(x)\, dx\right)^{1/p'} < \infty, \tag{30}$$

where the supremum is over all $J \in \mathbf{J}^n$. Here $0 \cdot \infty$ is taken to be zero.

PROOF. If (30) holds and for some $J \in \mathbf{J}^n$

$$\int_J w^q(x)\, dx = \infty,$$

then for any $J_1 \in \mathbf{J}^n$ with $J \supset J_1$

$$\int_{J_1} w^q(x)\, dx = \infty.$$

Consequently, by (30), $w(x) = \infty$ almost everywhere in R^n. If we assume that

$$\int_J w^q(x)\, dx = 0,$$

then $w(x) = 0$ almost everywhere in R^n, and, consequently, (29) holds.

Suppose now that $w \in \mathscr{A}^{q,p}(\mathbf{J}^n)$. For $p = q$ the sufficiency part of Theorem 3 follows from Theorem 1. By Corollary 2 to Theorem 2, for λ-almost all $(x, y) \in R^n \times Y$

$$H^{(1)}(f)(x) = \int_{R^n} f(t, y) \prod_{i=1}^{n} \frac{1}{x_i - t_i}\, dt = \int_{-\infty}^{\infty} \frac{dt_1}{x_1 - t_1} \cdots \int_{-\infty}^{\infty} \frac{f(t, y)}{x_n - t_n}\, dt_n. \tag{31}$$

According to Proposition 1, there exists a $c > 0$ such that

$$\left(\frac{1}{|I|}\int_I w^q(x_1,\dots,x_i,\dots,x_n)\,dx_i\right)^{1/q}\left(\frac{1}{|I|}\int_I w^{-p'}(x_1,\dots,x_i,\dots,x_n)\,dx_i\right)^{1/p'} < c$$

for an arbitrary one-dimensional interval I and a point

$$\tilde{x}^{(i)} = (x_1,\dots,x_{i-1},x_{i+1},\dots,x_n) \in R^{n-1}.$$

Therefore, (29) is obtained by using Theorem B n times in (31).

Now let $1 < p < q < \infty$. For $n = 1$ inequality (29) follows immediately from a theorem of Muckenhoupt and Wheedon [10]. Indeed, by the generalized Minkowski inequality and the cited theorem,

$$\int_{R^1}\left(\int_Y\left|\int_{R^1} f(t,y)\frac{\operatorname{sgn}(x-t)}{|x-t|^\alpha}\,dt\right|^\theta d\nu(y)\right)^{q/\theta} w^q(x)\,dx$$
$$\leqslant \int_{R^1}\left(\int_{R^1}\frac{\|f(t)\|}{|x-t|^\alpha}\,dt\right)^q w^q(x)\,dx \leqslant b\int_{R^1}\|f(t)\|^p w^p(t)\,dt.$$

We now assume that (29) holds for $n = k-1$ and prove it for $n = k$. Let $\bar{x} = (x_1,\dots,x_{k-1})$ and $\bar{t} = (t_1,\dots,t_{k-1})$. Our assumption and Minkowski's inequality allow us to write the following chain of inequalities:

$$\left(\int_{R^1}\int_{R^{k-1}}\left\|\int_{R^{k-1}}\frac{d\bar{t}}{\prod_{i=1}^{k-1}|x_i-t_i|^\alpha}\left(\int_{R^1}\frac{f(\bar{t},t_k)}{|x_k-t_k|^\alpha}\,dt_k\right)\right\|^q w^q(\bar{x},x_k)\,d\bar{x}\,dx_k\right)^{1/q}$$
$$\leqslant b\left(\int_{R'}\left(\int_{R^{k-1}}\left\|\int_{R^1}\frac{f(\bar{x},t_k)}{|x_k-t_k|^\alpha}\,dt_k\right\|^p w^p(\bar{x},x_k)\,d\bar{x}\right)^{p/q} dx_k\right)^{(p/q)(1/p)}$$
$$\leqslant b\left(\int_{R^{k-1}}\left(\int_{R^1}\left\|\int_{R^1}\frac{f(\bar{x},x_k)}{|x_k-t_k|^\alpha}\right\|^q w^q(\bar{x},x_k)\,dx_k\right)^{p/q} d\bar{x}\right)^{1/p}$$
$$\leqslant b\left(\int_{R^k}\|f(t)\|^p w^p(t)\,dt\right)^{1/p}.$$

This tells us that the sufficiency part of the theorem is valid for any n.

We now proceed to the proof that condition (30) is necessary for (29) to hold. It suffices to carry out the proof for scalar-valued functions.

Obviously, if

$$\int_J w^{-p'}(x)\,dx = 0 \quad\text{or}\quad \int_J w^q(x)\,dx = 0$$

for some parallelepiped $J \in \mathbf{J}^n$, then (30) holds.

Consider now the parallelepipeds $J \in \mathbf{J}^n$ such that

$$0 < \int_J w^{-p'}(x)\,dx < \infty \quad\text{and}\quad \int_J w^q(x)\,dx > 0.$$

Let

$$J = [a_1-h_1,a_1;\cdots;a_n-h_n,a_n],$$
$$J_1 = [a_1,a_1+h_1;\cdots;a_n,a_n+h_n], \qquad h_j > 0,\ j = 1,\dots,n.$$

It is clear that for a nonnegative f vanishing outside J

$$|H^{(\alpha)}(f)(x)| \geqslant b|J|^{-\alpha}\int_J f(t)\,dt\chi_{J_1}(x).$$

By (29),

$$|J|^{-\alpha}\left(\int_{J_1} w^q(x)\,dx\right)^{1/q}\left(\int_J f(y)\,dy\right) \leqslant b\left(\int_{R^n} |f(x)|^p w^p(x)\,dx\right)^{1/p}. \quad (32)$$

In (32) assume that

$$f(x) = \begin{cases} 1 & \text{for } x \in J, \\ 0 & \text{for } x \notin J. \end{cases}$$

Then

$$\left(|J_1|^{-1}\int_{J_1} w^q(x)\,dx\right)^{1/q} \leqslant b\left(|J|^{-1}\int_J w^p(x)\,dx\right)^{1/p}. \quad (33)$$

For $p < q$ we apply Hölder's inequality on the right-hand side of (33), and get

$$\left(|J_1|^{-1}\int_{J_1} w^q(x)\,dx\right)^{1/q} \leqslant b\left(|J|^{-1}\int_J w^q(x)\,dx\right)^{1/q} \quad (34)$$

with some constant $b > 0$ independent of J and J_1. Analogous arguments give us that

$$\left(|J|^{-1}\int_J w^q(x)\,dx\right)^{1/q} \leqslant b\left(|J|^{-1}\int_{J_1} w^q(x)\,dx\right)^{1/2}, \quad (35)$$

where the constant b is independent of J and J_1.

In (32) we now let

$$f(x) = \begin{cases} w^{-p'}(x) & \text{for } x \in J, \\ 0 & \text{for } x \notin J. \end{cases}$$

Then

$$|J|^{-(1/q+1/p')}\left(\int_{J_1} w^q(x)\,dx\right)^{1/q}\left(\int_J w^{-p'}(x)\,dx\right)^{1/p'} \leqslant b\left(\int_J w^{-pp'+p}(x)\,dx\right)^{1/p}.$$

Hence,

$$\left(|J|^{-1}\int_{J_1} w^q(x)\,dx\right)^{1/q}\left(|J|^{-1}\int_J w^{-p'}(x)\,dx\right)^{1/p'} \leqslant b. \quad (36)$$

According to (35) and (36),

$$\left(|J|^{-1}\int_J w^q(x)\,dx\right)^{1/q}\left(|J|^{-1}\int_J w^{-p'}(x)\,dx\right)^{1/p'} \leqslant b$$

with some $b > 0$ independent of J.

Assume now that for some parallelepiped $J \in \mathbf{J}^n$

$$\int_J w^{-p'}(x)\,dx = \infty.$$

Then there exists a nonnegative function $g \in L^p(\mathbf{J})$ such that

$$\int_J \frac{g(x)}{w(x)} dx = \infty.$$

Let

$$h(x) = \frac{fg(x)}{w(x)} \chi_J(x)$$

and let J_1 be attached to a vertex of the parallelepiped J in the same way as above. For $x \in J_1$

$$\left| \int_J \prod_{k=1}^n \frac{h(t)}{(x_k - t_k)^\alpha} dt \right| \geqslant b|J|^{-\alpha} \int_J h(t)\, dt.$$

Consequently, $|H^{(\alpha)}(f)(x)| = \infty$ for $x \in J_1$. According to the conditions of the theorem,

$$\left(\int_{J_1} |H^{(\alpha)}(h)(x)|^q w^q(x)\, dx \right)^{1/q} \leqslant b \left(\int_J |g(x)|^p\, dx \right)^{1/p}.$$

Therefore, $w(x) = 0$ almost everywhere in J_1, and

$$\int_{J_1} w^{-p'}(x)\, dx = \infty.$$

From this, analogous arguments lead to the conclusion that $w(x) = 0$ almost everywhere in J. Continuing this process, we arrive at the fact that $w(x) = 0$ almost everywhere in R^n and, consequently, (30) holds.

This proves the theorem.

4. The spaces $L^{p,\theta}_{w,m}(R^n)$

Let S be the Schwartz space of test functions, and S' the space of continuous linear functionals over S. The elements of S' are called *generalized functions over S*.

For $\varphi \in S$ the Fourier transform $\hat{\varphi}$ is defined by

$$\hat{\varphi}(\lambda) = (2\pi)^{-n/2} \int_{R^n} \varphi(x) e^{-i\lambda x}\, dx.$$

The inverse Fourier transform is given as follows:

$$\overline{\varphi}(x) = (2\pi)^{-n/2} \int_{R^n} \varphi(\lambda) e^{i\lambda x}\, d\lambda.$$

Fourier transformation determines a topological isomorphism of the space S onto itself.

The Fourier transforms of a generalized function f are defined by

$$(\hat{f}, \varphi) = (f, \hat{\varphi}) \quad \text{and} \quad (\overline{f}, \varphi) = (f, \overline{\varphi}), \qquad \varphi \in S.$$

Let $(m_{j,i})_{-\infty}^{\infty}$ $(j = 1, \ldots, n)$ be two-sided sequences of positive numbers such that there is an $\varepsilon > 0$ for which

$$m_{j,i+1} \geqslant (1 + \varepsilon) m_{j,i}, \tag{37}$$

for any $j = 1,\dots,n$ and $i = 0, \pm 1,\dots$. For all possible integer n-tuples $(s_1,\dots,s_n)$, $s_j \neq 0$ ($j = 1,\dots,n$), we consider the sets of the form

$$\left(-m_{1,s_1+1}, -m_{1,s_1}\right] \cup \left[m_{1,s_1}, m_{1,s+1}\right) \cdots \left(-m_{n,s_n+1}, -m_{n,s_n}\right] \cup \left[m_{n,s_n}, m_{n,s_n+1}\right). \tag{38}$$

This gives us a partition of $R^n \setminus \{0\}$ which will be denoted by $(U_k)_0^\infty$ in what follows. For any $\varphi \in S(R^n)$ let

$$\varphi_k(x) = \overline{\chi U_k \hat{\varphi}}(x) \qquad (k = \pm 1, \pm 2, \dots). \tag{39}$$

Suppose that $1 < p$, $\theta < \infty$ and let w: $R^n \to R^1$ be a nonnegative measurable function. If the sequences $(m_{j,i})_{i=-\infty}^\infty$ and the function w are chosen so that for any $\varphi \in S(R^n)$

$$\left|\varphi, \mathscr{L}_{w,m}^{p,\theta}(R^n)\right| = \left(\int_{R^n} \left(\sum_{k=-\infty}^{\infty} |\varphi_k(x)|^\theta\right)^{p/\theta} w^p(x)\, dx\right)^{1/p} < \infty, \tag{40}$$

then the completion of $S(R^n)$ in the norm (40) will be the definition of the space $\mathscr{L}_{w,m}^{p,\theta}(R^n)$.

Obviously, if the sequences $(m_{j,i})_i$ ($j = 1,\dots,n$) satisfy condition (37), then for any numbers p and θ with $1 < p$, $\theta < \infty$, any function $\varphi \in S(R^n)$, and any $w \in \mathscr{A}^p(\mathbf{J}^n)$

$$\left|\varphi, \mathscr{L}_{w,m}^{p,\theta}(R^n)\right| < \infty.$$

We give a proof for $n = 1$ (cf. [2], Appendix, §I). In this case $j = 1$, and we set $m_{1,i} = m_i$ ($i = 0, \pm 1,\dots$). Without loss of generality we can consider a function $\varphi \in S(R^1)$ whose Fourier transform $\hat{\varphi}$ vanishes for $\lambda < 0$. Then

$$\begin{aligned}
\left|\varphi, \mathscr{L}_{w,m}^{p,\theta}(R^1)\right| &= \left(\int_{-\infty}^{\infty} \left(\sum_{k=-\infty}^{\infty} |\varphi_k(x)|^\theta\right)^{p/\theta} w^p(x)\, dx\right)^{1/p} \\
&\leqslant \left(\int_{-\infty}^{\infty} \left(\sum_{k=-\infty}^{\infty} |\varphi_k(x)|\right)^{p} w^p(x)\, dx\right)^{1/p} \\
&\leqslant \left(\int_{|x|<1} \left(\sum_{k=-\infty}^{0} |\varphi_k(x)|\right)^{p} w^p(x)\, dx\right)^{1/p} \\
&\quad + \left(\int_{|x|>1} \left(\sum_{k=-\infty}^{0} |\varphi_k(x)|\right)^{p} w^p(x)\, dx\right)^{1/p} \\
&\quad + \left(\int_{|x|<1} \left(\sum_{k=1}^{\infty} |\varphi_k(x)|^p\right) w^p(x)\, dx\right) \\
&\quad + \left(\int_{|x|>1} \left(\sum_{k=1}^{\infty} |\varphi_k(x)|\right)^{p} w^p(x)\, dx\right)^{1/p} \\
&= V_1 + V_2 + V_3 + V_4.
\end{aligned} \tag{41}$$

Note first that under our conditions

$$\sum_{l=1}^{\infty} m_{-l}^{\alpha} < \infty \tag{42}$$

for any $\alpha > 0$. Indeed, $m_{-l}/m_{-l+1} \leqslant (1+\varepsilon)^{-1}$ $(l = 1, 2, \ldots)$ by assumption. Therefore, $m_{-l} \leqslant (1+\varepsilon)^{-l} m_{-l}$ $(l = 1, 2, \ldots)$ and, consequently,

$$\sum_{l=1}^{\infty} m_{-l}^{\alpha} \leqslant \sum_{l=1}^{\infty} (1+\varepsilon)^{-\alpha l} < \infty.$$

It is easy to see that

$$|\varphi_k(x)| \leqslant \sqrt{2\pi} \int_{m_k}^{m_{k+1}} |\hat{\varphi}(\lambda)|\, d\lambda \leqslant b m_{k+1}.$$

Consequently,

$$V_1 \leqslant b \sum_{k=-\infty}^{0} m_{k+1} \left(\int_{|x|<1} w^p(x)\, dx \right)^{1/p}. \tag{43}$$

By a property of the weight function and (42), the right-hand side of (43) is finite.

We use an estimate of $\varphi_k(x)$ (see, for example, [2]):

$$|\varphi(x)| \leqslant \frac{|\hat{\varphi}(m_{k+1}) - \hat{\varphi}(m_k)|}{m_{k+1}} \frac{m_{k+1}}{x} + \hat{\varphi}(m_k) \frac{|e^{i m_k x} - 1|}{|x|} + \frac{1}{|x|} \int_{m_k}^{m_{k+1}} |\hat{\varphi}(\lambda)|\, d\lambda,$$

which follows easily from integration by parts. This gives us that

$$\begin{aligned} V_2 &\leqslant b \left(\left(\int_{|x|>1} \left(\sum_{k=-\infty}^{0} m_{k+1} \right)^p \frac{w^p(x)}{|x|^p}\, dx \right)^{1/p} \right. \\ &\quad \left. + \left(\int_{|x|>1} \left(\sum_{k=-\infty}^{0} |e^{i m_k x} - 1|^p \frac{w^p(x)}{|x|^p} \right) dx \right)^{1/p} \right) \\ &= b\left(V_2^{(1)} + V_2^{(2)} \right). \end{aligned}$$

By Proposition 5,

$$\int_{|x|>1} \frac{w^p(x)}{|x|^p}\, dx < \infty;$$

therefore, $V_2^{(1)}$ is finite.

Let us now estimate $V_2^{(2)}$. We have that

$$V_2^{(2)} \leqslant \sum_{k=-\infty}^{0} \left(\int_{|x|>1} \frac{|e^{i m_k x} - 1|^p w^p(x)}{|x|^p}\, dx \right)^{1/p}.$$

By Proposition 4, there exists a number $\eta > 0$ such that $w^p \in \mathscr{A}_{p-\eta}(\mathbf{I}^1)$, and so (see Proposition 3)

$$\int_{|x|>1} \frac{w^p(x)}{|x|^{p-\eta}}\, dx < \infty.$$

Moreover,

$$V_2^{(2)} \leqslant b \sum_{k=-\infty}^{0} \left(\int_{|x|>1} \frac{|\sin m_k(x/2)|^{\eta} w^p(x)}{|x|^p} \, dx \right)^{1/p}$$

$$\leqslant b \sum_{k=-\infty}^{0} m_k^{\eta/p} \left(\int_{|x|>1} \frac{w^p(x)}{|x|^{p-\eta}} \, dx \right)^{1/p}. \tag{45}$$

According to (42), this implies that $V_2^{(2)}$ is finite. Further,

$$\int_{m_k}^{m_{k+1}} |\hat{\varphi}(\lambda)| \, d\lambda \leqslant b \int_{m_k}^{m_{k+1}} \lambda^{-2} \, d\lambda \leqslant b m_k^{-1},$$

since $\hat{\varphi} \in S(R^n)$ and $|\hat{\varphi}(\lambda)| \leqslant b|\lambda|^{-2}$. Consequently, V_3 satisfies

$$V_3 \leqslant b \sum_{k=1}^{\infty} m_k^{-1} \left(\int_{-1}^{1} w^p(x) \, dx \right)^{1/p}. \tag{46}$$

By the condition that $m_{k+1} \geqslant (1+\varepsilon) m_k$,

$$\sum_{k=1}^{\infty} m_k^{-1} < \infty.$$

Therefore, $V_3 < \infty$. Integrating by parts and using the estimates $|\hat{\varphi}(\lambda)| \leqslant b|\lambda|^{-1}$ and $\hat{\varphi}'(\lambda)| \leqslant b|\lambda|^{-2}$, we get that

$$V_4 \leqslant \left(\int_{|x|>1} \left(\sum_{k=1}^{\infty} |\hat{\varphi}(m_{k+1})| + |\hat{\varphi}(m_k)| \right)^p \frac{w^p(x)}{|x|^p} \, dx \right)^{1/p}$$

$$+ \left(\int_{|x|>1} w^p(x) \left(\sum_{k=1}^{\infty} \int_{m_k}^{m_{k+1}} |\hat{\varphi}(\lambda)| \, d\lambda \right)^p dx \right)^{1/p}$$

$$\leqslant b \sum_{k=1}^{\infty} m_k^{-1} \left(\int_{|x|>1} \frac{w^p(x)}{|x|^p} \, dx \right)^{1/p} < \infty. \tag{47}$$

According to (43)–(47), we now conclude that

$$\left| \varphi, \mathscr{L}_{w,m}^{p,\theta}(R^1) \right| < \infty, \qquad \varphi \in S(R^1).$$

For $1 < p, \theta < \infty$ the spaces $\mathscr{L}_{w,m}^{p,\theta}(R^n)$ are Banach spaces, and each $f \in \mathscr{L}_{w,m}^{p,\theta}$ can be regarded as an element of the space S'. These facts can be proved just as done in the case without a weight in [2], so we omit the proofs. For $\theta = 2$ the norm (40) is equivalent to the norm of φ in $L_w^p(R^n)$. This follows from the next theorem, which is a weighted analogue of the Littlewood-Paley theorem (see [7], Chapter IV) on "dissection" of Fourier transforms.

THEOREM C [23]. *If* $1 < p < \infty$, $w \in \mathscr{A}^p(\mathbf{J}^n)$, *and the conditions* (37) *hold, then there exist positive constants* b_1 *and* b_2 *such that for any* $\varphi \in S(R^n)$

$$b_1 \|\varphi\|_{p,w} \leqslant \left\| \left(\sum_{k=-\infty}^{\infty} |\varphi_{U_k}(x)|^2 \right)^{1/2} \right\|_{p,w} \leqslant b_2 \|\varphi\|_{p,w}. \tag{48}$$

The fact that $\mathscr{L}_w^{p,\theta},m$ is continuously imbedded in $L_w^p(R^n)$ for $1 < \theta < 2$ follows from (48).

§5. Theorems on multipliers in $\mathscr{L}^{p,\theta}_{w,m}(R^n)$

Let E_i ($i = 1, 2$) be Banach spaces in which S is dense in the sense of the norm in E_i. Assume that each element in E_i can be regarded as an element of S'.

A bounded measurable function $\Phi\colon R^n \to R^1$ will be called an (E_1, E_2)-*multiplier* if for any function $\varphi \in S$

$$\|\overline{\Phi\hat{\varphi}}\|_{E_2} \leqslant c\|\varphi\|_{E_1},$$

where the constant $c > 0$ does not depend on φ.

Clearly, $\Phi \in S'$. If Φ is an (E_1, E_2)-multiplier, then the convolution of the generalized function $\overline{\Phi}$ with an f in E_1 is defined by

$$\overline{\Phi} * f = F,$$

where F is the limit of the sequence $(\overline{\Phi\hat{\varphi}_k})_k$ in the norm of E_2, with $(\varphi_k)_k$ satisfying the conditions that $\varphi_k \in S$ ($k = 1, 2, \ldots$) and $\|\varphi_k - f\|_{E_1} \to 0$ as $k \to \infty$. This makes it possible to define the product $\Phi\hat{f}$ for $f \in E_1$ by the equality $\Phi\hat{f} = \hat{F}$.

The multiplier method is a very useful tool in constructing an imbedding theory for spaces of smooth functions of several variables. Here an important role is played by tests providing sufficient conditions for a measurable function Φ to be a multiplier in $L^p(R^n)$ for all p with $1 < p < \infty$. The first theorem of this type was proved in an article by Marcinkiewicz (see [24]). The theory of multipliers of these types was subsequently developed in work of Mikhlin, Hörmander, Lizorkin, and others (see the survey of Chapter 1 in [3]).

In the present section we prove, for the weighted spaces $\mathscr{L}^{p,\theta}_{w,m}(R^n)$, multiplier theorems of the same type as Lizorkin's theorems. We further develop a method (in concept going back to Marcinkiewicz) on representation of an operator under investigation in the form of a composition of certain elementary transformations. Here a complete description is given for the class of weight functions for which all the Φ in a large class of measurable functions are $(\mathscr{L}^{p,\theta}_{w,m}, \mathscr{L}^{q,\theta}_{w,m})$-multipliers.

Suppose that the sequences $(m_{j,k})^{\infty}_{-\infty}$ ($j = 1, \ldots, n$) satisfy condition (37) and, as above, the definition of the set U_k.

THEOREM 4. *Suppose that* $1 < p \leqslant q < \infty$, $1 < \theta < \infty$, *and* $w \in \mathscr{A}^{a,p}(\mathbf{J}^n)$. *Let the function* $\Phi\colon R^n \to R^1$ *be defined by*

$$\Phi(\lambda) = \int_{-\infty}^{\lambda_1} \cdots \int_{-\infty}^{\lambda_n} \frac{d\mu_k}{(x_1 - \lambda_1)^{\beta} \cdots (x_n - \lambda_n)^{\beta}}, \quad \lambda \in U_k, k = 0, \pm 1, \ldots, \tag{49}$$

where $\beta = 1/p - 1/q$, *and the family* $(\mu_k)_k$ *of Borel measures satisfies the condition*

$$\int_{R^n} |d\mu_k| \leqslant M < \infty. \tag{50}$$

Then the operator T_Φ *defined in terms of Fourier transforms by the formula*

$$\hat{T}_\Phi(f)(\lambda) = \Phi(\lambda)\hat{f}(\lambda) \tag{51}$$

is a bounded transformation from $\mathscr{L}^{p,\theta}_{w,m}(R^n)$ *to* $\mathscr{L}^{q,\theta}_{w,m}(R^n)$, *and there exists a constant* $b > 0$ *independent of* f *and* Φ *such that*

$$|T_\Phi(f), \mathscr{L}^{q,\theta}_{w,m}(R^n)| \leqslant bM|f, \mathscr{L}^{p,\theta}_{w,m}(R^n)|.$$

Before proceeding to the proof we introduce some notation and prove some preliminary lemmas (cf. [1]).

For $r \geqslant 0$ and $\lambda = (\lambda_1, \ldots, \lambda_n) \in R^n$ let

$$\lambda_+^{-r} = \prod_{j=1}^{n} (\lambda_j)_+^{-r}, \quad \text{where} \quad (\lambda_j)_+^{-r} = \begin{cases} \lambda_j^{-r} & \text{for } \lambda_j > 0, \\ 0 & \text{for } \lambda_j \leqslant 0. \end{cases}$$

Let $\mathbf{S}(R^n)$ be the space of vector-valued functions $\varphi = (\varphi_j)_1^\infty$ with $\varphi_j \in S$ ($j = 1, \ldots, n$).

Consider the family of operators V_α^t, $\alpha \in [0, 1]$, $t \in R^n$, defined on $\mathbf{S}(R^n)$ as follows. Let $\varphi = (\varphi_j)_1^\infty \in \mathbf{S}(R^n)$. Then V_α^t: $\varphi \to \psi$, where $\psi = (\psi_j)_1^\infty$ and $\hat{\psi}(\lambda) = (\lambda - t)^{\alpha - 1} \hat{\varphi}_j(\lambda)$.

LEMMA 2. *Suppose that* $1 < p \leqslant q < \infty$, $1 < \theta < \infty$ *and* $w \in \mathscr{A}^{p,q}(\mathbf{J}^n)$. *The transformation* V_α^t, $\alpha = 1 - 1/p + 1/q$, *on* $\mathbf{S}$ *generates a bounded transformation of* $L_w^p[R^n, l^\theta]$ *into* $L_w^q[R^n, l^\theta]$ *when it is closed in* $L_w^p[R^n, l^\theta]$. *Moreover, the norm of this operator does not depend on* t.

Boundedness of V_α^0 in $L_w^p[R^n, l^\theta]$ is equivalent to boundedness of the operator $H^{(\alpha)}$ in the same space (see [1] in the case without a weight). Therefore, Lemma 2 follows from Theorem 3. The fact that for any $t \in R^n$ the norm of V_α^t coincides with that of V_α^0 follows from the facts that to translation of the argument of the Fourier transform by t there corresponds multiplication by e^{itx}.

LEMMA 3. *Suppose that* $1 < p \leqslant p < \infty$, $1 < \theta < \infty$, $w \in \mathscr{A}^{q,p}(\mathbf{J}^n)$, *and that the measurable function* Φ *on* R^n *is representable in the form*

$$\Phi(\lambda) = \int_{-\infty}^{\lambda_1} \cdots \int_{-\infty}^{\lambda_n} \frac{d\mu}{(x_1 - \lambda_1)^\beta \cdots (x_n - \lambda_n)^\beta}, \qquad \beta = \frac{1}{p} - \frac{1}{q}, \tag{52}$$

where μ *is a finite Borel measure on* R^n. *Then the transformation defined in terms of Fourier transforms by*

$$\hat{T}(\varphi)(\lambda) = \Phi(\lambda)\hat{\varphi}(\lambda), \qquad \lambda \in S,$$

is a bounded transformation from $L_w^p[R^n, l^\theta]$ *to* $L_w^q[R^n, l^\theta]$.

LEMMA 4. *Suppose that* $1 < p \leqslant q < \infty$, $1 < \theta < \infty$, *and* $w \in \mathscr{A}^{q,p}(\mathbf{J}^n)$. *Assume that*

$$\Phi_j(\lambda) = s\int_{-\infty}^{\lambda_1} \cdots \int_{-\infty}^{\lambda_n} \frac{d\mu_j}{(t_1 - \lambda_1)^\beta \cdots (t_n - \lambda_n)^\beta}, \qquad \beta = \frac{1}{p} - \frac{1}{q},$$

where $(\mu_j)_1^\infty$ *is a family of finite measures such that*

$$\int_{R^n} |d\mu_j| \leqslant M < \infty.$$

Further, assume that $(\mathscr{P}_N)_1^\infty$ *is a sequence of operators acting on the vector-valued functions in* $\mathbf{S}(R^n)$ *as follows: for* $F = (F_j)_1^\infty \in S(R^n)$ *let* $\mathscr{P}_N(F) = (g_j^{(N)})_1^\infty$, *where*

$$\hat{g}_j^{(N)}(\lambda) = \begin{cases} \Phi_j(\lambda)\hat{F}_j(\lambda) & \text{for } j \leqslant N, \\ \hat{F}_j(\lambda) & \text{for } j > N. \end{cases}$$

Then

$$|\mathscr{P}_N(F), L^p_w[R^n, l^\theta]| \leqslant bM|F, L^p_w[R^n, l^\theta]|,$$

where the constant b is independent of F, Φ, and N.

Lemmas 3 and 4 can be proved with the help of Lemma 2 just as in the case without a weight (see [1]), so the proofs will not be given here.

LEMMA 5. *Suppose that $1 < p \leqslant q < \infty$, $1 < \theta < \infty$ and $w \in \mathscr{A}^{q,p}(\mathbf{J}^n)$. Assume that the function Φ_j ($j = 1, 2, \ldots$) can be represented in the form*

$$\Phi_j(\lambda) = \int_{-\infty}^{\lambda_1} \cdots \int_{-\infty}^{\lambda_n} \frac{d\mu_j}{(t_1 - \lambda_1)^\beta \cdots (t_n - \lambda_n)^\beta}, \qquad \beta = \frac{1}{p} - \frac{1}{q}, \tag{53}$$

where $(\mu_j)_1^\infty$ is a family of finite measures such that

$$\int_{R^n} |d\mu_j| \leqslant M < \infty \qquad (j = 1, 2, \ldots).$$

Further, suppose that the transformation $\mathscr{P}$ is defined as follows on the functions $f = (f_j)_1^\infty \in \mathbf{S}(R^n) \cap L^p_w[R^n, l^\theta]$:

$$\mathscr{P}(f) = \left(\overline{\Phi_j \hat{f}_j}\right)_{j=1}^\infty.$$

Then $\mathscr{P}$ can be extended to a bounded transformation from $L^p_w[R^n, l^\theta]$ to $L^q_w[R^n, l^\theta]$. Moreover,

$$|\mathscr{P}(f), L^q_w[R^n, l^\theta]| \leqslant bM|f, L^p_w[R^n, l^\theta]|, \tag{54}$$

where the constant b is independent of f and Φ.

PROOF. First of all we show that under the conditions of the theorem the limit $\lim_{N\to\infty} \mathscr{P}_N(F) = \mathscr{P}(F)$ exists in the sense of convergence in the norm of $L^p_w[R^n, l^\theta]$ for any $F \in S \cap L^p_w[R^n, l^\theta]$.

Let

$$\mathscr{P}_{N_1}(F) - \mathscr{P}_N(F) = (\psi_j)_{j=1}^\infty, \qquad N_1 > N. \tag{55}$$

By the definition of the operator $\mathscr{P}_N$ and Lemma 4,

$$|\mathscr{P}_N(F) - \mathscr{P}_{N_1}(F), L^q_w[R^n, l^\theta]| \leqslant bM \left\| \left(\sum_{j=N_1+1}^{N} |F_j(x)|^\theta \right)^{1/\theta} \right\|_{p,w}. \tag{56}$$

Then the right-hand side of (56) tends to zero; hence $\mathscr{P}_N(F)$ has a limit in the mean in $L^q_w[R^n, l^\theta]$. Suppose, further, that

$$\lim_{N\to\infty} |\mathscr{P}_N(F) - h, L^q_w[R^n, l^\theta]| = 0$$

for a given $F \in S(R^n) \cap L^p_w[R^n, l^\theta]$. We prove that $h(x) = \mathscr{P}(F)(x)$, i.e., $\hat{h}_j(\lambda) = \Phi_j(\lambda)\hat{F}_j(\lambda)$ for any $j = 1, 2, \ldots$.

As above, let $\mathscr{P}_N(F) = (g_j^{(N)})_{j=1}^\infty$, where

$$\hat{g}_j^{(N)}(\lambda) = \begin{cases} \Phi(\lambda)\hat{F}_j(\lambda) & \text{for } j \leqslant N, \\ \hat{F}_j(\lambda) & \text{for } j > N. \end{cases}$$

Since $h = (h_j)_j \in L^q_w[R^n, l^\theta]$, it follows that $h_j \in L^q_w(R^n)$; hence, it can be regarded as an element of S'.

We have that $(\hat{h}_j, \varphi) = (h_j, \hat{\varphi})$. Our purpose is to prove that for any $j = 1, 2, \ldots$

$$\lim_{N\to\infty} \left(g_j^{(N)}, \hat{\varphi}\right) = \left(h_j, \hat{\varphi}\right). \tag{57}$$

Estimating the difference of these two integrals, we get

$$\begin{aligned}
&\int_{R^n} \left|g_j^{(N)}(\lambda) - h_j(\lambda)s\right| |\hat{\varphi}(\lambda)|\, d\lambda \\
&\quad \leqslant \int_{R^n} \frac{\left|g_j^{(N)}(\lambda) - h_j(\lambda)\right|(1 + |\lambda|^n)}{\prod_{j=1}^n (1 + |\lambda_j|)} |\hat{\varphi}(\lambda)|\, d\lambda \\
&\quad \leqslant \sup_{\lambda\in R^n} (1 + |\lambda|^n)|\hat{\varphi}(\lambda)| \int_{R^n} \frac{\left|g_j^{(N)}(\lambda) - h_j(\lambda)\right|}{\prod_{j=1}^n \left(1 + |\lambda_j|\right)}\, d\lambda \\
&\quad \leqslant b \int_{R^n} \frac{\left|g_j^{(N)}(\lambda) - h_j(\lambda)\right|}{\prod_{j=1}^n \left(1 + |\lambda_j|\right)}\, d\lambda.
\end{aligned} \tag{58}$$

Let us now show that the right-hand side of (58) tends to zero as $N \to \infty$ for each $j = 1, 2, \ldots$. According to Hölder's inequality,

$$\int_{R^n} \frac{\left|g_j^{(N)}(\lambda) - h_j(\lambda)\right|}{\prod_{j=1}^n \left(1 + |\lambda_j|\right)}\, d\lambda \leqslant \left\|g_j^{(N)} - h_j\right\|_{q,w} \left(\int_{R^n} \frac{w^{-q'}(\lambda)}{\prod_{j=1}^n \left(1 + |\lambda_j|\right)^{q'}}\, d\lambda\right)^{1/q'}. \tag{59}$$

The first factor on the right-hand side tends to zero, because $\lim_{N\to\infty} \mathscr{P}_n(F) = 0$ in the sense of convergence in the mean in $L^q_w[R^n, l^\theta]$. On the other hand, the condition $w \in \mathscr{A}^{q,p}(\mathbf{J}^n)$ implies that $1/w \in \mathscr{A}^{q',p'}(\mathbf{J}^n)$ and thus $1/w \in \mathscr{A}^{q'}(\mathbf{J}^n)$. Hence, the second factor on the right-hand side of (59) is finite. Therefore, (59) implies that

$$\lim_{N\to\infty} \int_{R^n} g_j^{(N)}(\lambda)\hat{\varphi}(\lambda)\, d\lambda = \int_{R^n} h_j(\lambda)\hat{\varphi}(\lambda)\, d\lambda \qquad (j = 1, 2, \ldots).$$

Further,

$$\left(\hat{h}_j, \varphi\right) = \left(h_j, \hat{\varphi}\right) = \lim_{N\to\infty} \left(\hat{g}_j^{(N)}, \varphi\right) = \left(\Phi_j \hat{F}_j, \varphi\right).$$

Consequently, $\hat{h}_j = \Phi_j \hat{F}_j$ and $\lim_{N\to\infty} \mathscr{P}_N(F) = \mathscr{P}(F)$ in the sense of convergence in the mean in $L^q_w[R^n, l^\theta]$.

The lemma is proved.

The proof of Theorem 4 is concluded with the help of the lemmas proved above in the same way as in the case without a weight in [1].

THEOREM 5. *Suppose that* $1 < p \leqslant q < \infty$, *and* $w \in \mathscr{A}^{q,p}(\mathbf{J}^n)$. *Assume that* Φ *is continuous outside the coordinate planes and has there continuous derivatives*

$$\partial^k \Phi / \partial\lambda_1^{k_1} \cdots \partial\lambda_n^{k_n}, \qquad 0 \leqslant k_1 + \cdots + k_n = k \leqslant n,\ k_j = 0, 1.$$

Moreover, assume that

$$\left|\lambda_1^{k_1+\beta_1} \cdots \lambda_n^{k_n+\beta_n} \partial^k \Phi / \partial\lambda_1^{k_1} \cdots \partial\lambda_n^{k_n}\right| \leqslant M, \qquad \beta = 1/p - 1/q. \tag{60}$$

Then Φ *is an* $(\mathscr{L}^{p,\theta}_{w,m}(R^n), \mathscr{L}^{q,\theta}_{w,m}(R^n))$*-multiplier.*

Theorem 5 is a corollary to Theorem 4; it suffices only to note that a function Φ satisfying (60) can be represented in the form (49) (see [1], §4).

We now dwell on multipliers in $L_w^p(R^n)$. Below we give a description of the class of all weight functions w such that all the operators

$$T_\Phi : f \to \overline{\Phi \hat{f}}$$

with Φ representable in the form (49) act boundedly from $L_w^p(R^n)$ to $L_w^q(R^n)$.

The family of all measurable functions Φ: $R^n \to R^1$ representable in the form (49) is denoted by G. For each $\Phi \in G$ there is a family $(\mu_k)_k$ of finite measures such that Φ is representable in the form (49). Each $\Phi \in G$ generates an operator T_Φ with domain containing $S(R^n)$ by means of (51).

Let $1 < p \leqslant q < \infty$, and let w: $R^n \to R^1$ be a measurable function such that

$$\int_{R^n} \frac{w^q(x)}{\prod_{k=1}^n (1 + |x_k|)^q} \, dx < \infty.$$

This condition ensures the inclusion $S \subset L_w^p \cap L_w^q$.

THEOREM 6. *Suppose that* $1 < p \leqslant q < \infty$, *and conditions* (49) *and* (50) *hold. The operators* T_Φ, $\Phi \in G$, *act boundedly from* $L_w^p(R^n)$ *to* $L_w^q(R^n)$ *if and only if* (2) *holds.*

PROOF. The sufficiency part of the theorem follows from Theorems 4 and C.

As noted above (see Lemma 2), Φ can be chosen so that T_Φ takes the form of the operator $H^{(\infty)}$. For the rest it suffices to refer to Theorem 3.

COROLLARY. *Suppose that the measurable function* Φ: $R^n \to R^1$ *satisfies the conditions of Theorem* 4. *Then* T_Φ *acts boundedly from* $L_w^p(R^n)$ *to* $L_w^q(R^n)$ *if* $1 < p \leqslant q < \infty$ *and* $w \in \mathscr{A}^{q,p}(\mathbf{J}^n)$.

This corollary generalizes a multiplier theorem proved in [23] for $p = q$.

§6. The spaces $\mathscr{L}_{w,m}^{p,\theta}(R^n)$. An imbedding theorem

In this section we introduce weighted spaces of anisotropic Bessel potentials, and give imbedding theorems for these spaces.

Let $r = (r_1, \ldots, r_n)$, $r_j > 0$ $(j = 1, \ldots, n)$. Assume that $1 < p < \infty$, $w \in \mathscr{A}^p(\mathbf{J}^n)$ and $1 < \theta < \infty$. The symbol G_r will denote the anisotropic Bessel-Macdonald kernel (see [3]).

DEFINITION. We say that $f \in \mathscr{L}_{w,m}^{p,\theta,r}(R^n)$ if

$$f = G_r * g, \qquad g \in \mathscr{L}_{w,m}^{p,\theta}(R^n).$$

By definition,

$$\left|f, \mathscr{L}_{w,m}^{p,\theta,r}(R^n)\right| = \left|g, \mathscr{L}_{w,m}^{p,\theta}(R^n)\right|.$$

These spaces were introduced and studied by Lizorkin [3] in the case $w(x) \equiv 1$. They generalize the Liouville spaces, which for their part form a natural extension of the Sobolev classes W_p^l to noninteger exponents.

For the spaces $\mathscr{L}_{w,m}^{p,\theta,r}$ the multiplier method can be used to prove imbedding theorems establishing a connection between different differentiability properties of functions of several variables in different metrics, as well as a connection between differentiability properties of a function considered on spaces of different dimensions.

THEOREM 7. *Suppose that* $1 < p < q < \infty$, $1 \leqslant s \leqslant n$ *and* $l = (l_1,\ldots,l_n)$, *the* l_j *being integers. If* $w \in \mathscr{A}^{q,p}(\mathbf{J}^s)$ *and*

$$\kappa(l) = 1 - \sum_{j=1}^{s} \frac{1}{r_j}\left(\frac{1}{p} - \frac{1}{q}\right) - \frac{1}{p}\sum_{j=s+1}^{n} \frac{1}{r_j} - \sum_{j=1}^{n} \frac{l_j}{r_j} > 0,$$

then for any $f \in \mathscr{L}^{p,\theta,r}_{w,m}(R^n)$ *the function*

$$\Psi(x_1,\ldots,x_s) = \mathscr{D}^l f(x_1,\ldots,x_s, x^0_{s+1},\ldots,x^0_n)$$

belongs to the space $\mathscr{L}^{q,\theta,\nu}_{w,m}(R^s)$, $\nu = \kappa r$, *for any* $x^0_{s+1},\ldots,x^0_n$. *Moreover,*

$$\left|\Psi,\, \mathscr{L}^{q,\theta,\nu}_{w,m}(R^s)\right| \leqslant b\left|f,\, \mathscr{L}^{p,\theta,r}_{w,m}(R^n)\right|,$$

This theorem can be proved by the scheme given in [2], where one should use Theorem 4 on multipliers and the following lemma.

LEMMA 5. *Let*

$$\|x\| = \left(\sum_{j=1}^{n} |x_j|^{2\alpha_j}\right)^{1/2}, \qquad \alpha_j > 0 \ (j = 1,\ldots,n).$$

Assume that $1 \leqslant s \leqslant n$, $1 < p < q < \infty$, *and* $w\colon R^s \to R^1$ *belongs to the class* $\mathscr{A}^{q,p}(\mathbf{J}^s)$. *If*

$$\mu = \frac{1}{p'}\sum_{j=1}^{n} \frac{1}{\alpha_j} + \frac{1}{q}\sum_{j=1}^{s} \frac{1}{\alpha_j},$$

then for any $f \in L^p_w(R^n)$ *and any* $(x_{s+1},\ldots,x_n)$ *the function*

$$h(x) = \int_{R^n} \frac{f(y)}{\|x-y\|^{\mu}}\,dy$$

belongs to $L^q_w(R^s)$, *and*

$$\|h\|_{L^q_w(R^s)} \leqslant b\|f\|_{L^p_w(R^n)},$$

where the constant b *is independent of* f *and* $(x_{s+1},\ldots,x_n) \in R^{n-s}$.

PROOF. Let $y' = (y_1,\ldots,y_s)$ and $y'' \in (y_{s+1},\ldots,y_n)$. We show that for $f \in L^p_w(R^n)$ and $g \in L^{q'}_{1/w}(R^n)$, $1 < p < q < \infty$, we have

$$\left|\int_{R^s} dx \int_{R^n} \frac{g(x)f(y)}{\|x-y\|^{\mu}}\,dy\right| \leqslant b\|f\|_{L^p_w(R^n)}\cdot\|g\|_{L^{q'}_{1/w}(R^s)}, \tag{61}$$

where the constant b is independent of f and g. By Hölder's inequality,

$$\int_{R^{n-s}} \frac{|f(y)|}{\|x-y\|^{\mu}}\,dy'' \leqslant \|f\|_{L^p(R^{n-s})}\left(\int_{R^{n-s}} \frac{dy''}{\|x-y\|^{\mu p'}}\right)^{1/p'}.$$

Since

$$\int_{R^{n-s}} \frac{dy''}{\|x-y\|^{\mu p'}} \leqslant b \prod_{j=1}^{s} |x_j - y_j|^{1/p-q-1}$$

with an absolute constant $b > 0$, we can write the inequality

$$\int_{R^s} dx \int_{R^n} \frac{|g(x)f(y)|}{\|x-y\|^{\mu}} dy \leqslant b \int_{R^s} \int_{R^s} \frac{|g(x)| \left(\int_{R^{n-s}} |f(y)| dy'' \right)^{1/p} dx\, dy'}{\prod_{j=1}^{s} |x_j - u_j|^{1/p'+1/q}}.$$

If we use Theorem 3 on the right-hand side of (62), we get (61). The lemma follows from (61).

Lemma 5 was established previously by Il′ in [25] for the case $w(x) \equiv 1$.

BIBLIOGRAPHY

1. P. I. Lizorkin, *On multipliers of Fourier integrals in the spaces $L_{p,\theta'}$*, Trudy Mat. Inst. Steklov. **89** (1967), 231–248; English transl. in Proc. Steklov Inst. Math. **89** (1967).

2. ______, *Properties of functions in the space $\Lambda^r_{p,\theta'}$*, Trudy Mat. Inst. Steklov. **131** (1974), 158–194; English transl. in Proc. Steklov Inst. Math. **131** (1974)

3. ______, *Generalized Liouville differentiation and the multiplier method in imbedding theory for classes of differentiable functions*, Trudy Mat. Inst. Steklov. **105** (1969), 89–167; English transl. in Proc. Steklov Inst. Math. **105** (1969).

4. Hans Triebel, *Fourier analysis and function spaces*, Teubner, Leipzig, 1977.

5. G. A. Kalyabin, *Imbedding theorems for generalized Besov and Liouville spaces*, Dokl. Akad. Nauk SSSR **232** (1977), 1245–1248; English transl. in Soviet Math. Dokl. **18** (1977).

6. M. A. Gol′dman, *A description of traces for certain function spaces*, Trudy Mat. Inst. Steklov. **150** (1979), 99–127; English transl. in Proc. Steklov Inst. Math. **1981**, no. 4 (150).

7. Elias M. Stein, *Singular integrals and differentiability properties of functions*, Princeton Univ. Press, Princeton, N. J., 1970.

8. Benjamin Muckenhoupt, *Weighted norm inequalities for the Hardy maximal function*, Trans. Amer. Math. Soc. **165** (1972), 207–226.

9. Richard Hunt, Benjamin Muckenhoupt and Richard Wheeden, *Weighted norm inequalities for the conjugate function and Hilbert transform*, Trans. Amer. Math. Soc. **176** (1973), 227–251.

10. Benjamin Muckenhoupt and Richard L. Wheeden, *Weighted norm inequalities for fractional integrals*, Trans. Amer. Math. Soc. **192** (1974), 261–274.

11. R. R. Coifman and C. Fefferman, *Weighted norm inequalities for maximal functions and singular integrals*, Studia Math. **51** (1974), 241–1250.

12. V. M. Kokilashvili, *Weight inequalities for maximal functions with respect to regular Vitali families*, Trudy Tbiliss. Mat. Inst. Razmadze Akad. Nauk Gruzin. SSR **66** (1980), 36–45. (Russian)

13. ______, *Maximal inequalities and multipliers in weighted Lizorkin-Triebel spaces*, Dokl. Akad. Nauk SSSR **239** (1978), 42–45; English transl. in Soviet Math. Dokl. **19** (1978).

14. ______, *Maximal functions in weighted spaces*, Trudy Tbiliss. Mat. Inst. Razmadze Akad. Nauk Gruzin. SSR **65** (1980), 110–121. (Russian)

15. C. Fefferman and E. M. Stein, *Some maximal inequalities*, Amer. J. Math. **93** (1971), 107–115.

16. V. M. Kokilashvili, *On singular integrals and multipliers of Fourier transforms*, Trudy Tbiliss. Mat. Inst. Razmadze Akad. Nauk Gruzin. SSSR **43** (1976), 38–61. (Russian)

17. A. P. Calderón, *Commutators of singular integral operators*, Proc. Nat. Acad. Sci. U.S.A. **53** (1965), 1092–1099.

18. R. Bajsanski [Bajšanski] and R. Coifman, *On singular integrals*, Singular Integrals (Proc. Sympos. Pure Math., vol. 10), Amer. Math. Soc., Providence, R. I., 1967, pp. 1–17.

19. R. Coifman and Y. Meyer, *Commutateurs d'integrales singulières et opèrateurs multilinéaires*, Ann. Inst. Fourier (Grenoble) **28** (1978), fasc. 3, 177–202.

20. K. Sokół-Sokołowski, *On trigonometric series conjugate to Fourier series in two variables*, Fund. Math. **33** (1945), 166–182.

21. A. Zygmund, *On the boundary values of functions of several complex variables*. I, Fund. Math. **36** (1949), 207–235.

22. M. Cotlar, *Some generalizations of the Hardy-Littlewood maximal theorem*, Rev. Mat. Cuyana **1** (155), 85–104.

23. Douglas S. Kurtz, *Littlewood-Paley and multiplier theorems on weighted L^p spaces*, Trans. Amer. Math. Soc. **259** (1980), 235–263.

24. Józef Marcinkiewicz, *Collected papers*, PWN, Warsaw, 1964.

25. V. P. Il'in, *Some integral inequalities and their application in the theory of differentiable functions of several variables*, Mat. Sb. **54 (96)** (1961), 331–380. (Russian)

Translated by H. H. McFADEN

Труды Матем. Инст. Стеклов.
161(1983)

Proceedings of the Steklov
Institute of Mathematics
1984, Issue 3

ON THE QUESTION OF POLYNOMIAL TRACES

UDC 517.518.235

L. D. KUDRYAVTSEV

ABSTRACT. Conditions are given for the trace of a function of several variables on a hyperplane to be a polynomial, in terms of the strong divergence of weighted integrals of powers of the moduli of differences of the function
Bibliography: 4 titles.

Under certain conditions on the weight, a function defined on an n-dimensinal parallelepiped and belonging to the corresponding weighted space has only algebraic polynomials as its traces on the faces of the parallelepiped (see [1]–[3]). In this article we give conditions ensuring that a trace is a polynomial and containing such previously known conditions as a special case. For simplicity the main theorem will be stated and proved only for the case of functions defined on an infinite n-dimensional strip. The case of functions defined on an arbitrary n-dimensional parallelepiped is studied according to the same scheme. The question of polynomial traces of functions on hyperplanes of dimension less than $n-1$ will also be considered briefly. The trace of a function is understood in the sense of convergence on any compact set in the L_p-metric.

Let N be the set of natural numbers, R the set of real numbers, R^n the n-dimensional space of points $x=(x_1,\dots,x_n)$, $x_i\in R$, $i=1,\dots,$ $n\in N$, $Q_a^n=\{x\colon |x_i|\leqslant a,\ i=1,\dots,n\}$, $a\in R$, $r\in N$, $h_1\in R$, $Q^n_{a+r|h_1|}=\{x\colon |x_1|<a+r|h_1|, |x_j|\leqslant a, j=1,\dots,n\}$, $E\subset R^n$, $u\colon E\to R$, $|u; L_p(E)|=(\int_E|u(x)|^p\,dx)^{1/p}$, $p\in R$, $p\geqslant 1$, $x=(x^{(n-1)},x_n)$, $x^{(n-1)}=(x_1,x^{(n-2)})=(x_1,\dots,x_{n-1})$ and $x^{(n-2)}=(x_1,\dots,x_{n-1})$.

For a function $u\colon Q^n_{a+r|h_1|}\to R$ let

$$\Delta^r_{h_1}u(x)=\sum_{k=0}^r(-1)^{k+r}C_r^k u\big(x_1+kh_1,x^{(n-2)},x_n\big)$$

denote its rth-order difference with respect to the variable x_1 at a point $x\in Q_a^n$. It is known that if $u(x)$ has an rth-order generalized derivative ${}^ru/\partial x_1^r$ on $Q^n_{q+(r+m)|h_1|}$ $(m\in N)$ (see [4], §4.1), then

$$\left|\Delta^{r+m}_{h_1}u(x); L_p(Q_a^n)\right|\leqslant 2^m|h_1|^r\left|\partial^r u/\partial x_1^r; L_p\left(Q^n_{a+(r+m)|h_1|}\right)\right|. \tag{1}$$

1980 *Mathematics Subject Classification*. Primary 46E35, 26C99.

Let $R^{(n-1)} = \{x\colon\ x_n = 0\}$, $\overset{+}{R}{}^n_1 = \{x\colon\ 0 < x_n < 1\} = R^{n-1} \times (0,1)$. $Q_a^{n-1} = \{x^{(n-1)}\colon |x_i| \leqslant a,\ i = 1,\ldots,n-1\}$ and $\overset{+}{R} = \{t \in R\colon t \geqslant 0\}$, and let μ be Lebesgue measure.

Definition. Let $\varphi\colon Q_a^{n-1} \times (0,1) \to \overset{+}{R}$. The integral

$$\int_{Q_a^{n-1}\times(0,1)} \varphi(x)\,dx = \int_0^1 dx_n \int_{Q_a^{n-1}} \varphi\left(x^{(n-1)}, x_n\right) dx^{(n-1)}$$

is said to be *strongly divergent near the hyperplane* R^{n-1} if for any $\alpha > 0$, any $\beta > 0$, and any set $E_{x_n} \subset Q_a^{n-1}$ such that

$$\mu E_{x_n} \geqslant \alpha, \qquad x_n \in (0,\beta),\quad \beta < 1, \tag{2}$$

it is true that

$$\int_0^\beta dx_n \int_{E_{x_n}} \varphi\left(x^{(n-1)}, x_n\right) dx^{(n-1)} = +\infty. \tag{3}$$

An example of an integral that is strongly divergent near the hyperplane R^{n-1} is an integral

$$\int_0^1 dx_n \int_{Q_a^{n-1}} \varphi\left(x^{(n-1)}, x_n\right) dx^{(n-1)},$$

such that for any $\beta \in (0,1)$

$$\int_0^\beta \operatorname*{ess\,inf}_{Q_a^{n-1}} \varphi\left(x^{(n-1)}, x_n\right) dx_n = +\infty,$$

in particular, an integral such that $\varphi(x^{(n-1)}, x_n) = \varphi(x_n)$ and $\int_0^\beta \varphi(x_n)\,dx_n = +\infty$. Indeed, for any system of sets E_{x_n} satisfying (2) we have in this case

$$\begin{aligned}\int_0^\beta dx_n \int_{Q_a^{n-1}} \varphi\left(x^{(n-1)}, x_n\right) dx_n &\geqslant \int_0^\beta \operatorname*{ess\,inf}_{Q_a^{n-1}} \varphi\left(x^{(n-1)}, x_n\right) dx_n \int_{Q_a^{n-1}} dx^{(n-1)} \\ &\geqslant \int_0^\beta \operatorname*{ess\,inf}_{Q_a^{n-1}} \varphi\left(x^{(n-1)}, x_n\right) dx_n \int_{E_{x_n}} dx^{(n-1)} \\ &\geqslant \alpha \int_0^\beta \operatorname*{ess\,inf}_{Q_a^{n-1}} \varphi\left(x^{(n-1)}, x_n\right) dx_n = +\infty.\end{aligned}$$

Remark 1. There exist functions $\varphi(x^{(n-1)}, x_n)$ such that for all $x^{(n-1)} \in Q_a^{n-1}$ and all $\beta \in (0,1)$ but the integral $\int_0^1 dx_n \int_{Q_a^{n-1}} \varphi(x^{(n-1)}, x_n)\,dx^{(n-1)}$ does not diverge strongly near the hyperplane R^{n-1}.

We give an example confirming this assertion. For simplicity we confine ourselves to the case $n = 2$; a corresponding example can be constructed similarly for any positive integer n.

Example 1. Let (Figure 1)

$$P = \{(x,y)\colon 0 < x < 1,\ 0 < y < 1\},$$

$$\Pi_k = \left\{(x,y)\colon 0 < x < 1,\ -\frac{3}{2^{k+3}}x + \frac{7}{2^{k+3}} < y < -\frac{3}{2^{k+3}}x + \frac{1}{2^k}\right\},$$

$$k = 0, 1, \ldots.$$

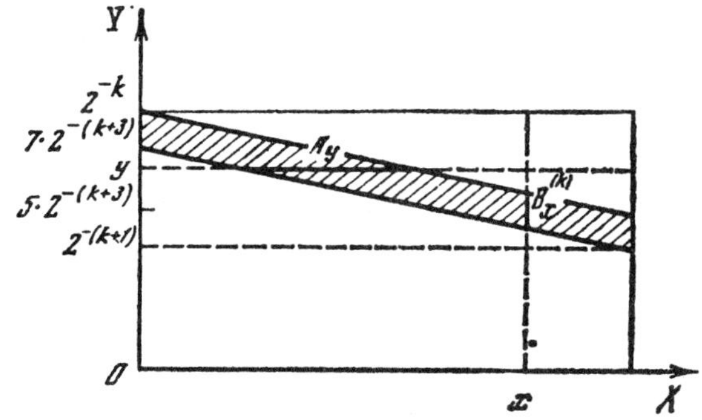

FIGURE 1

The function $\varphi(x, y)$ is defined as follows: $\varphi(x, y) = 0$ on $P \setminus \bigcup_0^\infty \Pi_k$, and $\varphi(x, y) = 2^k$ for $(x, y) \in \Pi_k$. Then for all $x \in (0,1)$ and any $k = 0, 1, \ldots$ we set $B_x^{(k)} = \{y\colon (x, y) \in \Pi_k\}$ and get that the length $\mu B_x^{(k)}$ of the segment $B_k^{(k)}$ is equal to $1/2^{k+3}$, and so

$$\int_0^{1/2^k} \varphi(x, y)\, dy = \sum_{j=k}^{+\infty} 2^j \mu B_x^{(j)} = +\infty,$$

i.e., (4) holds for φ.

However, if for each $y \in (0,1)$ we set

$$A_y = \left\{x\colon (x, y) \in P \cap \bigcup_{k=0}^{\infty} \Pi_k\right\}, \qquad E_y = \left\{x\colon (x, y) \in P \Big\backslash \bigcup_{k=0}^{\infty} \Pi_k\right\},$$

then for the segment A_y we have $A_y \cup E_y = (0,1)$, $\mu A_y \leqslant 1/3$ (see Figure 1) and, consequently, $\mu E_y \geqslant 2/3$, i.e., condition (2) holds for $\alpha = 2/3$. But $\varphi(x, y) = 0$ on E_y; therefore,

$$\int_0^1 dy \int_{E_y} \varphi(x, y)\, dx = 0,$$

i.e., condition (3) does not hold.

LEMMA. *Suppose that* $f\colon Q_a^{n-1} \times (0,1) \to \overset{+}{R}$, $\varphi\colon Q_a^{n-1} \times (0,1) \to \overset{+}{R}$, *and:*

1°. $\int_0^1 dx_n \int_{Q_a^{n-1}} \varphi(x^{(n-1)}, x_n) f(x^{(n-1)}, x_n)\, dx^{(n-1)} < +\infty$;

2°. $\int_0^1 dx_n \int_{Q_a^{n-1}} \varphi(x^{(n-1)}, x_n)\, dx^{(n-1)}$ *is strongly divergent near the hyperplane* R^{n-1}; *and*

3°. *the function* $f(x^{(n-1)}, x_n)$ *has a trace* $f(x^{(n-1)}, 0)$ *on the cube* Q_a^{n-1}.

Then the trace $f(^{(n-1)}, 0)$ *is equal to zero almost everywhere on* Q_a^{n-1}.

The existence of the trace of f indicated in the lemma means that there exist a function $f^*\colon Q_a^{n-1} \times (0,1) \to R$ equivalent to f in the sense of the n-dimensional measurable function $g\colon Q_a^{n-1} \to R$, called the trace of f on Q_a^{n-1}: $f(x^{(n-1)}, 0) = g(x^{(n-1)})$, such that

$$\lim_{x_n \to 0} \int_{Q_a^{n-1}} \left|f^*\left(x^{(n-1)}, x_n\right) - g\left(x^{(n-1)}\right)\right|^p dx^{(n-1)} = 0. \tag{5}$$

Suppose that $g(x^{(n-1)}) \neq 0$ on a subset of positive measure in Q_a^{n-1}. Let

$$A^{(0)} = \left\{x^{(n-1)} \in Q_a^{n-1}\colon g\left(x^{(n-1)}\right) = 0\right\},$$
$$A^{(m)} = \left\{x^{(n-1)} \in Q_a^{n-1}\colon g\left(x^{(n-1)}\right) > 1/m\right\}. \tag{6}$$

Since $Q_a^{n-1} = \bigcup_0^\infty A^{(m)}$ and $\mu A^{(0)} < \mu Q_a^{n-1}$, there exists an $m_0 \neq 0$ such that $\mu A^{(m_0)} > 0$. Let

$$B_{x_n} = \left\{ x^{(n-1)} \in A^{(m_0)} : \left| f^*\left(x^{(n-1)}, x_n\right) - g\left(x^{(n-1)}\right) \right| > 1/2m_0 \right\}.$$

Then, observing that

$$\begin{aligned} &\left| f^*\left(x^{(n-1)}, x_n\right) - g\left(x^{(n-1)}\right); L_p\left(Q_a^{n-1}\right) \right| \\ &\qquad \leqslant \left| f^*\left(x^{(n-1)}, x_n\right) - g\left(x^{(n-1)}\right); L_p\left(B_{x_n}\right) \right| \geqslant \left(\mu B_{x_n}\right)^{1/p} / 2m_0, \end{aligned}$$

we have from (5) that

$$\lim_{x_n \to 0} \mu B_{x_n} = 0$$

(this is the well-known fact that convergence in the mean implies convergence in measure). Therefore, there exists a $\delta > 0$ such that for all $x_n \in (0, \delta)$

$$\mu B_{x_n} < \mu A^{(m_0)}/2.$$

If $E_{x_n} = A^{(m_0)} \setminus B_{x_n}$, then for any $x_n \in (0, \delta)$

$$\mu E_{x_n} > \mu A^{(m_0)}/2 \tag{7}$$

and for any $x^{(n-1)} \in E_{x_n}$

$$\left| f^*\left(x^{(n-1)}, x_n\right) - g\left(x^{(n-1)}\right) \right| \leqslant 1/2m_0$$

and, consequently,

$$\left| g\left(x^{(n-1)}\right) \right| - \left| f^*\left(x^{(n-1)}, x_n\right) \right| \leqslant \left| f^*\left(x^{(n-1)}, x_n\right) - g\left(x^{(n-1)}\right) \right| \leqslant 1/2m_0,$$

from which it follows (see (6) with $m = m_0$) that for all $x^{(n-1)} \in E_{x_n}$

$$\left| f^*\left(x^{(n-1)}, x_n\right) \right| \geqslant \left| g\left(x^{(n-1)}\right) \right| - 1/2m_0 \geqslant 1/m_0 - 1/2m_0 = 1/2m_0.$$

Using this inequality, we now have from (7) and the strong divergence of the integral $\int_0^1 dx_n \int_{Q_a^{n-1}} \varphi(x^{(n-1)}, x_n)\, dx^{(n-1)}$ that

$$\begin{aligned} &\int_0^1 dx_n \int_{Q_a^{n-1}} \varphi\left(x^{(n-1)}, x_n\right) f\left(x^{(n-1)}, x_n\right) dx^{(n-1)} \\ &\qquad \geqslant \int_0^\delta dx_n \int_{E_{x_n}} \varphi\left(x^{(n-1)}, x_n\right) f\left(x^{(n-1)}, x_n\right) dx^{(n-1)} \\ &\qquad \geqslant \frac{1}{2m_0} \int_0^\delta dx_n \int_{E_{x_n}} \varphi\left(x^{(n-1)}, x_n\right) dx^{(n-1)} = +\infty, \end{aligned}$$

but this contradicts the condition 1° in the lemma.

REMARK 2. If in the premises of the lemma the condition of strong divergence of the integral $\int_0^1 dx_n \int_{Q_a^{n-1}} \varphi(x^{(n-1)}, x_n)\, dx^{(n-1)}$ near the hyperplane R^{n-1} is replaced by the weaker condition of divergence of the integral $\int_0^\beta \varphi(x^{(n-1)}, x_n)\, dx_n$ for all $\beta \in (0, 1)$ and all $x^{(n-1)} \in Q_a^{n-1}$, then the assertion of the lemma ceases to be true. We construct an example confirming this.

EXAMPLE 2. As in Example 1, let $P = \{(x, y): 0 < x < 1, 0 < y < 1\}$, and let π_k be defined now by the following equalities (see Figure 2):

$$\Pi_k = \left\{(x, y): 0 < x < 1, -\left(\frac{1}{2^{k+1}} - \frac{1}{2^{2k}}\right)x + \frac{1}{2^k} - \frac{1}{2^{2k}} < y < -\left(\frac{1}{2^{k+1}} - \frac{1}{2^{2k}}\right)x + \frac{1}{2^k}\right\}.$$

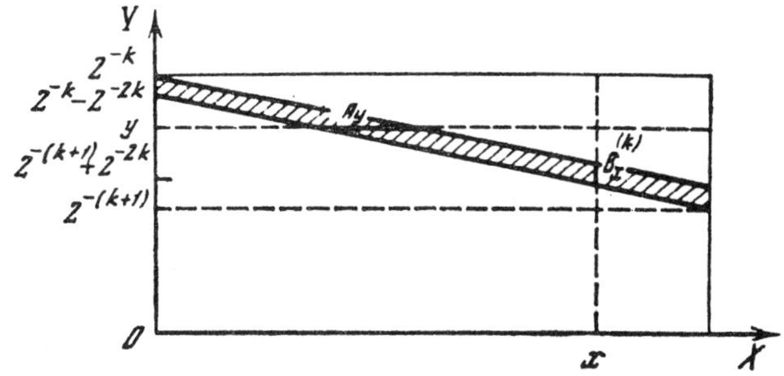

FIGURE 2

Here the construction of the sets Π_k differs from that in Example 1 in that the length (measure) of the segment

$$A_y = \left\{x: (x, y) \in P \cap \bigcap_{k=0}^{\infty} \Pi_k\right\}$$

with

$$y \in [1/2^{k+1}, 1/2^k] \tag{8}$$

satisfies the inequality

$$\mu A_y \leqslant 1/(2^{k-1} - 1), \tag{9}$$

from which (since $y \to 0$ implies that $k \to +\infty$ under condition (8)) it follows that

$$\lim_{y \to 0} \mu A_y = 0. \tag{10}$$

For a particular $x \in (0, 1)$ let

$$B_x^{(k)} = \{y: (x, y) \in \Pi_k\}, \qquad k = 0, 1, \ldots.$$

Obviously, $B_k^{(k)}$ is a segment of length $1/2^{2k}$:

$$\mu B_x^{(k)} = 1/2^{2k}. \tag{11}$$

Now let the functions $\varphi: P \to \overset{+}{R}$ and $f: P \to \overset{+}{R}$ be defined as follows:

$$\varphi(x, y) = \begin{cases} 0 & \text{if } (x, y) \in P \setminus \bigcup_{k=0}^{+\infty} \Pi_k, \\ 2^{2k} & \text{if } (x, y) \in P \cap \Pi_k, k = 0, 1, \ldots, \end{cases}$$

$$f(x, y) = \begin{cases} 0 & \text{if } (x, y) \in P \cap \Pi_k, k = 0, 1, \ldots, \\ 1 & \text{if } (x, y) \in P \setminus \bigcup_{k=0}^{+\infty} \Pi_k. \end{cases}$$

For all $(x, y) \in P$ we have that $\varphi(x, y) f(x, y) = 0$; hence,

$$\iint_P \varphi(x, y) f(x, y)\, dx\, dy = 0. \tag{12}$$

By (11), for all $x \in (0,1)$

$$\int_0^1 \varphi(x, y)\, dy = \sum_{k=0}^{+\infty} 2^{2k} \mu B_x^{(k)} = +\infty; \tag{13}$$

and by (9), for all $y \in (0,1)$

$$\int_0^1 |f(x, y) - 1|^p\, dx = \int_{A_y} dx = \mu A_y \leqslant \frac{1}{2^{k-1} - 1},$$

which implies that

$$\lim_{y \to 0} \int_0^1 |f(x, y) - 1|^p\, dx = 0. \tag{14}$$

Thus, despite the fact that the integral $\int_0^1 \varphi(x, y)\, dy$ diverges for all $x \in (0,1)$ (see (13)), the function $f(x, y)$ has a trace $f(x, 0) \equiv 1$ $(x \in (0,1))$ by (14), and

$$\iint_P \varphi(x, y) f(x, y)\, dx\, dy < +\infty$$

(see (12)).

We now determine conditions under which a function has a trace that is a polynomial in one of the variables. For simplicity let us consider the case of functions defined in an infinite strip.

THEOREM 1. *Suppose that* $u\colon R^{n-1} \times (0,1) \to R$, $\varphi\colon R^{n-1} \times (0,1) \times (0,1) \to \overset{+}{R}$, *and*:

1) *for any* $a > 0$

$$\int_0^1 dh_1 \int_0^1 dx_n \int_{Q_a^{n-1}} \varphi\left(x^{(n-1)}, x_n; h_1\right) \left|\Delta_{h_1}^r u\left(x^{(n-1)}, x_n\right)\right|^p dx^{(n-1)} < +\infty;$$

2) *for any* $a > 0$ *and almost all* $h \in (0,1)$ *the integral*

$$\int_0^1 dx_n \int_{Q_a^{n-1}} \varphi\left(x^{(n-1)}, x_n, h\right) dx^{(n-1)}$$

is strongly divergent near the hyperplane R^{n-1}; *and*

3) *for any* $a > 0$ *the trace* $u(x^{(n-1)}, 0)$ *of the function* u *exists on the cube* Q_a^{n-1} *in the* $L_p(Q_a^{n-1})$*-metric.*

Then $u(x^{(n-1)}, 0)$ *can be modified on a set of* $(n-1)$*-dimensional measure zero in such a way that the function obtained (which obviously is also the trace of* u *on* Q_a^{n-1}*) is a polynomial of degree at most* $r - 1$ *in the variable* x_1.

PROOF. It follows from the first condition that for almost all $h_1 \in (0,1)$

$$\int_0^1 \int_{Q_a^{n-1}} \varphi\left(x^{(n-1)}, x_n, h_1\right) \left|\Delta_{h_1}^r u\left(x^{(n-1)}, x_n\right)\right|^p dx^{(n-1)} < +\infty. \tag{15}$$

Thus, both the second condition of the theorem and inequality (15) hold for almost all $h_1 \in (0,1)$. Choose one such h_1 and fix it. Fix also some arbitrary $a > 0$. The existence of the trace $g(x^{(n-1)})$ of u on Q_a^{n-1} means that there exists a function $u^*\colon Q_a^{n-1} \times (0,1) \to R$ equivalent to u on $Q_a^{n-1} \times (0,1)$ and such that

$$\lim_{x_n \to 0} \left|u^*\left(x^{(n-1)}, x_n\right) - g\left(x^{(n-1)}\right); L_p\left(Q_a^{n-1}\right)\right| = 0. \tag{16}$$

Since

$$\begin{aligned}&\left|\Delta^r_{h_1}u^*(x^{(n-1)},x_n)-\Delta^r_{h_1}g(x^{(n-1)});L_p(Q^{n-1}_a)\right|\\&\quad=\left|\Delta^r_{h_1}\left[u^*(x^{(n-1)},x_n)-g(x^{(n-1)})\right];L_p(Q^{n-1}_a)\right|\\&\quad\leqslant 2^r\left|u^*(x^{(n-1)},x_n)-g(x^{(n-1)});L_p\left(Q^{n-1}_{a+r|h_1|}\right)\right|,\end{aligned}$$

it follows from the validity of (16) for any $a > 0$ that

$$\lim_{x_n\to 0}\left|\Delta^r_{h_1}u^*(x^{(n-1)},x_n)-\Delta^r_{h_1}g(x^{(n-1)});L_p(Q^{n-1}_a)\right|=0,$$

i.e., for any $a > 0$ the function $\Delta^r_{h_1}u(x^{(n-1)}, x_n)$ also has a trace on Q^{n-1}_a in the sense of the L_p-metric, and $\Delta^r_{h_1}g(x^{(n-1)})$ is this trace. Consequently, $|\Delta^r_{h_1}g(x^{(n-1)})|$ is the trace of $|\Delta^r_{h_1}u(x^{(n-1)}, x_n)|$ on Q^{n-1}_a. We apply the lemma proved above to the function

$$f(x^{(n-1)},x_n)=\left|\Delta^r_{h_1}u(x^{(n-1)},x_n)\right|$$

on the set $Q^{n-1}_a \times (0,1)$ for the values of h_1 and a fixed in the way indicated above. This is possible by (15) and the second condition of the theorem. Then we get for almost all $x^{(n-1)} \in Q^{n-1}_a$ and almost all $h \in (0,1)$ that

$$\left|\Delta^r_{h_1}g(x^{(n-1)})\right|=0.$$

This gives us (see [2]) that $g(x^{(n-1)})$ is a polynomial of degree at most $r-1$ in the variable x_1.

REMARK 3. It follows from the theorem that if u has a generalized derivative $\partial^r u/\partial x_1^r$ and if for all $a > 0$ and all $b > 0$

$$\int_0^1\varphi(x_n)\,dx_n\int_{Q^{n-1}_a}\left|\frac{\partial^r u(x^{(n-1)},x_n)}{\partial x_1^r}\right|^p dx^{(n-1)}<+\infty, \tag{17}$$

$$\int_0^b\varphi(x_n)\,dx_n=+\infty, \tag{18}$$

then the first and second conditions of the theorem hold, and, consequently, if the function $u(x^{(n-1)}, x_n)$ has, moreover, a trace on Q^{n-1}, then it is a polynomial of degree at most $r-1$ (this result is contained in Tandit's article [3] with inessential additional restrictions).

Indeed, with the help of (1) we get from (17) that

$$\begin{aligned}&\int_0^1 dh_1\int_0^1\varphi(x_n)\,dx_n\int_{Q^{n-1}_a}\left|\Delta^r_{h_1}u(x^{(n-1)},x_n)\right|^p dx^{(n-1)}\\&\quad\leqslant\int_0^1|h_1|^{rp}\,dh_1\int_0^1\varphi(x_n)\,dx_n\int_{Q^{n-1}_a}\left|\frac{\partial^r u(x^{(n-1)},x_n)}{\partial x_1^r}\right|^p dx^{(n-1)}\\&\quad=\frac{|h_1|^{rp+1}}{rp+1}\int_0^1\varphi(x_n)\,dx_n\int_{Q^{n-1}_a}\left|\frac{\partial^r u(x^{(n-1)},x_n)}{\partial x_1^r}\right|^p dx^{(n-1)}<+\infty.\end{aligned}$$

If now $\alpha > 0$, $\beta > 0$, $E_{x_n} \subset Q^{n-1}_a$ and $\mu E_{x_n} > \alpha$, $x_n \in (0,\beta)$, then by (18)

$$\int_0^\beta dx_n\int_{E_{x_n}}\varphi(x_n)\,dx^{(n-1)}=\mu E_{x_n}\int_0^\beta\varphi(x_n)\,dx_n\geqslant\alpha\int_0^\beta\varphi(x_n)\,dx_n=+\infty.$$

The test obtained for a trace on the hyperplane R^{n-1} to be a polynomial is easily generalized to the case of traces of functions on hyperplanes of any dimension.

We introduce the necessary notation and definitions. Let $R^m = \{x\colon x_{m+1} = \cdots = x_n = 0\}$, $1 \leqslant m \leqslant n-1$, $x^{(m)} = (x_1,\ldots,x_m)$, $x^{(n-m)} = (x_{m+1},\ldots,x_n)$ and $x = (x^{(m)}, x^{(n-m)})$. Consider the spherical coordinates

$$x^{(n-m)} = (\rho, \theta), \qquad 0 \leqslant \rho < +\infty, \theta = (\theta_1,\ldots,\theta_{n-m-1}) \in S^{n-m-1}, \tag{19}$$

in the space R^{n-m} of points $x^{(n-m)}$, where $S^{n-m-1} = \{x^{(n-m)}\colon \rho = 1\}$ is the unit $(n-m-1)$-dimensional sphere in R^{n-m}, and $\rho = \sqrt{x_{m+1}^2 + \cdots + x_n^2}$. Let

$$Q_a^m = \left\{x^{(m)}\colon |x_i| \leqslant a, i = 1,\ldots,m\right\}, \qquad 0 < a < \infty,$$
$$P_a^n = \left\{x\colon x^{(m)} \in Q_a^m, \rho \leqslant 1\right\}.$$

DEFINITION 2. Let $\varphi\colon P_a^n \to \overset{+}{R}$. The integral $\int_{P_a^n} \varphi(x)\,dx$ is said to be *strongly divergent near the hyperplane* R^m if the integral

$$\int_0^1 \rho^{n-m-1}\,d\rho \int_{S^{n-m-1}} d\theta \int_{Q_a^m} \varphi(x^{(m)}, \rho, \theta)\,dx^{(m)}$$

is strongly divergent near $\rho = 0$.

THEOREM 2. *Suppose that* $u\colon P_a^n \to R$, $\varphi\colon P_a^n \times (0,1) \to \overset{+}{R}$, *and*:

1) *for any finite* $a > 0$

$$\int_0^1 dh_1 \int_{P_a^n} \varphi(x, h_1)\left|\Delta_{h_1}^r u(x)\right|^p dx < +\infty;$$

2) *for almost all* $h \in (0,1)$ *and any finite* $a > 0$ *the integral*

$$\int_{P_a^n} \varphi(x, h)\,dx$$

is strongly divergent near the hyperplane R^m; *and*

3) *for any* $a > 0$ *the function* u *has a trace* $u(x^{(m)}, 0)$ *on the cube* Q_a^m *in the* L_p*-metric.*

Then there exists a function which is a polynomial of degree at most $r-1$ *in* x_1 *and is equivalent to the trace* $u(x^{(m)}, 0)$ *in the sense of the m-dimensional measure.*

To prove this, pass to the spherical coordinates (19) in R^{n-m}, and then Theorem 2 follows from Theorem 1.

BIBLIOGRAPHY

1. G. N. Yakovlev, *On traces of functions whose derivatives are integrable with some weight*, Imbedding Theorems and Their Applications (Proc. Sympos., Baku, 1966), "Nauka", Moscow, 1970, pp. 225–233. (Russian)

2. L. D. Kudryavtsev, *On stabilization of functions at infinity and at a hyperplane*, Trudy Mat. Inst. Steklov. **134** (1975), 124–141; English transl. in Proc. Steklov Inst. Math. **134** (1977).

3. B. V. Tandit, *Boundary properties of functions in weighted spaces*, Differentsial'nye Uravneniya **15** (1979), 492–506; English transl. in Differential Equations **15** (1979).

4. S. M. Nikol'skiĭ, *Approximation of functions of several variables and embedding theorems*, 2nd rev. aug. ed., "Nauka", Moscow, 1977; English transl. of 1st ed., Springer-Verlag, 1974.

Translated by H. H. MCFADEN

Труды Матем. Инст. Стеклов.
161(1983)

Proceedings of the Steklov
Institute of Mathematics
1984, Issue 3

COERCIVE PROPERTIES OF AN ELLIPTIC EQUATION WITH DEGENERACY AND A GENERALIZED RIGHT-HAND SIDE

UDC 517.518.235

P. I. LIZORKIN AND S. M. NIKOL'SKIĬ

ABSTRACT. The first boundary value problem is considered for a differential equation

$$Lu \equiv \sum_{|k|,\,|l| \leqslant r} (-1)^{|l|} D^l\left(a_{kl}(x) D^k u\right) = F(x), \qquad x \in \Omega, \tag{1}$$

which is degenerate on the boundary $\partial\Omega \equiv \Lambda$ of a bounded domain $\Omega \subset R^n$. The degeneracy is characterized by the condition

$$\sum_{|k|,\,|l| \leqslant r} a_{kl}(x)\xi_k\xi_l \geqslant \frac{\kappa}{\rho(x)^{2\alpha}} \sum_{|k|=r} \xi_k^2. \tag{2}$$

Moreover, the coefficients a_{kl} are assumed to be symmetric and measurable functions in Ω such that for some integer $\gamma \geqslant 0$

$$\left|D^\lambda a_{k\lambda}(x)\right| \leqslant M^2/[\rho(x)]^{2(r+\alpha)-|k|-|l|+|\lambda|} \qquad (|\lambda| \leqslant \gamma). \tag{3}$$

The parameters α are subject to the constraints

$$-r < \alpha < -r + \tfrac{1}{2}, \qquad s_0 = \left[r + \alpha - \tfrac{1}{2}\right] \geqslant r/2.$$

The variational method is used to find a generalized solution U of (1) with finite norm

$$\int\left(\rho^{2\alpha} \sum_{|k|=r} \left|D^k u\right|^2 + |u|^2\right) dx,$$

and satisfying on Γ the conditions

$$u\big|_\Gamma = \varphi_0, \qquad \frac{\partial u}{\partial n}\bigg|_\Gamma = \varphi_1, \ldots, \frac{\partial^{s_0-1} u}{\partial n^{s_0-1}}\bigg|_\Gamma = \varphi_{s_0-1},$$

where the functions φ_j belong to the corresponding Besov spaces. The homogeneous boundary value problem is studied first. A theorem on existence and uniqueness of a solution U is obtained for $\gamma = 0$. Theorems are proved on improved smoothness of the solution when $\gamma > 0$. Coercive estimates are given for the solution, along with isomorphism theorems. These results are carried over to the case of the nonhomogeneous boundary value problem under certain additional requirements on the coefficients. The main attention is focussed on the case $0 \leqslant \gamma < r$, when the boundary part of the equation is a generalized function in an appropriate class (the case $\gamma \geqslant 2$ was treated in a previous article by the authors).

Bibliography: 9 titles.

1980 *Mathematics Subject Classification*. Primary 35D05, 35J50, 35J55; Secondary 35B65, 46E35.

§1. Introduction

This article is a continuation of our joint article [5]. We consider the differential equation

$$LU = F(x) \qquad (x \in \Omega), \tag{1}$$

$$LU = \sum_{|k|,\,|l| \leqslant r} (-1)^{|l|} D^{(l)}\big(a_{kl}(x) D^k U(x)\big), \tag{2}$$

where $k = (k_1, \ldots, k_n)$, $l = (l_1, \ldots, l_n)$, $|k| = \Sigma_1^n k_j$, $|l| = \Sigma_1^n l_j$, and

$$D^{(k)}\Psi = \Psi^{(k)} = \frac{\partial^{|k|}\Psi}{\partial x_1^{k_1} \cdots \partial x_n^{k_n}}.$$

Here $\Omega \subset R_n$ is a bounded domain with sufficiently smooth $(n-1)$-dimensional boundary Γ, and the $a_{kl}(x) = a_{lk}(x)$ are measurable functions on Ω indexed by pairs (k, l) of nonnegative integer vectors k and l and satisfying the inequality

$$\sum_{|k|,\,|l| \leqslant r} a_{kl}(x)\xi_k\xi_l \geqslant \frac{\kappa}{\rho(x)^{2\alpha}} \sum_{|k| = r} \xi_k^2, \tag{3}$$

where ξ_k and ξ_l are real variables indexed by the indicated vectors k and l, $\rho(x)$ is the distance from $x \in \Omega$ to Γ, κ is a constant not depending on x, ξ_k, nor ξ_l, and α is a particular real number. The inequality (3) shows that LU is an elliptic operator of order $2r$ with degeneracy on the boundary Γ.

We give an arbitrary nonnegative integer γ and assume that the coefficients a_{kl} have continuous partial derivatives $a_{kl}^{(\lambda)}(x)$ through order γ that satisfy the inequalities

$$\big|a_{kl}^{(\lambda)}(x)\big| \leqslant M^2/\rho(x)^{2(r+\alpha)-|k|-|l|+|\lambda|} \qquad (|\lambda| \leqslant \gamma), \tag{4}$$

where M does not depend on x.

We also impose differential properties (depending on the given γ) on the function $F(x)$ and study the corresponding properties of a solution (see §2) that are characterized by membership in a certain weighted class of functions.

As in [4], we solve the boundary value problem of the first kind for the differential equation (1) and study the differential properties of a solution in dependence on the properties of the right-hand side in the case of homogeneous boundary conditions, and also in dependence on the properties of the boundary functions in the case of nonhomogeneous boundary conditions. These properties improve as γ increases. The case $\gamma \geqslant r$ was completely solved for $r = 1$ in Nikol'skiĭ's article [7], and for $r > 1$ in our joint article [5]. The main goal of the present article is to study the differential properties of a solution of (1) when $0 \leqslant \gamma < r$. In this case $F(x)$ is a generalized function (distribution) in general.

In §2 the problem is formulated, and the main result, Theorem 1, is stated. The direct part of Theorem 1 is proved there.

§§3–7 deal with the converse part of Theorem 1.

In §3 we prove the existence of a solution for the boundary value problem (1) (the case $\gamma = 0$).

In §4 we prove a needed lemma for a classical elliptic equation (without degeneracy). It must be regarded as known (see the bibliography in [5]), but the method of proof may nevertheless be of interest.

§5 deals with generalizations of L_p-spaces to the case of systems of functions. They are needed in §6, where a general coercive estimate is derived for a solution of

(1). Finally, the converse part of Theorem 1 is proved in §7 for $0 < \gamma \leqslant r$.

For nonnegative A and B we write $A \ll B$ if $A \leqslant cB$, where $c > 0$ is a constant not depending on A nor B, and we write $A \sim B$ if $A \ll B \ll A$.

If norms $\|f\|_{E_1}$ and $\|f\|_{E_2}$ are introduced in the same linear space of functions f and

$$\|f\|_{E_1} \sim \|f\|_{E_2}, \tag{5}$$

then we say that they are equivalent and write $E_1 = E_2$.

REMARK. We shall take the liberty of writing

$$\|f\|_{E_1} = \|f\|_{E_2} \tag{6}$$

in place of (5).

The symbol $W^r_{2,\alpha}(\Omega)$ denotes the weighted class of functions $f(x)$ defined on Ω, with the norm

$$\|f\|_{W^r_{2,\alpha}(\Omega)} = \|f\|_{L_2(\Omega)} + \|f\|_{w^r_{2,\alpha}(\Omega)}, \quad \|f\|_{w^r_{2,\alpha}(\Omega)} = \left(\int_\Omega \sum_{|k|=r} \left| \frac{f^{(k)}(x)}{\rho(x)^\alpha} \right|^2 dx \right)^{1/2}.$$

Let $C_0^\infty(\Omega)$ denote the set of infinitely differentiable functions in Ω with compact support. This set, with the standard topology of a space of "test functions", is denoted by $\mathscr{D}(\Omega) = \mathscr{D}$, and the collection of generalized functions over $\mathscr{D}$ is denoted by $\mathscr{D}'(\Omega) = \mathscr{D}'$.

§2. Formulation of the problem. Main theorem

2.1. In this article we assume that r is a given natural number, α is a real number such that

$$0 < r + \alpha - \tfrac{1}{2} < r, \tag{1}$$

where $r + \alpha - \frac{1}{2}$ is not an integer, and s_0 is a natural such that

$$r + \alpha - \tfrac{1}{2} < s_0 < r + \alpha + \tfrac{1}{2}. \tag{2}$$

There is only one case (§6) when the conditions (1) are replaced by other conditions.

The most far-reaching results obtained here require the additional condition

$$2s_0 \geqslant r. \tag{3}$$

It follows from (1) and (2) (see [5], Chapter 10) that a function $f \in W^r_{2,\alpha}(\Omega)$ has the s_0 boundary functions

$$\left.\frac{\partial^s f}{\partial \nu^s}\right|_\Gamma = \varphi_s \in L_2(\Gamma), \qquad s = 0,\ldots,s_0 - 1, \Gamma \in C^{(s_0+1)}, \tag{4}$$

where ν is the outer normal to Γ. By definition, the boundary Γ of a domain Ω belongs to $C^{(\mu)}$ ($\Gamma \in C^{(\mu)}$) if it is described locally by functions μ times continuously differentiable.

If $\Phi(x) \in W^r_{2,\alpha}(\Omega)$ and $\Gamma \in C^{(s_0+1)}$, then $\mathfrak{M}_\Phi$ denotes the class of functions in $W^r_{2,\alpha}(\Omega)$ having the same boundary functions as Φ:

$$\mathfrak{M}_\Phi = \left\{ f : f \in W^r_{2,\alpha}(\Omega);\ \left.\frac{\partial^s f}{\partial \nu^s}\right|_\Gamma = \left.\frac{\partial^s \Phi}{\partial \nu^s}\right|_\Gamma, s = 0,\ldots,s_0 - 1 \right\}.$$

In particular, $\mathfrak{M}_\Phi = \mathfrak{M}_0$ if $\Phi(x) \equiv 0$.[1]

---

[1] We denote the class $\mathfrak{M}_0$ also by $\overset{0}{W}_{2,\alpha}(\Omega)$.

We remark that under the conditions (1), (2), and (3)

$$\|f\|_{L_2(\Omega)} \leqslant \|f\|_{w^r_{2,\alpha}(\Omega)}, \quad \forall f \in \overset{0}{W}{}^r_{2,\alpha}(\Omega), \Gamma \in C^{(s_0+1)}$$

(see [4] and [6], Chapter 10); therefore,

$$\|f\|_{\overset{0}{W}{}^r_{2,\alpha}(\Omega)} \sim \|f\|_{w^r_{2,\alpha}(\Omega)}, \quad \forall f \in \overset{0}{W}{}^r_2(\Omega), \Gamma \in C^{(s_0+1)}. \tag{5}$$

The symbol $\|f\|_{\overset{0}{W}{}^r_{2,\alpha}(\Omega)}$ denotes the norm of a function $f \in \overset{0}{W}{}^r_{2,\alpha}(\Omega)$ in the $W^r_{2,\alpha}(\Omega)$-metric.

We have the well-known inequalities

$$\|f^{(k)}/\rho^\beta\|_{L_2(\Omega)} \leqslant c_1\|f\|_{W^r_{2,\beta-r+|k|}} \leqslant c_2\|f\|_{W^r_{2,\alpha}(\Omega)},$$
$$\beta < 1/2, \beta - (r - |k|) \leqslant \alpha, \Gamma \in C^1 \tag{6}$$

(see [6], Chapter 10). A particular consequence of (6) is the imbedding

$$W^{r+\gamma}_{2,\alpha-\gamma}(\Omega) \to W^r_{2,\alpha}(\Omega) \quad (\gamma > 0, \alpha < \tfrac{1}{2}), \qquad \Gamma \in C^1, \tag{7}$$

which expresses that

$$W^{r+\gamma}_{2,\alpha-\gamma}(\Omega) \subset W^r_{2,\alpha}(\Omega) \quad \text{and} \quad \|f\|_{W^r_{2,\alpha}(\Omega)} \ll \|f\|_{W^{r+\gamma}_{2,\alpha-\gamma}(\Omega)}.$$

Under conditions (1) and (2) this property implies that an arbitrary function $f \in W^{r+\gamma}_{2,\alpha-\gamma}(\Omega)$ has the s_0 boundary functions (4).

2.2. Along with the weighted space $W^r_{2,\alpha}(\Omega)$ we consider the space $V^r_{2,\alpha}(\Omega)$ with "compatible weights" in which the norm is

$$\|f\|_{V^r_{2,\alpha}(\Omega)} = \sum_{|k|\leqslant r} \left\| \frac{D^k f}{\rho^{r+\alpha-|k|}} \right\|_{L_2(\Omega)}. \tag{8}$$

This space was considered in [2]. It is closely related to $\overset{0}{W}{}^r_{2,\alpha}(\Omega)$. Namely, under conditions (1) and (2) we have the equivalence relation

$$V^r_{2,\alpha}(\Omega) \sim \overset{0}{W}{}^r_{2,\alpha}(\Omega), \tag{9}$$

and under conditions (1)–(3) the relations

$$V^r_{2,\alpha}(\Omega) \sim \overset{0}{W}{}^r_{2,\alpha}(\Omega) \sim w^r_{2,\alpha}(\Omega), \quad \forall f \in \overset{0}{W}{}^r_2(\Omega) \tag{10}$$

(which follow immediately from (5) and (9)).

The relation (9) follows on the one hand from an inequality that is valid for $f \in \overset{0}{W}{}^r_{2,\alpha}(\Omega)$:

$$\|f\|_{V^r_{2,\alpha}(\Omega)} \ll \|f\|_{\overset{0}{W}{}^r_{2,\alpha}(\Omega)} \tag{11}$$

(see [5], Lemma 2). On the other hand, if $f \in V^r_{2,\alpha}(\Omega)$, then

$$\|f\|_{V^r_{2,\alpha}(\Omega)} = \sum_{|k|\leqslant r} \left\| \frac{f^{(k)}}{\rho^{r+\alpha-|k|}} \right\|_{L_2(\Omega)} \geqslant \left\| \frac{f}{\rho^\alpha} \right\| + \sum_{|k|=r} \left\| \frac{f^{(k)}}{\rho^\alpha} \right\|_{L_2(\Omega)} \gg \|f\|_{W^r_{2,\alpha}(\Omega)}, \tag{12}$$

because under conditions (1) and (2)

$$\|f/\rho^\alpha\|_{L_2(\Omega)} + \|f\|_{w^r_{2,\alpha}(\Omega)} \sim \|f\|_{L_2(\Omega)} + \|f\|_{w^r_{2,\alpha}(\Omega)} = \|f\|_{W^r_{2,\alpha}(\Omega)}$$

(see [6], Chapter 10). Consequently, $f \in W^r_{2,\alpha}(\Omega)$ and f has the s_0 boundary functions

$$\left.\frac{\partial^s f}{\partial \nu^s}\right|_\Gamma = \varphi_s \qquad (s = 0,\ldots,s_0 - 1)$$

in view of (1) and (2).

We note the important result that the class $C_0^\infty(\Omega)$ of infinitely differentiable functions with compact support in Ω is dense in the space $V^r_{2,\alpha}(\Omega)$.(²)

Let $v_l(x)$, $l = 1,2,\ldots$, be compactly supported infinitely differentiable functions ($v_l \in V^r_{2,\alpha}(\Omega)$) approximating f in the $V^r_{2,\alpha}(\Omega)$-metric, and a fortiori (by (12)) in the $W^r_{2,\alpha}(\Omega)$-metric. We have (explanation below)

$$\|f - v_l\|_{W^r_{2,\alpha}(\Omega)} \gg \left(\int_\Gamma \sum_{s=0}^{s_0-1} \varphi_s^2 \, d\Gamma\right)^{1/2} \qquad (l = 1,2,\ldots).$$

This well-known inequality (see [5], §10.1, (13)) is true for (1) and (2); it must be taken into account that the v_l are compactly supported functions. Its left-hand side tends to zero as $l \to \infty$; consequently, $\varphi_s \equiv 0$ on Γ $(s = 0,\ldots,s_0 - 1)$, which shows that $f \in \overset{0}{W}{}^r_{2,\alpha}(\Omega)$, and instead of (12) we can write

$$\|f\|_{V^r_{2,\alpha}(\Omega)} \geqslant \|f\|_{\overset{0}{W}{}^r_{2,\alpha}(\Omega)}, \qquad \forall f \in V^r_{2,\alpha}(\Omega), \tag{13}$$

which together with (11) leads to (9).

Observe that the following lemma is a particular consequence of the above arguments.

LEMMA 1. *The set $C_0^\infty(\Omega)$ is dense in $\mathfrak{M}_0 = \overset{0}{W}{}^r_{2,\alpha}(\Omega)$. Moreover, M_0 coincides with the completion of $C_0^\infty(\Omega)$ in the norm of the space $W^r_{2,\alpha}(\Omega)$, $-r + \frac{1}{2} < \alpha < \frac{1}{2}$, $-\alpha + \frac{1}{2} \neq 1,\ldots,r$.*

This fact was proved in [9] for domains with infinitely differentiable boundary, and in Besov's article [1] (in this volume) for domains with Lipschitz boundary.

We consider (bounded) linear functionals $F(v)$ on $V^\mu_{2,\alpha}(\Omega)$ $(\mu = 0,1,\ldots)$, written formally as integrals

$$F(v) = \int_\Omega F(x) v(x)\, dx.$$

The functionals F form the space $\overset{*}{V}{}^\mu_{2,\alpha}(\Omega)$ dual to $V^\mu_{2,\alpha}(\Omega)$, with the norm

$$\|F\|_{\overset{*}{V}{}^\mu_{2,\alpha}(\Omega)} = \sup_{\|v\|_{V^\mu_{2,\alpha}(\Omega)} \leqslant 1} F(v).$$

It will be seen below that $\overset{*}{V}{}^r_{2,\alpha}(\Omega)$ is a subspace of the space $\mathscr{D}'(\Omega) = \mathscr{D}'$ of generalized functions on Ω (over the test space $\mathscr{D}(\Omega) = \mathscr{D}$ of infinitely differentiable functions with compact support in Ω).

(²) See the Appendix.

2.3. Finally, we introduce the class $H_{r,\gamma}$. This class is the space of functions $F(x) \in \mathscr{D}'$ such that the following norm makes sense and is finite:

$$\|F\|_{H_{r,\gamma}} = \begin{cases} \|F\|_{\overset{*}{V}{}^{r-\gamma}_{2,\alpha+\gamma}(\Omega)} = \sup\limits_{\|v\|_{V^{\mu-\gamma}_{2,\alpha+\gamma}(\Omega)} \leqslant 1} F(v) & (0 \leqslant \gamma \leqslant r) \\ \left(\sum_{|\lambda| \leqslant \gamma - r} \|F^{(\lambda)} \rho^{r+\alpha+|\lambda|}\|_{L_2(\Omega)} \right)^{1/2} & (\gamma \geqslant r). \end{cases} \tag{14}$$

If $\gamma < r$, then F is in general a generalized function ($F \in \mathscr{D}'$; see Remark 6 in §5.3 for an explanation). On the basis of what was stated in §2.2, equality (14) can also be written for $\gamma = 0$ (under the conditions (1), (2), and $\Gamma \in C^{(s_0+1)}$) also as

$$\|F\|_{H_{r,\gamma}} = \sup_{\|v\|_{V^{\mu}_{2,\alpha}(\Omega)} \leqslant 1} F(v) = \sup_{\|v\|_{\overset{0}{W}{}^{r}_{2,\alpha}(\Omega)} \leqslant 1} F(v). \tag{15}$$

The two expressions on the right-hand side in (14) coincide if $\gamma = r$, because then

$$\|v\|_{V^0_{2,r+\alpha}(\Omega)} = \|v/\rho^{r+\alpha}\|_{L_2(\Omega)}, \tag{16}$$

and a linear functional on the space of functions with the norm (16) can be written in the form

$$F(v) = \int_\Omega F(x) v(x)\, dx,$$

where (see §5 below)

$$\|F(x)\rho^{r+\alpha}\|_{L_2(\Omega)} < \infty.$$

2.4. We consider the operator L (see §1, (1)) on functions U in $\overset{0}{W}{}^{r+\gamma}_{2,\alpha-\gamma}(\Omega)$ or $W^{r+\gamma}_{2,\alpha-\gamma}(\Omega)$, and automatically assume that its coefficients $a_{kl}(x)$ have continuous partial derivatives through order γ that satisfy (3) and (4) in §1. If $\gamma \geqslant r$, then LU is understood in the usual sense; if $\gamma < r$, in the generalized sense.

In our considerations the functions $a_{kl}(x)$ are measurable and locally bounded, and the functions $U(x)$ belong locally to W_2^r, so the product $a_{kl}(x)U^{(k)}(x)$ is locally integrable, and the integral

$$\int_\Omega \sum_{|k|, |l| \leqslant r} a_{kl}(x) U^{(k)}(x) v^{(l)}(x)\, dx = F(v), \quad \forall v \in D, \tag{17}$$

is a linear functional (distribution) over $\mathscr{D}$ ($F \in \mathscr{D}'$). Each function U belonging locally to W_2^r can also be regarded as belonging to $\mathscr{D}'$ ($U \in \mathscr{D}'$).

Thus, to each function U belonging locally to W_2^r there corresponds (by (17), which is true for all $v \in \mathscr{D}$) a generalized function F ($F \in \mathscr{D}'$). This determines the operator $LU = F$.

Now we can state the main theorem.

2.5. THEOREM 1. *Under conditions* (3) *and* (4) *in* §1, *along with conditions* (1), (2), *and* (3) *and the condition* $\Gamma \in C^{(s_0+1)}$, *the operator* L (*see* §1, (1)) *maps the class* $\overset{0}{W}{}^{r+\gamma}_{2,\alpha-\gamma}(\Omega)$ *onto the class* $H_{r,\gamma}(\Omega)$ *in a one-to-one and continuous fashion*:

$$L\left(\overset{0}{W}{}^{r+\gamma}_{2,\alpha-\gamma}\right) = H_{r,\gamma}(\Omega). \tag{18}$$

Thus, for any nonnegative integer γ *and any function* $U \in \overset{0}{W}{}^{r+\gamma}_{2,\alpha-\gamma}(\Omega)$ *the corresponding function*

$$F(x) = LU \tag{19}$$

belongs to the class $H_{r,\gamma}$, *and*

$$\|F\|_{H_{r,\gamma}} \leqslant c\|U\|_{\overset{0}{W}{}^{r+\gamma}_{2,\alpha-\gamma}(\Omega)}. \tag{20}$$

Conversely, to each function $F \in H_{r,\gamma}$ *there corresponds a unique function* $U \in \overset{0}{W}{}^{r}_{2,\alpha}(\Omega)$ *satisfying* (19) *and the estimate*

$$\|U\|_{\overset{0}{W}{}^{r+\gamma}_{\alpha-\gamma}(\Omega)} \leqslant c_1\|F\|_{H_{r,\gamma}}. \tag{21}$$

The constants c *in* (20) *and* c_1 *in* (21) *are independent of* U *and* F, *respectively.*

2.6. PROOF OF THE DIRECT ASSERTION IN THEOREM 1 FOR $0 \leqslant \gamma \leqslant r$.(3) Let γ be an integer such that $0 \leqslant \gamma \leqslant r$, and let $U(x) \in \overset{0}{W}{}^{r+\gamma}_{2,\alpha-\gamma}(\Omega)$. We have agreed (see §1, (4)) that in this case the coefficients $a_{kl}(x)$ automatically have continuous partial derivatives $a_{kl}^{(\lambda)}$ through order γ on Ω and that the inequalities (4) in §1 are satisfied. This shows that in the terms of the sum (17) corresponding to vectors l with $|l| > r-\gamma$ we can transfer some of the partial derivatives of orders $|l|-(r-\gamma)$ ($\leqslant \gamma$!) over from $U^{(l)}$ to the adjacent factor. With this in mind, we write

$$\int_\Omega \sum_{|k|,|l|\leqslant r} a_{kl}U^{(k)}v^{(l)}\,dx = \sum_{\substack{|k|,|l|\leqslant r\\ |l|\leqslant r-\gamma}} \int_\Omega a_{kl}U^{(k)}v^{(l)}\,dx + \sum_{\substack{|k|,|l|\leqslant r\\ |l|>r-\gamma}} \int_\Omega a_{kl}U^{(k)}v^{(l)}\,dx. \tag{22}$$

The terms of the first sum on the right-hand side of (22) can be estimated as follows (explanation below):

$$\left|\int_\Omega a_{kl}U^{|k|}v^{|l|}\,dx\right| \ll \int_\Omega \left|\frac{U^{(k)}}{\rho^{r+\alpha-|k|}}\frac{v^{(l)}}{\rho^{r+\alpha-|l|}}\right| dx \leqslant \left\|\frac{U^{(k)}}{\rho^{r+\alpha-|k|}}\right\|_{L_2(\Omega)}\left\|\frac{v^{(l)}}{\rho^{r+\alpha-|l|}}\right\|_{L_2(\Omega)}$$

$$\ll \|U\|_{\overset{0}{W}{}^{r}_{2,\alpha}(\Omega)}\|v\|_{V^{r-\gamma}_{2,\alpha+\gamma}(\Omega)} \ll \|U\|_{\overset{0}{W}{}^{r+\gamma}_{2,\alpha-\gamma}(\Omega)}\|v\|_{V^{r-\gamma}_{2,\alpha+\gamma}(\Omega)}. \tag{23}$$

The first inequality in this chain follows from the condition (4) in §1; the third is true by (10) and the fact that $|l| \leqslant r-\gamma$; the last one follows from (7) if we take into account that $\alpha < 1/2$.

(3) Theorem 1 was proved in [5] for $\gamma \geqslant r$.

For any term in the second sum in (22) we have $|l| > r - \gamma$. If $l^0 \leqslant l$, $|l^0| = r - \gamma$, and $s = l - l^0$, then

$$\left|\int_\Omega a_{kl} U^{(k)} v^{(l)}\,dx\right| = \left|\int_\Omega \left(a_{kl}U^{(k)}\right)^{(s)} v^{(l^0)}\,dx\right| \leqslant \sum_{0\leqslant\lambda\leqslant s} c_s^\lambda \int_\Omega \left|a_{kl}^{(\lambda)} U^{(k+s-\lambda)} v^{(l^0)}\right| dx$$

$$\ll \sum_{0\leqslant\lambda\leqslant s} \int_\Omega \left| \frac{U^{(k+s-\lambda)}}{\rho^{2r+\alpha-\gamma-|k|-|l|+|\lambda|}} \frac{v^{(l^0)}}{\rho^{r+\alpha-(r-\gamma)}}\right| dx$$

$$\leqslant \sum_{0\leqslant\lambda\leqslant s} \left\| \frac{U^{(k+s-\lambda)}}{\rho^{r+\alpha-|s|+|\lambda|-|k|}}\right\|_{L_2(\Omega)} \|v\|_{V^{r-\gamma}_{r,\alpha+\gamma}(\Omega)}$$

$$= \sum_{0\leqslant\mu\leqslant r+\gamma} \left\| \frac{U^{(\mu)}}{\rho^{(r+\gamma)+(\alpha-\gamma)-|\mu|}}\right\|_{L_2(\Omega)} \|v\|_{V^{r-\gamma}_{r,\alpha+\gamma}(\Omega)}$$

$$\leqslant \|U\|_{\overset{0}{W}{}^{r+\gamma}_{\alpha-\gamma}(\Omega)} \|v\|_{V^{r-\gamma}_{2,\alpha+\gamma}(\Omega)}. \tag{24}$$

In the first relation in this chain (the equality) it is legal to transfer $s = l - l^0$ derivatives over to $(a_{kl}U^{(k)})$ because $|s| \leqslant \gamma$ and (by assumption) a_{kl} has continuous derivatives through order γ; the function $U^{(k)}$ can also be differentiated s times, yielding a locally integrable function (in fact, $U \in W^{r+\gamma}_{\alpha-\gamma}$, and $|k| + |s| \leqslant r + \gamma$). In the second relation the Leibniz formula was used, in the third the inequality (4) in §1 was used, in the fourth it was necessary to take into account that $|l| - (r - \gamma) = |s|$, and in the fifth we set $\mu = k + s - \lambda$. Since $|\mu| = |k| + |s - \lambda| \leqslant r + \gamma$, $\alpha - \gamma < \frac{1}{2}$, and $(r + \gamma) + (\alpha - \gamma) - \frac{1}{2} = r + \alpha - \frac{1}{2}$ is not a positive integer, the last inequality in (24) follows from (11), where it is only necessary to replace r and α by $r + \gamma$ and $\alpha - \gamma$, respectively. It follows from (23) and (24) that the linear functional (17) of $v \in V^{r-\gamma}_{r,\alpha}(\Omega)$ satisfies the inequality

$$\left|\int_\Omega \sum_{|k|,\,|l|\leqslant r} a_{kl} U^{(k)} v^{(l)}\,dx\right| \ll \|U\|_{\overset{0}{W}{}^{r+\gamma}_{\alpha-\gamma}(\Omega)} \|v\|_{V^{r-\gamma}_{r,\alpha+\gamma}(\Omega)}.$$

Thus, it has been proved that each function $U \in \overset{0}{W}{}^{r+\gamma}_{\alpha-\gamma}(\Omega)$ generates a linear functional $F = LU$ on $V^{r-\gamma}_{r,\alpha+\gamma}(\Omega)$, namely

$$F(v) = \int_\Omega \sum_{|k|,\,|l|\leqslant r} a_{kl}(x) U^{(k)}(x) v^{(l)}(x)\,dx,$$

whose norm is majorized by the number $\|U\|_{\overset{0}{W}{}^{r+\gamma}_{\alpha-\gamma}(\Omega)}$.

As explained above, this proves property (18) and the inequality (2) accompanying it.

§3. The theorem on existence of a solution (the case $\gamma = 0$)

3.1. The aim in this section is to prove the converse assertion in Theorem 1 when $\gamma = 0$. Specifically, we establish

ASSERTION 1. *Under the conditions* (*see* (3) *and* (4) *in* §1)

$$|a_{kl}(x)| \leqslant c/\rho(x)^{2(r+\alpha)-|k|-|l|}, \tag{1}$$

$$\sum_{|k|,\,|l|\leqslant r} a_{kl}(x)\xi_k\xi_l \geqslant \frac{\kappa}{\rho(x)^{2\alpha}} \sum_{|k|=r} \xi_k^2, \tag{2}$$

and the condition that s_0 *is a natural number and*

$$0 < r + \alpha - \tfrac{1}{2} < r, \quad r + \alpha - \tfrac{1}{2} < s_0 < r + \alpha + \tfrac{1}{2}, \quad F \in C^{(s_0+1)}, \tag{3}$$

$$2s_0 \geqslant r, \tag{4}$$

there exists for any linear functional $F = F(v)$ *on the space of functions* $v \in C_0^\infty(\Omega)$ *with norm (see* §2.2)

$$\|v\|_{V_{2,\alpha}^r(\Omega)} = \left(\sum_{|k| \leqslant r} \left\| \frac{v^{(k)}}{\rho^{r+\alpha-|k|}} \right\|_{L_2(\Omega)}^2 \right)^{1/2} \tag{5}$$

(i.e., for $F \in H_{r,0}$*) a unique function* $U \in \overset{0}{W}{}_\alpha^r(\Omega) = \mathfrak{M}_0$ *such that*

$$LU = F, \tag{6}$$

or, in more detail,

$$E(U, v) = \int_\Omega \sum_{|k|,|l| \leqslant r} a_{kl}(x) U^{(k)}(x) v^{(l)}(x)\, dx = F(v), \quad \forall v \in C_0^\infty(\Omega). \tag{6'}$$

Moreover,

$$\|U\|_{\overset{0}{W}{}_{2,\alpha}^r(\Omega)} \leqslant c \sup_{\|v\|_{V_{2,\alpha}^r(\Omega)} \leqslant 1} F(v). \tag{7}$$

REMARK 1. A function U having property (6′) is called a generalized solution of (6) in the class $\overset{0}{W}{}_{2,\alpha}^r(\Omega) = \mathfrak{M}_0$.

REMARK 2. As will be clear from the proof of Assertion 3.1, (6′) can be replaced by the equivalent property

$$E(U, v) = \int_\Omega \sum_{|k|,|l| \leqslant r} a_{kl}(x) U^{(k)}(x) v^{(l)}(x)\, dx = F(v), \quad \forall v \in \overset{0}{W}{}_{2,\alpha}^r(\Omega), \tag{6''}$$

where it is then necessary to assume that the linear functional $F(v)$ is extended by continuity in the sense of the $W_{2,\alpha}^r(\Omega)$-metric from $C_0^\infty(\Omega)$ to $\overset{0}{W}{}_{2,\alpha}^r(\Omega) \sim V_{2,\alpha}^r(\Omega)$.

REMARK 3. Assertion 3.1 is obviously true for $\alpha = 0$, and then $s_0 = r$. In this case the operator L is elliptic (without degeneracy). But the lower-order coefficients $a_{kl}(x)$ $(|k| + |l| < 2r, |k|, |l| \leqslant r)$ can be unbounded (see (1)). Since $\mathfrak{M}_0 = \overset{0}{W}{}_{2,\alpha}^r(\Omega) = \overset{0}{W}{}_2^r(\Omega)$ is the usual class of Sobolev functions with zero boundary values for $\alpha = 0$, the inequality

$$\|U\|_{\overset{0}{W}{}_2^r(\Omega)} \leqslant c \sup_{\|v\|_{\overset{0}{W}{}_2^r(\Omega)} \leqslant 1} F(v) = c\|F\|_{W_2^{-r}(\Omega)}$$

holds for a generalized solution of (6) in the class $\mathfrak{M}_0$ (see §4 below).

This says that Assertion 3.1 generalizes the well-known classical result, which was established under the assumption that inequality (2) holds for $\alpha = 0$ and the coefficients a_{kl} are bounded: in fact, if $|a_{kl}(x)| \leqslant c$, inequality (1) is thereby satisfied.

3.2. Proof of Assertion 1. The bilinear form

$$E(\varphi,\psi)=\int_\Omega \sum_{|k|,|l|\leqslant r} a_{kl}(x)\varphi^{(k)}(x)\psi^{(l)}(x)\,dx \qquad (\varphi,\psi\in\mathfrak{M}_0)$$

can be regarded as an inner product on $\mathfrak{M}_0$, as follows from the properties

1) $E(\varphi,\psi)=E(\psi,\varphi)$,
2) $E(\lambda\varphi+\mu\psi,\chi)=\lambda E(\varphi,\chi)+\mu E(\psi,\chi)$,
3) $E(\varphi,\varphi)\geqslant 0$,
4) if $E(\varphi,\varphi)=0$ then $\varphi=0$.

The existence of $E(\varphi,\psi)$ for $\varphi,\psi\in\mathfrak{M}_0$ follows from the inequalities (see (9) in §2)

$$\int_\Omega |a_{kl}\varphi^{(k)}\psi^{(l)}|\,dx \leqslant \int_\Omega \left|\frac{\varphi^{(k)}}{\rho^{r+\alpha-|k|}}\frac{\psi^{(l)}}{\rho^{r+\alpha-|l|}}\right|dx$$
$$\leqslant \left\|\frac{\varphi^{(k)}}{\rho^{r+\alpha-|k|}}\right\|_{L_2(\Omega)}\left\|\frac{\psi^{(l)}}{\rho^{r+\alpha-|l|}}\right\|_{L_2(\Omega)}\leqslant \|\varphi\|_{\overset{0}{W}{}^r_{2,\alpha}(\Omega)}\|\psi\|_{\overset{0}{W}{}^r_{2,\alpha}(\Omega)}. \quad (8)$$

Properties 3) and 4) can be obtained from the inequalities (explanation below)

$$E(\varphi,\varphi)=\int_\Omega \sum_{|k|,|l|\leqslant r} a_{kl}\varphi^{(k)}\varphi^{(l)}\,dx\geqslant \kappa\int_\Omega\sum_{|k|=r}\left(\frac{\varphi^{(k)}}{\rho^\alpha}\right)^2dx$$
$$=\kappa\|\varphi\|_{w^r_{2,\alpha}(\Omega)}\gg \|\varphi\|_{W^r_{2,\alpha}(\Omega)}\geqslant 0. \quad (9)$$

The second relation in (9) follows from (2); see (10) in §2 concerning the next to last relation. It is closely connected with the condition $2s_0\geqslant r$.

The space $\mathfrak{M}_0$ with the norm $E(f)^{1/2}=E(f,f)^{1/2}$ is complete, because it is complete in the sense of the norm $w^r_{2,\alpha}(\Omega)\sim \overset{0}{W}{}^r_{2,\alpha}(\Omega)$ and because of the inequalities

$$\kappa\|\varphi\|^2_{w^r_{2,\alpha}(\Omega)}\leqslant E(\varphi)\leqslant c\|\varphi\|^2_{w^r_{2,\alpha}(\Omega)}. \quad (10)$$

The first of these follows from (2), and the second from (8).

Let us take an arbitrary (bounded) functional $F(v)$ on $V^r_{2,\alpha}(\Omega)\cap C_0^\infty(\Omega)$ and extend it by continuity to $V^r_{2,\alpha}(\Omega)\sim\mathfrak{M}_0$. On the basis of a property of the inner product in $\mathfrak{M}_0$ there exists a unique function $U\in\mathfrak{M}_0$ such that

$$E(U,v)=F(v),\quad \forall v\in\mathfrak{M}_0. \quad (11)$$

But then (explanation below)

$$\|U\|_{\overset{0}{W}{}^r_2(\Omega)}\ll\|U\|_{w^r_2(\Omega)}\ll E(U)^{1/2}\ll\|F\|_{\overset{*}{V}_{2,\alpha}(\Omega)}=\sup_{\|v\|_{V^r_{2,\alpha}(\Omega)}\leqslant 1}F(v), \quad (12)$$

and we obtain (7).

See the next to last inequality in (9) about the first inequality in (12), and see (10) about the second inequality; the third follows from (11).

There cannot be a function $U_1\in\mathfrak{M}_0$ satisfying (6′) and different from the function $U\in\mathfrak{M}_0$ found, for if that were so, then we would have that

$$E(U-U_1,v)=0,\quad \forall v\in C_0^\infty(\Omega).$$

But since the inner product is continuous in v, and $C_0^\infty(\Omega)$ is dense in $\overset{0}{W}{}^r_{2,\alpha}(\Omega)$, this equality would then be satisfied for all $v \in \overset{0}{W}{}^r_{2,\alpha}(\Omega)$, implying $U = U_1$.

Thus, there exists a unique function $U \in \overset{0}{W}{}^r_{2,\alpha}(\Omega)$ such that $E(U, v) = F(v)$ for all $v \in C_0^\infty(\Omega)$ or, what is the same, for all $v \in \overset{0}{W}{}^r_{2,\alpha}(\Omega)$. This function can be obtained as the minimum

$$\min_{f \in \mathfrak{M}_0} [E(f, f) - 2F(f)] = E(U, U) - 2F(U)$$

by arguing just as in the corresponding place in Theorem 1 of [5]. The converse assertion of Theorem 1 is proved for $\gamma = 0$.

§4. A lemma on improved smoothness

Lemma 1 below relates in essence to the theory of nondegenerate elliptic equations and exists in various forms in the literature. We present it in the form we need and accompany it by a proof that may be of independent interest.

4.1. The Sobolev class $W_2^r(\Omega)$ generalizes in a natural way to negative integers $-r$. As before, $\mathscr{D} = \mathscr{D}(\Omega)$ is the test space of infinitely differentiable functions with compact support in Ω, and $\mathscr{D}' = \mathscr{D}'(\Omega)$ is the space of generalized functions over $\mathscr{D}$.

By definition, a function $f \in \mathscr{D}'$ belongs to the class $W_2^{-r}(\Omega)$ if the norm

$$\|f\|_{W_2^{-r}(\Omega)} = \sup_{\|v\|_{W_2^r(\Omega)} \leqslant 1} \int_\Omega f(x) v(x)\, dx \tag{1}$$

makes sense for it and is finite for all $v \in \mathscr{D}$. Recall that the integral in (1) stands for the value of the functional f on an element $v \in C_0^\infty(\Omega)$.

For a domain Ω with boundary $\Gamma \in C^{(r+1)}$ the set $\mathscr{D}$ is dense in $\overset{0}{W}{}^r_2(\Omega)$ (see [3]); therefore, the functional $\int_\Omega fv\, dx$ can be extended by continuity to all possible functions $v \in \overset{0}{W}{}^r_2(\Omega)$. Moreover, the norm of the extended functional remains unchanged.

We mention an inequality for a domain Ω with smooth boundary, namely

$$\|f\|_{W_2^{r_1}(\Omega)} \leqslant c \|f\|_{W_2^{r_2}(\Omega)} \qquad (r_1 < r_2), \tag{2}$$

which is true for any real r_1 and r_2, where the constant c depends on r_1, r_2, and Ω, but not on f.

The inequality (2) is well known for $0 \leqslant r_1 < r_2$. But if $r_1 < r_2 \leqslant 0$, then

$$\|v\|_{W_2^{-r_2}(\Omega)} \leqslant c \|v\|_{W_2^{-r_1}(\Omega)},$$

where c does not depend on $v \in \mathscr{D}$. Therefore,

$$\sup_{\|w\|_{W_2^{-r_1}(\Omega)} \leqslant 1} \int fw\, dx \leqslant \sup_{\|w\|_{W_2^{-r_2}(\Omega)} \leqslant c} \int fw\, dx = c \sup_{\|w\|_{W_2^{-r_2}(\Omega)} \leqslant 1} \int fw\, dx,$$

and we have obtained (2). The case $r_1 < 0 < r_2$ can be reduced to the two cases considered.

4.2. LEMMA 1. *Let σ and σ' be two concentric balls, with $\sigma \subset \bar{\sigma} \subset \sigma'$. Let $F(x) \in \mathscr{D}'(\sigma')$, and let the $a_{kl}(x) = a_{kl}(x)$, $x \in \Omega$, be measurable functions indexed by pairs of nonnegative integer vectors k, l and satisfying for a given nonnegative integer γ the following conditions*:

$$|a_{kl}^{(\lambda)}(x)| \leqslant K, \qquad |\lambda| \leqslant \gamma, \qquad |k|, |l| \leqslant r; \tag{3}$$

$$F(x) \in W_2^{\gamma - r}(\sigma') \equiv (W_2^{r-\gamma}(\sigma'))^* \quad \text{if } \gamma \leqslant r; \tag{4}$$

$$F(x) \in W^{\gamma - r}(\sigma') \quad \text{if } \gamma \geqslant r; \tag{5}$$

$$\int_{\sigma'} \sum_{|k|,|l|=r} a_{kl}(x) f^{(k)}(x) f^{(l)}(x)\,dx \geqslant \kappa \int_{\sigma'} \sum_{|k|=r} f^{(k)}(x)^2\,dx \tag{6}$$

for all f such that these integrals make sense. Further, let $U(x) \in W_2^r(\sigma')$ satisfy the condition

$$\int_{\sigma'} \Big\{ \sum_{|k|,|l|\leqslant r} a_{kl}(x) U^{(k)} v^{(l)}(x) - F(x) v(x) \Big\}\,dx = 0, \quad \forall v \in C_0^\infty(\sigma') \tag{7}$$

(*see Remark* 1).

Then in fact U belongs to $W_2^{r+\gamma}(\sigma)$ and satisfies the inequality

$$\|U\|_{W_2^{r+\gamma}(\sigma)} \leqslant c\Big(\|U\|_{W_2^r(\sigma')} + \|F\|_{W_2^{\gamma-r}(\sigma')}\Big). \tag{8}$$

This lemma was proved in §2 of our article [5] for $\gamma \geqslant r$. We focus attention on a proof of it when $0 \leqslant \gamma < r$. Recall that in the case $0 < \gamma \leqslant r$ the symbol

$$\int_{\sigma'} F(x) v(x)\,dx$$

denotes here the value of the functional $F \in \overset{*}{V}{}_2^{\gamma-r}(\sigma')$ at an element $v \in \overset{0}{W}{}_2^{r-\gamma}(\sigma')$.

REMARK 4. The validity of (7) for all $v \in C_0^\infty(\sigma')$ implies its validity for all $v \in \overset{0}{W}{}_2^r(\sigma')$ if it is assumed that the functional $F(v)$ is extended by continuity in the $W_2^{r-\gamma}(\sigma')$-metric to $\overset{0}{W}{}_2^{r-\gamma}(\sigma')$. This extension is possible and then is automatically unique. In fact, $F(v)$ is (by a condition of the lemma) bounded at $v \in C_0^\infty(\sigma')$ with the norm $\|v\|_{\overset{0}{W}{}_2^{r-\gamma}(\sigma')}$; therefore, it can be extended by continuity in the $W_2^{r-\gamma}(\sigma')$-metric to $\overset{0}{W}{}_2^{r-\gamma}(\sigma')$. But $\overset{0}{W}{}_2^{r-\gamma}(\sigma') \supset \overset{0}{W}{}_2^r(\sigma')$, and so the extended functional $F(v)$ is bounded on $\overset{0}{W}{}_2^r(\sigma')$, and it is then bounded (continuous) on $\overset{0}{W}{}_2^r(\sigma')$, as is clear from the inequalities

$$\begin{aligned} |F(v_1) - F(v_2)| = |F(v_1 - v_2)| &\leqslant M\|v_1 - v_2\|_{\overset{0}{W}{}_2^{r-\gamma}(\sigma')} \\ &\leqslant cM\|v_1 - v_2\|_{\overset{0}{W}{}_2^r(\sigma')}, \quad \forall v_1, v_2 \in \overset{0}{W}{}_2^r(\sigma). \end{aligned}$$

If now $v \in \overset{0}{W}{}_2^r(\Omega)$, $v_\mu \in \mathcal{D}\,(\mu = 1, 2, \ldots)$, and $\|v_\mu - v\|_{\overset{0}{W}{}_2^r(\sigma')} \to 0$ as $\mu \to \infty$, then

$$\left| \int_{\sigma'} \sum_{|k|, |l| \leqslant r} a_{kl} U^{(k)} v^{(l)}\, dx - F(v) \right|$$

$$= \left| \int_{\sigma} \sum_{|k|, |l| \geqslant r} a_{kl} U^{(k)} (v - v_\mu)^{(l)}\, dx - F(v - v_\mu) \right|$$

$$\ll \sum_{|k|, |l| \leqslant r} \|U^{(k)}\|_{L_2(\sigma)} \left\| (v - v_\mu)^{(l)} \right\|_{L_2(\sigma')} + cM \|v - v_\mu\|_{W_2^r(\sigma')}$$

$$\ll \sum_{|k|, |l| \leqslant r} \|U\|_{\overset{0}{W}{}_2^r(\sigma')} \|v - v_\mu\|_{\overset{0}{W}{}_2^r(\sigma')} + cM \|v - v_\mu\|_{W_2^r(\sigma')} \to 0, \qquad \mu \to \infty,$$

i.e., the first expression in this chain is equal to zero.

The proof of the lemma does not differ on the whole from the proof of a similar lemma in [5], where the case $\gamma \geqslant r$ was completely taken care of, and a rough estimate with $\|F\|_{L_2(\Omega)}$ instead of $\|F\|_{W_2^{\gamma - r}(\sigma')}$ was obtained in the case $\gamma < r$. But when $\gamma < r$

$$\|F\|_{W_2^{\gamma - r}(\sigma')} \ll \|F\|_{W_2^0(\sigma')} = \|F\|_{L_2(\sigma')};$$

see (2).

Let $\eta(x) \in C^{r+\gamma}(R_n)$ be a function with compact support in σ' and such that: $\eta \equiv 1$ on σ, $0 \leqslant \eta(x) \leqslant 1$ $(x \in R_n)$, and $|\eta^{(\tau)}(x)|^2 \leqslant c\eta(x)$ for $|\tau| \leqslant r + \gamma$, where c does not depend on x.

We consider functions $v \in \mathfrak{M}_0(\sigma')$ with compact support in σ'. For sufficiently small h such functions satisfy the inequality (which follows from (7))

$$\int_\Omega \left\{ \sum_{|k|, |l| \leqslant r} \Delta_h^p \left(a_{kl} U^{(k)}(x) \right) v^{(l)} - F(x) \Delta_{-h}^p v(x) \right\} dx = 0,$$

where $p = (p_1, \ldots, p_n)$, $|p_n| = \gamma$ is a nonnegative integer vector, $hp = (hp_1, \ldots, hp_n)$ is the corresponding increment at the point $x = (x_1, \ldots, x_n)$, and $\Delta^p F$ is the pth-order difference of F at the point x with step h. If

$$v(x) = \eta(x) \Delta^p U(x) / h^{|p|}, \qquad |p| = \gamma,$$

then $v \in \mathfrak{M}_0(\sigma')$ for sufficiently small h. Substituting this function in (7), we get, after transposing some terms from the left-hand side to the right-hand side, that

$$I = \int_{\sigma'} \sum_{|k|, |l| = r} a_{kl}(x + ph) \left(\frac{\Delta^p U}{h^{|p|}} \right)^{(k)} \left(\frac{\Delta^p U}{h^{|p|}} \right)^{(l)} \eta\, dx$$

$$= -{\sum}' I_{k, l, \lambda, \tau} + \int_{\sigma'} F \frac{\Delta_{-h}^p}{h^{|p|}} \left(\frac{\Delta^p U}{h^{|p|}} \eta \right) dx. \tag{9}$$

By (6),

$$I \geqslant \kappa \int_{\sigma'} \sum_{|k| = r} \left(\frac{\Delta^p U^{(k)}}{h^{|p|}} \right)^2 \eta\, dx = X^2. \tag{10}$$

It was shown in [5] that the terms in the sum Σ' can be estimated above as follows:

$$|I_{k, l, \lambda, \tau}| \ll \|U\|_{W_2^{r+\gamma-1}(\sigma')} \left(X + \|U\|_{W_2^{r+\gamma-1}(\sigma')} \right). \tag{11}$$

4.3. As for the last number on the right-hand side of (9), for $\gamma < r$ we can now get a sharper estimate of it than was obtained in [5].

Recalling that the indicated term is a value of the functional $F \in W_2^{\gamma-r}(\sigma')$, we write

$$\left| \int_{\sigma'} F \frac{1}{h^p} \Delta_{-h}^p \left(\frac{\Delta_h^p U}{h^{|p|}} \eta \right) dx \right| \leqslant \|F\|_{W_2^{\gamma-r}(\sigma')} \|w_h\|_{W_2^{r-\gamma}(\sigma')}, \tag{12}$$

where the function $w_h(x)$ (which has compact support in σ') is defined by

$$w_h(h) = \frac{1}{h^{|p|}} \Delta_{-h}^p \left(\frac{\Delta^p U}{h^{|p|}} \eta \right).$$

Any partial derivative $w_h^{(\lambda)}(x)$ with $|\lambda| \leqslant r - \gamma - 1$ can be estimated as follows. Assume that the function U has derivatives up to order $r + \gamma - 1$. Using Proposition A2 in [5], we get that

$$T_\lambda \equiv \left(\int_{\sigma'} \left| w_h^\lambda(x) \right|^2 dx \right)^{1/2} = \left(\int_{\sigma'} \left| \frac{1}{h^{|p|}} \Delta_{-h}^p \left(\frac{\Delta_h^p U}{h^{|p|}} \eta \right)^{(\lambda)} \right|^2 dx \right)^{1/2}$$

$$\leqslant \left(\int_{\sigma'} \left| \left(\frac{\Delta_h^p U}{h^{|p|}} \eta \right)^{(\lambda+p)} \right|^2 dx \right)^{1/2}.$$

It is now clear that T_λ is majorized by the norm $\|U\|_{W_2^{r+\gamma-1}(\sigma')}$ (use the Leibniz formula, the boundedness of the derivatives of η, and again Proposition A2 from [5]).

Unlike in the preceding case, an additional term of the form

$$\left(\int_{\sigma'} \left(\frac{\Delta_h^p U^{\lambda+p}}{h^{|p|}} \right)^2 \eta^2 \, dx \right)^{1/2} \ll \left(\int_{\sigma'} \left(\frac{\Delta^p U^k}{h^{|p|}} \right)^2 \eta \, dx \right)^{1/2}$$

which can be estimated in terms of X appears when a derivative $w_h^{(\lambda)}(x)$ with $|\lambda| = r - \gamma$ is estimated (we have used the estimate $\eta^2 \ll \eta$ and have set $p + \lambda = k$, $|k| = r$).

Thus, in this case T_λ is majorized by the quantity $(X + \|U\|_{W_2^{r+\gamma-1}(\sigma')})$.

Beginning from the foregoing, we finally arrive at the inequality

$$\|w_h\|_{W_2^{r-\gamma}(\sigma')} \ll X + \|U\|_{W_2^{r+\gamma-1}(\sigma')}, \tag{13}$$

which upon substitution in (12) completes the estimate of the term of interest to us.

It now follows from (9)–(11) and (13) that

$$X^2 \ll \left(\|U\|_{W_2^{r+\gamma-1}(\sigma')} + \|F\|_{W_2^{\gamma-r}(\sigma')} \right) X + \|U\|_{W_2^{r+\gamma-1}(\sigma')} \left(\|U\|_{W_2^{r+\gamma-1}(\sigma')} + \|F\|_{W_2^{\gamma-r}(\sigma')} \right).$$

But then ($\eta(x) = 1$ on σ, and $\eta(x) \geqslant 0$!)

$$\left(\kappa \int_\sigma \sum_{|k|=r} \left(\frac{\Delta^p U^{(k)}}{h^{|p|}} \right)^2 dx \right)^{1/2} \leqslant X \leqslant c \left(\|U\|_{W_2^{r+\gamma-1}(\sigma')} + \|F\|_{W_2^{\gamma-r}(\sigma')} \right),$$

where c does not depend on h nor the expression in parentheses on the right-hand side.

In this case we have (see Corollary A1 in [5]) the inequality

$$\left(\int_\sigma \sum_{|k|=r} [U^{(\alpha+p)}(x)]^2\,dx\right)^{1/2} \leqslant c_1\left(\|U\|_{W_2^{r+\gamma-1}(\sigma')} + \|F\|_{W_2^{\gamma-r}(\sigma')}\right),$$

from which, using the fact that p is an arbitrary nonnegative integer vector with $|p| = \gamma$, we get

$$\|U\|_{W_2^{r+\gamma}(\sigma)} \ll \|U\|_{W_2^{r+\gamma-1}(\sigma')} + \|F\|_{W_2^{\gamma-r}(\sigma')}. \tag{14}$$

We now prove inequality (8) for $0 \leqslant \gamma \leqslant r$. It is trivial for $\gamma = 0$. For $0 < \gamma \leqslant r$ we change the notation σ' to σ_0 and σ to σ_γ and construct a strictly increasing system of concentric balls

$$\sigma = \sigma_\gamma \subset \sigma_{\gamma-1} \subset \cdots \subset \sigma_1 \subset \sigma_0 = \sigma'.$$

On the basis of (14) and (2),

$$\|U\|_{W_2^{r+\gamma}(\sigma_\gamma)} \ll \|U\|_{W_2^{r+\gamma-1}(\sigma_{\gamma-1})} + \|F\|_{W_2^{\gamma-r}(\sigma_{\gamma-1})} \ll \|U\|_{W_2^{r+\gamma-2}(\sigma_{\gamma-2})} + \|F\|_{W_2^{\gamma-r}(\sigma_{\gamma-2})}$$

$$\ll \cdots \ll \|U\|_{W_2^{r+1}(\sigma_1)} + \|F\|_{W_2^{\gamma-r}(\sigma_1)} \ll \|U\|_{W_2^{r}(\sigma_0)} + \|F\|_{W_2^{\gamma-r}(\sigma_0)},$$

and we have inequality (8) for $0 \leqslant \gamma \leqslant r$.

§5. Linear functionals on weighted function spaces

Some facts about weighted spaces of vector-valued functions are given in the beginning of the section (proof are also presented for completeness). These facts are used to prove the central result in the section (Theorem 2) on representing a linear functional on a weighted space of smooth functions.

5.1. Let $L_{p,\beta}$ be the space of functions φ with finite norm

$$\|\varphi/\beta^\beta\|_p\|\varphi(x)/\rho^\beta(x)\|_{L_p(\Omega)}, \qquad 1 < p < \infty,$$

and let $F(\varphi)$ be a linear functional on $L_{p,\beta}$. We introduce the new functional

$$\Phi(\Psi) = F(\rho^\beta\Psi), \qquad \Psi \in L_p(\Omega).$$

This functional is linear on $L_p(\Omega)$ and is bounded:

$$|\Phi(\Psi)| = |F(\rho^\beta\Psi)| \leqslant \|F\|\cdot\|\rho^\beta\Psi/\rho^\beta\|_p \leqslant \|F\|\cdot\|\Psi\|_p.$$

But then there exists a unique function $\lambda(x) \in L_{p'}(\Omega)$ $(1/p + 1/p' = 1)$ such that

$$\Phi(\Psi) = \int_\Omega \lambda(x)\psi(x)\,dx, \quad \forall\Psi \in L_p,$$

or

$$F(\rho^\beta\Psi) = \int_\Omega \lambda(x)\Psi(x)\,dx, \quad \forall\Psi \in L_p.$$

Setting $\rho^\beta\Psi = \varphi(x) \in L_{p,\beta}$, we get that

$$F(\varphi) = \int_\Omega \frac{\lambda(x)}{\rho(x)^\beta}\varphi(x)\,dx = \int_\Omega \mu(x)\varphi(x)\,dx, \quad \forall\varphi \in L_{p,\beta}.$$

The function $\mu(x) = \lambda(x)/\rho(x)^\beta$ has the property

$$\|\mu\rho^\beta\|_{p'} = \|\lambda(x)\|_{p'} < \infty.$$

Thus, to every linear functional $F(\varphi)$, $\varphi \in L_{p,\beta}$, there corresponds a (unique) function $\mu(x) \in L_{p',-\beta}$ with the properties

$$F(\varphi) = \int_\Omega \mu(x)\varphi(x)\,dx, \quad \forall\varphi \in L_{p,\beta}, \qquad \|F\| = \|\mu\rho^\beta\|_{p'}.$$

It is obviously that, conversely, the quantity

$$\int_\Omega \mu(x)\varphi(x)\,dx = F(\varphi), \qquad \mu \in L_{p',-\beta},$$

is a linear functional defined on $L_{p,\beta}$ and with norm

$$\|F\| = \sup_{\|\varphi\|_{L_{p,\beta}} \leqslant 1} F(\varphi) = \|\mu\rho^\beta\|_{p'}.$$

Thus, $L_{p',-\boldsymbol{\beta}}$ is the space dual to $L_{p,\beta}(\Omega)$; that is,

$$\left(L_{p,\beta}(\Omega)\right)^* = L_{p',-\beta}, \qquad 1 < p < \infty, 1/p + 1/p' = 1.$$

5.2. We now introduce the vector $\boldsymbol{\beta} = (\beta_1,\ldots,\beta_N)$, where $\beta_1,\ldots,\beta_N$ are real numbers, and we let $L_{p,\boldsymbol{\beta}}$ stand for the space of vector-valued functions $\boldsymbol{\varphi} = (\varphi_1,\ldots,\varphi_N)$ defined on Ω and with norm

$$\|\boldsymbol{\varphi}\|_{p,\boldsymbol{\beta}} = \left(\sum_{\nu=1}^N \left\|\frac{\varphi_\nu}{\rho^{\beta_\nu}}\right\|_p^p\right)^{1/p} < \infty.$$

If $\boldsymbol{\mu}(x) = (\mu_1(x),\ldots,\mu_N(x)) \in L_{p',-\boldsymbol{\beta}}$, then the expression

$$F(\boldsymbol{\varphi}) = \int_\Omega \boldsymbol{\mu}(x)\boldsymbol{\varphi}(x)\,dx = \sum_{j=1}^N \int_\Omega \mu_j(x)\varphi_j(x)\,dx \tag{1}$$

is a linear functional on $L_{p,\boldsymbol{\beta}}$ with norm

$$\|F\| = \sup_{\|\boldsymbol{\varphi}\|_{p,\boldsymbol{\beta}} \leqslant 1} F(\boldsymbol{\varphi}) = \left(\sum_1^N \|\mu_\nu\rho^{\beta_\nu}\|_{p'}^{p'}\right)^{1/p'}. \tag{2}$$

Indeed, $F(\boldsymbol{\varphi})$ makes sense for any system $\boldsymbol{\varphi} \in L_{p,\boldsymbol{\beta}}$ and depends linearly on $\boldsymbol{\varphi}$; and

$$|F(\boldsymbol{\varphi})| \leqslant \sum_{j=1}^N \int_\Omega \left|\mu_j\rho^{\beta_j}\frac{\varphi_j}{\rho^{\beta_j}}\right| dx \leqslant \sum_{j=1}^N \|\mu_j\rho^{\beta_j}\|_{p'}\left\|\frac{\varphi_j}{\rho^{\beta_j}}\right\|_p$$

$$\leqslant \left(\sum_{j=1}^N \|\mu_j\rho^{\beta_j}\|_{p'}^{p'}\right)^{1/p'}\left(\sum_{j=1}^N \left\|\frac{\varphi_j}{\rho^{\beta_j}}\right\|_p^p\right)^{1/p} = \|\boldsymbol{\mu}\|_{p',-\boldsymbol{\beta}}\|\boldsymbol{\varphi}\|_{p,\boldsymbol{\beta}}.$$

Equality here is attained at the function $\boldsymbol{\varphi} = (\varphi_1,\ldots,\varphi_N)$ with components

$$\varphi_j = \rho^{\beta_j}(x)\left|\rho^{\beta_j}(x)\mu_j(x)\right|^{p'/p} \Big/ \left(\textstyle\sum_1^N \int_\Omega |\mu_\nu\rho^{\beta_\nu}|^{p'}\,dx\right)^{1/p} \qquad (j = 1,\ldots,N).$$

In fact,

$$\|\varphi\|_{p,\beta} = \left(\sum_1^N \left\| \frac{\varphi_\nu}{\rho^{\beta_\nu}} \right\|_p^p \right)^{1/p} = 1,$$

$$F(\varphi) = \sum_{\nu=1}^N \int_\Omega \mu_\nu \varphi_\nu \, dx = \frac{1}{\left(\sum_1^N \int_\Omega |\mu_\nu, \rho^{\beta_\nu}|^{p'} dx\right)^{1/p}} \sum_{\nu=1}^N \int_\Omega |\mu_\nu \rho^{\beta_\nu}|^{p'} dx$$

$$= \left(\sum_{\nu=1}^N \int_\Omega |\mu_\nu \rho^{\beta_\nu}|^{p'} dx \right)^{1/p'} = \|\mu\|_{p',-\beta}.$$

Conversely, an arbitrary (bounded) linear functional $F(\varphi)$ on $L_{p,\beta}$ can be represented as a sum (1) with a uniquely determined function $\mu(x) = (\mu_1(x), \ldots, \mu_N(x))$ in $L_{p',-\beta}$.

Indeed, a linear functional $F(\varphi)$ on the space $L_{p,\beta}$ of vector-valued functions $\varphi = (\varphi_1, \ldots, \varphi_N)$ generates for each $j = 1, \ldots, N$ a linear functional $F_j(\varphi_j)$ which can be written as an integral

$$F_j(\varphi_j) = \int_\Omega \mu_j(x) \varphi_j(x) \, dx. \tag{3}$$

But then for any system of functions φ

$$F(\varphi) = \sum_{j=1}^N F_j(\varphi_j) = \sum_{j=1}^N \int_\Omega \mu_j(x) \varphi_j(x) \, dx, \tag{4}$$

where

$$\left\| \mu_j(x) \rho(x)^{\beta_j} \right\|_{p'} < \infty \qquad (j = 1, \ldots, N). \tag{5}$$

5.3. We introduce the class $V_{2,\beta}^r(\Omega)$ of functions $f(x)$ on Ω, with the norm

$$\|f\|_{V_{2,\beta}^r} = \left(\sum_{|k| \leqslant r} \left\| \frac{f^{(k)}(x)}{\rho(x)^{\beta_k}} \right\|_{L_2(\Omega)}^2 \right)^{1/2}, \tag{6}$$

where the β_k are given real numbers indexed by nonnegative integer vectors k.

REMARK 5. It can be assumed that $V_{2,\beta}^r$ is the class of infinitely differentiable functions f with compact support in Ω ($f \in \mathcal{D}$), equipped with the norm (6). It is just this case that will be used subsequently in §6.

Obviously, $V_{2,\beta}^r$ can be regarded as a subspace of the space $L_{2,\beta}$ of systems of functions

$$\varphi(x) = \{\varphi_k(x)\} \qquad (|k| \leqslant r), \tag{7}$$

with the norm

$$\|\varphi\|_{2,\beta} = \left(\sum_{|k| \leqslant r} \left\| \frac{\varphi_k(x)}{\rho(x)^{\beta_k}} \right\|^2 \right)^{1/2}. \tag{8}$$

THEOREM 2. *Every linear functional $F(f)$ on $V_{2,\beta}^r$ can be represented in the form*

$$F(f) = \sum_{|k| \leqslant r} \int_\Omega \mu_k(x) f^{(k)}(x) \, dx \qquad (f \in V_{2,\beta}^r), \tag{9}$$

where the functions $\mu_k(x)$ *have the property*

$$\left\|\rho(x)^{\beta_k}\mu_k(x)\right\|_{L_2(\Omega)} < \infty \qquad (\|k\| \leqslant r). \tag{10}$$

Moreover,

$$\|F\| = \left(\sum_{|k|\leqslant r} \left\|\rho(x)^{\beta_k}\mu_k(x)\right\|^2_{L_2(\Omega)}\right)^{1/2}. \tag{11}$$

Conversely, if

$$\left(\sum_{|k|=r} \left\|\rho(x)^{\beta_k}\mu_k(x)\right\|^2_{L_2(\Omega)}\right)^{1/2} < \infty, \tag{12}$$

then the sum (9) *is a linear functional on* $V^r_{2,\beta}$ *with norm*

$$\|F\| = \sup_{\|f\|_{V^r_{2,\beta}}\leqslant 1} F(f) \leqslant \left(\sum_{|k|\leqslant r} \left\|\rho(x)^{\beta_k}\mu_k(x)\right\|^2_{L_2(\Omega)}\right)^{1/2}. \tag{13}$$

REMARK 6. If the functional (9) is considered only for functions $f \in \mathscr{D}$, then it can be regarded as a generalized function $F(x) \in \mathscr{D}'$ expressed explicitly by the formula

$$F(x) = \sum_{|k|\leqslant r} (-1)^{|k|}\mu_k^{(k)}(x),$$

where $\mu_k^{(k)}(x) \in \mathscr{D}'$ is the kth ($k = (k_1,\dots,k_n)$) derivative of the locally integrable function $\mu_k(x)$.

PROOF. We introduce the auxiliary space $L_{2,\beta}$ of systems of functions φ (see (7)), with the norm (8).

Each function $f \in V^r_{2,\beta}$ generates a definite system of functions $\{f^{(k)}(x)\}$, $|k| \leqslant r$, which can be regarded as an element φ of the space $L_{2,\beta}$, $\varphi = \{\varphi_k\}$, where $\varphi_k(x) = f^{(k)}(x)$, $|k| \leqslant r$, with norm (6)

$$\|\varphi\|_{L_2,\beta} = \left(\sum_{|k|\leqslant r} \left\|\frac{f^{(k)}}{\rho^{\beta_k}}\right\|^2_{L_2(\Omega)}\right)^{1/2} \equiv \left(\sum_{|k|\leqslant r} \left\|\frac{\varphi_k}{\rho^{\beta_k}}\right\|^2_{L_2(\Omega)}\right)^{1/2}.$$

In this sense $V^r_{2,\beta}$ is a subspace of $L_{2,\beta}$. By the Hahn-Banach theorem, the linear functional $F(f)$ on $V^r_{2,\beta}$ can be extended with preservation of norm to $L_{2,\beta}$, and then $F(\varphi)$ can (on the basis of statements in §5.2) be represented in the form

$$F(\varphi) = \sum_{|k|\leqslant r} \mu_k(x)\varphi_k(x),$$

and (11) holds. But then our linear functional $F(f)$, $f \in V^r_{2,\beta}$, can be written in the form (9), and its norm in the form (11).

On the other hand, it follows from (9) and (10) that

$$|F(f)| \leqslant \sum_{|k|\leqslant r} \left|\int_\Omega \mu_k\rho^{\beta_k}\frac{f^{(k)}}{\rho^{\beta_k}}\,dx\right| \leqslant \sum_{|k|\leqslant r} \left\|\mu_k\rho^{\beta_k}\right\|_{L_2(\Omega)}\left\|\frac{f^{(k)}}{\rho^{\beta_k}}\right\|_{L_2(\Omega)}$$

$$\leqslant \left(\sum_{|k|\leqslant r} \left\|\mu_k\rho^{\beta_k}\right\|^2_{L_2(\Omega)}\right)^{1/2}\left(\sum_{|k|\leqslant r} \left\|\frac{f^{(k)}}{\rho^{\beta_k}}\right\|^2_{L_2(\Omega)}\right)^{1/2},$$

and this shows that (9) is a bounded linear functional on $V^r_{2,\beta}$.

REMARK 7. The space $V^r_{2,\beta}(\Omega)$ with compatible weights we introduced in §2 is also contained among the spaces $V^r_{2,\beta}(\Omega)$. In this case the weight for the highest derivatives $D^k f$, $|k| = r$, is equal to $\rho^{-\beta}$ and varies with decreasing $|k|$ according to the law $D^k f \in L_{2,r+\beta-|k|}$. By Theorem 2, a bounded linear functional F on $V^r_{2,\beta}(\Omega)$ can be represented in the form

$$F(f) = \sum_{|k|\leqslant r} \int_\Omega \mu_k(x) f^{(k)}(x)\,dx, \tag{14}$$

where

$$\mu_k \in L_{2,-r-\beta+|k|}(\Omega), \qquad |k| \leqslant r.$$

In view of Remark 6 the functional can be written (formally) in the form

$$F(f) = \int_\Omega F(x) f(x)\,dx,$$

where

$$F(x) = \sum_{|k|\leqslant r} (-1)^{|k|} \mu_k^{(k)}(x) \in \mathscr{D}'(\Omega). \tag{15}$$

The collection of generalized functions $F(x)$ obtained in this way, with the norm

$$\|F\|_{(V^r_{2,\beta}(\Omega))^*} = \sum_{|k|\leqslant r} \|\rho^{r+\beta-|k|}\mu_k\|_{L_2(\Omega)},$$

forms the dual space $(V^r_{2,\beta}(\Omega))^*$, which is conveniently denoted by $V^{-r}_{2,-\beta}(\Omega)$.

The two-parameter scale of spaces

$$V^m_{2,\alpha}(\Omega) = \begin{cases} V^m_{2,\alpha}(\Omega) & \text{for } m \geqslant 0, \\ (V^{-m}_{2,-\alpha}(\Omega))^* & \text{for } m < 0 \end{cases}$$

is thus defined for any integers m and real numbers α.

Note that the space $H_{r,\gamma}(\Omega)$ (see (14) in §2.3) coincides with $V^{\gamma-r}_{2,-\alpha-\gamma}(\Omega)$. We use this in §7 in extending Theorem 1 to arbitrary values of α.

Note also the imbedding

$$\overset{*}{V}{}^{r-\gamma}_{2,\alpha+\gamma}(\Omega) \to V^r_{2,\alpha}(\Omega), \qquad 0 \leqslant \gamma \leqslant r, \tag{16}$$

which obviously follows from the imbedding

$$V^r_{2,\alpha}(\Omega) \to V^{r-\gamma}_{2,\alpha}(\Omega), \qquad 0 \leqslant \gamma \leqslant r.$$

§6. A general coercive estimate

This section deals with the following theorem.

THEOREM 3. *Suppose that $U(x) \in W^r_{2,\alpha}(\Omega)$ satisfies the equality*

$$\int_\Omega \Big\{ \sum_{|k|,|l|\leqslant r} a_{kl}(x) U^{(k)}(x) v^{(l)}(x) - F(x) v(x) \Big\}\,dx = 0, \quad \forall v \in \mathscr{D}, \tag{1}$$

let γ be a nonnegative number, and suppose that the $a_{kl}(x)$ and the function $F(x) \in \mathscr{D}'$, which is assumed to belong to $H_{r,\gamma}$ ($F \in H_{r,\gamma}$; see §2.3), *satisfy the conditions* (3) *and* (4) *in* §1.

Then in fact $U \in W^{r+\gamma}_2(\Omega)$ locally, and if the additional condition

$$\|U/\rho^{r+\alpha}\|_{L_2(\Omega)} < \infty \tag{2}$$

holds, then

$$\|U\|_{W^{r+\gamma}_{2,\alpha-\gamma}(\Omega)} \leqslant c\Big(\|U/\rho^{r+\alpha}\|_{L_2(\Omega)} + \|U\|_{w^r_{2,\alpha}(\Omega)} + \|F\|_{H_{r,\gamma}(\Omega)}\Big). \tag{3}$$

REMARK 8. In Theorem 3 the domain Ω is arbitrary, and the real number α can also be arbitrary, as long as condition (2) holds.

Theorem 3 was proved for $\gamma \geqslant r$ in our paper [5] (Theorem 3). Here we consider the case $0 \leqslant \gamma < r$.

PROOF. By a condition of the theorem, $F(x) \in \mathscr{D}'$ represents a linear functional (F, v) on the space $V^{r-\gamma}_{2,\alpha+\gamma}(\Omega)$ of functions $v \in \mathscr{D}$, with the norm

$$\|v\|_{V^{r-\gamma}_{2,\alpha+\gamma}(\Omega)} = \left(\sum_{|k|\leqslant r-\gamma}\left\|\frac{v^{(k)}}{\rho^{r+\alpha-|k|}}\right\|^2_{L_2(\Omega)}\right)^{1/2}.$$

By Theorem 2 (see Remark 6), there exist functions $\mu_k(x)$ ($|k| \leqslant r-\gamma$), such that

$$(F, v) = \sum_{|k|\leqslant r-\gamma}\int_\Omega \mu_k(x)v^{(k)}(x)\,dx,$$

$$\|F\|_{H_{r,\gamma}} = \|F\|_{\mathring{V}^{r-\gamma}_{2,\alpha+\gamma}(\Omega)} = \left(\sum_{|k|\leqslant r-\gamma}\|\rho^{r+\alpha-|k|}\mu_k(x)\|^2_{L_2(\Omega)}\right)^{1/2} < \infty.$$

Therefore, (1) can be rewritten as follows:

$$\int_\Omega \sum_{|k|,|l|\leqslant r} a_{kl}(x)U^{(k)}(x)v^{(l)}(x)\,dx = \int_\Omega \sum_{|l|\leqslant r-\gamma}\mu_l(x)v^{(l)}(x)\,dx, \quad \forall v \in \mathscr{D}. \tag{1'}$$

Let $x^0 \in \Omega$ be an arbitrary point, and denote by $I(x^0)$ the ball about it of radius $\rho(x^0)/2$. We confine ourselves to the functions $v \in \mathscr{D}\,(I(x^0)) \subset \mathscr{D}$, that are infinitely differentiable and compactly supported in $I(x^0)$. Then

$$\int_{I(x^0)} \sum_{|k|,|l|\leqslant r} a_{kl}(x)U^{(k)}(x)v^{(l)}(x)\,dx = \int_{I(x^0)} \sum_{|l|\leqslant r-\gamma}\mu_l(x)v^{(l)}(x)\,dx,$$
$$\forall v \in \mathscr{D}(I(x^0)). \tag{4}$$

In the integrals in (4) let us make the substitution

$$x = x^0 + (\rho(x^0)/2)(\xi - x^0) \qquad (\xi \in I(x^0)),$$

which implements a one-one-one continuously differentiable correspondence

$$I(x^0) \ni x \rightleftarrows \xi \in J(x^0),$$

where $J(x^0)$ is the ball of radius 1 about x^0 in the space of points $\xi = (\xi_1, \ldots, \xi_n)$.

We assume that

$$\Psi_*(\xi) = \Psi(x + (\rho(x^0)/2)(\xi - x^0)).$$

The equality (4), after multiplication by $[\rho(x^0)/2]^{2(r+\alpha)-n}$, becomes

$$\int_{J(x^0)} \sum_{|k|,|l|\leqslant r} b_{kl}(\xi)U^{(k)}_*(\xi)v^{(l)}_*(\xi)\,d\xi$$
$$= \int_{J(x^0)} \sum_{|l|\leqslant r-\gamma}\mu^*_l(\xi)v^{(l)}_*(\xi)\left[\frac{\rho(x^0)}{2}\right]^{2(r+\alpha)-|l|}d\xi, \tag{5}$$

where

$$b_{kl}(\xi) = a^*_{kl}(\xi)[\rho(x^0)/2]^{2(r+\alpha)-|k|-|l|}. \tag{6}$$

Here the $v_*(\xi)$ are arbitrary infinitely differentiable functions with compact support in the unit ball $J(x^0)$, and the $b_{kl}(\xi)$ have the properties

$$|b^{(\lambda)}_{kl}(\xi)| = |a^{(\lambda)}_{kl}(x)|[\rho(x^0)/2]^{|\lambda|+2(r+\alpha)-|k|-|l|} \leqslant 1, \quad \xi \in J(x^0), |\lambda| \leqslant \gamma, \tag{7}$$

$$\int_{J(x^0)} \sum_{|k|,|l|\leqslant r} b_{kl}(\xi) f^{(k)}_*(\xi) f^{(l)}_*(\xi)\, d\xi \geqslant \kappa \int_{J(x^0)} \sum_{|k|=r} [f^{(k)}_*(\xi)]^2\, d\xi \tag{8}$$

(see §4 in [5] for details). This shows that the operator of the same type as the operator in (1) of §1, but corresponding to the coefficients $b_{kl}(\xi)$, has an elliptic character (without degeneracy). Moreover,

$$\begin{aligned}
&\left|\int_{J(x^0)} \sum_{|l|\leqslant r-\gamma} \mu^*_l(\xi) v^{(l)}_*(\xi)\, d\xi\right| \left[\frac{\rho(x^0)}{2}\right]^{2(r+\alpha)-|l|} \\
&\quad \ll \sum_{|l|\leqslant r-\gamma} \|\mu^*_l(\xi)\rho(x^0)^{-|l|}\|_{L_2(J(x^0))} \|v^{(l)}_*(\xi)\|_{L_2(J(x^0))} \rho(x^0)^{2(r+\alpha)} \\
&\quad \ll \rho(x^0)^{2(r+\alpha)} \left(\sum_{|l|\leqslant r-\gamma} \|\mu^*_l(\xi)\rho(x^0)^{-|l|}\|^2_{L_2(J(x^0))} \right)^{1/2} \|v_*\|_{W^{r-\gamma}_2(J(x^0))},
\end{aligned} \tag{9}$$

because

$$\|v^{(l)}_*(\xi)\|_{L_2(\Omega)} \ll \|v_*\|_{W^{r-\gamma}_2(J(x^0))}, \qquad |l| \leqslant r-\gamma.$$

Inequality (9) shows that the linear functional $F_*(v_*)$ (of $v_* \in \mathscr{D}(J(x^0))$ on the right-hand side of (5) is bounded on the space of functions $v_* \in \mathscr{D}(J(x^0))$ for which the norm $\|v_*\|_{W^{r-\gamma}_2(J(x^0))}$ was introduced. Moreover, the norm of this functional can be estimated as follows:

$$\|F_*\|_{W^{\gamma-r}_2(J(x^0))} \ll \rho(x^0)^{2(r+\alpha)} \left(\sum_{|l|\leqslant r-\gamma} \|\mu^*_l \rho(x^0)^{-|l|}\|^2_{L_2(J(x^0))} \right)^{1/2}. \tag{10}$$

By (5), (7), (8), and (10), Lemma 1 gives us that in fact

$$U_*(\xi) \in W^{r+\gamma}_2(J^{1/2}(x^0)),$$

and

$$\begin{aligned}
&\sum_{|k|=r+\gamma} \left(\int_{J^{1/2}(x^0)} U^{(k)}_*(\xi)^2\, d\xi \right)^{1/2} \\
&\quad \ll \left(\int_{J(x^0)} U_*(\xi)^2\, d\xi \right)^{1/2} + \sum_{|k|=r} \left(\int_{J(x^0)} U^{(k)}_*(\xi)^2\, d\xi \right)^{1/2} \\
&\quad\quad + \rho(x^0)^{2(r+\alpha)} \left(\sum_{|l|\leqslant r-\gamma} \|\mu^*_l(\xi)\rho(x^0)^{-|l|}\|^2_{L_2(J(x^0))} \right)^{1/2}.
\end{aligned}$$

We divide this equality by $\rho(x^0)^{r+\alpha}$ and pass from ξ to x in the integrals (explanation below):

$$\sum_{|k|=r+\gamma} \rho(x^0)^{-\alpha+\gamma-n/2}\|U^{(k)}(x)\|_{L_2(I^{1/2}(x^0))}$$
$$\ll \rho(x^0)^{-r-\alpha-n/2}\|U(x)\|_{L_2(I(x^0))} + \sum_{|k|=r} \rho(x^0)^{-\alpha-n/2}\|U^{(k)}(x)\|_{L_2(I(x^0))}$$
$$+\rho(x^0)^{-n/2}\Big(\sum_{|l|\leqslant r-\gamma} \|\mu_l(x)\rho(x)^{r+\alpha-|l|}\|^2_{L_2(I(x^0))}\Big)^{1/2} \qquad (11)$$

(see §3 in [5]). The inequality (11) is preserved with a different (in general) constant in the relation $\ll$ when all its terms are squared and integrated over Ω. It now remains to use an inequality of Troisi (see [5], §3), namely,

$$\int_\Omega \left(\rho^{s-n/2}\|f(x)\|_{L_2(I(x^0))}\right)^2 dx \sim \int_\Omega \left(\rho^s(x)|f(x)|\right)^2 dx,$$

in order to get the inequality

$$\left(\sum_{|k|=r+\gamma}\int_\Omega\left(\frac{U^{(k)}(x)}{\rho(x)^{\alpha-\gamma}}\right)^2 dx\right)^{1/2}$$
$$\ll \left(\int_\Omega\left(\frac{U(x)}{\rho(x)^{r+\alpha}}\right)^2 dx\right)^{1/2} + \left(\sum_{|k|=r}\int_\Omega\left(\frac{U^{(k)}(x)}{\rho(x)^{\alpha}}\right)^2 dx\right)^{1/2}$$
$$+\left(\sum_{|l|\leqslant r-\gamma} \|\mu_l(x)\rho(x)^{r+\alpha-|l|}\|_{L_2(\Omega)}\right)^{1/2},$$

or (see (3))

$$\|U\|_{W^{r+\gamma}_{\alpha-\gamma}(\Omega)} \ll \|U/\rho^{r+\alpha}\|_{L_2(\Omega)} + \|U\|_{W^r_{2,\alpha}(\Omega)} + \|F\|_{H_{r,\gamma}}.$$

Theorem 3 is proved.

§7. Proof of the converse assertion in Theorem 1 $(0<\gamma\leqslant r)$

7.1. The conditions in Theorem 1 when $\gamma>0$ contain, in particular, the conditions in it when $\gamma=0$; hence the class $\mathfrak{M}_0 = \overset{0}{W}{}^r_{2,\alpha}(\Omega)$ contains (on the basis of §3.1) a unique generalized solution U ($U\in\mathfrak{M}_0$) of the equation

$$LU = F(x) \qquad (x\in\Omega) \qquad (1)$$

(see Remark 1 in §3.1) satisfying the inequality (see (7) in §3.1)

$$\|U\|_{\overset{0}{W}{}^r_\alpha(\Omega)} \ll \sup_{\|v\|_{V^r_{2,\alpha}(\Omega)}\leqslant 1} F(v).$$

Since

$$\|v\|_{V^{r-\gamma}_{2,\alpha+\gamma}(\Omega)} = \sum_{|k|\leqslant r-\gamma}\left\|\frac{v^{(k)}}{\rho^{r+\alpha-|k|}}\right\|_{L_2(\Omega)} \leqslant \sum_{|k|\leqslant r}\left\|\frac{v^{(k)}}{\rho^{r+\alpha-|k|}}\right\|_{L_2(\Omega)} \leqslant 1,$$

it follows that

$$\|U\|_{w^r_{2,\alpha}(\Omega)} \ll \|U\|_{\overset{0}{W}{}^r_{2,\alpha}(\Omega)} \ll \sup_{\|v\|_{V^{r-\gamma}_{2,\alpha+\gamma}(\Omega)} \leqslant 1} F(v) = \|F\|_{\overset{*}{V}{}^{r-\gamma}_{2,\alpha+\gamma}(\Omega)}. \tag{2}$$

Moreover, by the generalized Poincaré inequality (see Lemma 2 in §3 of [5]) and inequality (16) in §6,

$$\|U/\rho^{r+\alpha}\|_{L_2(\Omega)} \ll \|U\|_{\overset{0}{W}{}^r_{2,\alpha}(\Omega)} \ll \|F\|_{\overset{*}{V}{}^r_{2,\alpha}(\Omega)} \ll \|F\|_{\overset{*}{V}{}^{r-\gamma}_{2,\alpha+\gamma}(\Omega)}. \tag{3}$$

Thus, the generalized solution U of (1) in $\overset{0}{W}{}^r_\alpha(\Omega)$ satisfies the conditions of Theorem 3. But then, by Theorem 3, it belongs to the class $W^{r+\gamma}_{\alpha-\gamma}(\Omega)$ and thus to the class $\overset{0}{W}{}^{r+\gamma}_{\alpha-\gamma}(\Omega)$, and the inequality

$$\|U\|_{\overset{0}{W}{}^{r+\gamma}_{\alpha-\gamma}(\Omega)} \ll \|U/\rho^{r+\alpha}\|_{L_2(\Omega)} + \|U\|_{w^r_{2,\alpha}(\Omega)} + \|F\|_{H_{r,\gamma}(\Omega)}$$

is satisfied (see (14) in §2). By (2) and (3), this inequality implies

$$\|U\|_{W^{r+\gamma}_{\alpha-\gamma}(\Omega)} \ll \|F\|_{H_{r,\gamma}(\Omega)},$$

which is what we were required to prove.

Thus, for each function $F(x) \in H_{r,\gamma}(\Omega)$ $(0 \leqslant \gamma \leqslant r)$ there exists a function $U \in \overset{0}{W}{}^{r+\gamma}_{\alpha-\gamma}(\Omega)$ such that

$$E(U, v) = F(v), \quad \forall v \in \overset{0}{W}{}^r_\alpha(\Omega).$$

If there were another such function U_1, then we would have

$$E(U - U_1, v) = 0, \quad \forall v \in \overset{0}{W}{}^r_\alpha(\Omega).$$

But $U - U_1 \in \overset{0}{W}{}^{r+\gamma}_{\alpha-\gamma}(\Omega) \subset \overset{0}{W}{}^r_\alpha(\Omega)$, and $E(\varphi, \varphi)$ is the inner product in $\mathfrak{M}_0$ generating the norm $E(f)^{1/2}$, in which $\mathfrak{M}_0$ is complete. And then $U - U_1 \equiv 0$.

The proof of the converse assertion in Theorem 1 for $0 \leqslant \gamma \leqslant r$ is completed.

7.2. We remark that, under certain restrictions on the lower-order coefficients (see condition 4) below, Theorem 1 can be extended to all possible values of the parameter α in terms of the spaces $V^r_{2,\alpha}(\Omega)$.

THEOREM 4. *Suppose that the coefficients $a_{kl}(x)$ of the integro-differential form*

$$E(u, v) = \int_\Omega \sum_{|k|, |l| \leqslant r} a_{kl} D^k u D^l v \, dx$$

satisfy the following conditions:

1) *The $a_{kl}(x)$ are measurable in Ω, and $a_{kl}(x) = a_{lk}(x)$, $|k|, |l| \leqslant r$.*

2) $\sum_{|k|,|l|=r} a_{kl}(x)\xi_k\xi_l \geqslant (\kappa/\rho^{2\alpha}(x)) \sum_{|k|=r} \xi_k^2$, $\kappa > 0$.

3) $|D^\lambda a_{kl}(x)| \leqslant M^2/\rho^{2(r+\alpha)-|k|-|l|+|\lambda|}$, $|\lambda| \leqslant \gamma$, *for some nonnegative integer* γ.

4) $a_{00}(x) \geqslant c/\rho^{2(\alpha+r)}$, $c > 0$, *and the intermediate coefficients $a_{kl}(x)$, $0 < |k| + |l| < 2r$, are sufficiently small in modulus.*

Then there exists a unique solution $U \in V^r_{2,\alpha}(\Omega)$ of the equation

$$E(u, v) = \int_\Omega F(x) v(x) \, dx, \quad \forall v \in V^r_{2,\alpha}(\Omega),$$

where $F \in V^{\gamma-r}_{2,-(\alpha+\gamma)}(\Omega)$.

This solution actually belongs to the space $V_{2,\alpha-\gamma}^{r+\gamma}(\Omega)$ and satisfies the estimate

$$\|U\|_{V_{2,\alpha-\gamma}^{r+\gamma}(\Omega)} \leqslant c\|F\|_{V_{2,-(\alpha+\gamma)}^{\gamma-r}(\Omega)},$$

where the constant c does not depend on F. Moreover, the mapping implemented by the corresponding operator L, namely,

$$Lu = \sum(-1)^{|l|}D^l\left(a_{kl}D^k u\right),$$

is a topological isomorphism of $V_{2,\alpha-\gamma}^{r+\gamma}(\Omega)$ onto $V_{2,-(\alpha+\gamma)}^{\gamma-r}(\Omega)$.

§8. The nonhomogeneous boundary value problem

8.1. Below we extend Theorem 1 to the case where the solution U of the differential equation

$$LU = F(x) \qquad (x \in \Omega) \tag{1}$$

has nonzero boundary values in general.

All the conditions of Theorem 1 are assumed. Moreover, condition (4) in §1 must be replaced by the stronger condition

$$\left|a_{kl}^{(\lambda)}(x)\right| \leqslant c/\rho(x)^{2\alpha+|\lambda|}, \qquad |\lambda| \leqslant \gamma. \tag{2}$$

Let $\Phi(x) \in W_{\alpha-\gamma}^{r+\gamma}(\Omega)$ be arbitrary. It must be taken into account that

$$W_{\alpha-\gamma}^{(r+\gamma)}(\Omega) \to W_\alpha^r(\Omega)$$

(see (7) in §2). Therefore, Φ has the s_0 boundary functions

$$\left.\frac{\partial^s\Phi}{\partial\nu^s}\right|_\Gamma = \varphi_s \in L_2(\Gamma) \qquad (s = 0,\dots,s_0-1),$$

which, generally speaking, are not identically zero on Γ. Let $\mathfrak{M}_\Phi$ denote the class of functions $U \in W_{\alpha-\gamma}^{r+\gamma}(\Omega)$ with the same boundary functions as Φ.

THEOREM 5. *Under the conditions indicated above, the operator L maps the class $W_{2,\alpha-\gamma}^{r+\gamma}(\Omega)$ onto the class $H_{r,\gamma}(\Omega)$:*

$$L\left(W_{\alpha-\gamma}^{r+\gamma}(\Omega)\right) = H_{r,\alpha}(\Omega). \tag{3}$$

Moreover, for the functions $F = LU$

$$\|F\|_{H_{r,\gamma}(\Omega)} \leqslant c\|U\|_{W_{2,\alpha-\gamma}^{r+\gamma}(\Omega)}, \tag{4}$$

where the constant c does not depend on U.

Conversely, each function $F \in H_{r,\gamma}$ corresponds by means of (1) *to a unique function $U \in \mathfrak{M}_\Phi$, and*

$$\|U\|_{W_{2,\alpha-\gamma}^{r+\gamma}(\Omega)} \leqslant c_1\left(\|F\|_{H_{r,\gamma}} + \|\Phi\|_{W_{2,\alpha-\gamma}^{r+\gamma}(\Omega)}\right), \tag{5}$$

where the constant c_1 does not depend on F nor Φ.

REMARK 9. The assertions in Theorem 5 have already been proved for the case $\gamma \geqslant r$ (see §2 of [7] and Theorem 4 of [5]). The case $\gamma < r$ was also considered in [5], but it was assumed there that

$$\|F\|_{H_{r,\gamma}(\Omega)} = \|F\rho^{r+\alpha}\|_{L_2(\Omega)}. \tag{6}$$

With this convention the direct part of Theorem 5 does not go through. We have (see §§2.2 and 2.3)

$$\|F\|_{\overset{*}{V}{}^{r-\gamma}_{2,\alpha+\gamma}(\Omega)} = \sup_{\|v\|_{V^{r-\gamma}_{2,\alpha+\gamma}(\Omega)} \leqslant 1} F(v) \leqslant \sup_{\|v/\rho^{r+\alpha}\| \leqslant 1} F(v) = \|F(x)\rho^{r+\alpha}\|_{L_2(\Omega)}. \tag{7}$$

The inequality $\leqslant$ in (7) becomes an equality when $\gamma = r$ (see (16) in §2). However, this is not so when $\gamma < r$. It follows from (7) that if $F(x)$ has a finite norm (6), then it belongs to the class $\overset{*}{V}{}^{r-\gamma}_{2,\alpha+\gamma}(\Omega)$. But for $\gamma < r$ the functions $F(x) \in \overset{*}{V}{}^{r-\gamma}_{2,\alpha+\gamma}(\Omega)$ include truly generalized functions such that the norm (6) does not make sense. For example, for $\gamma = 0$ and

$$r + \alpha - n/2 > 0 \tag{8}$$

the δ-function is such a function if it is assumed that $0 = (0, \ldots, 0) \in \Omega$.

Indeed, under condition (8) (see Chapter 10 in [6]) we have

$$\left| \int_\Omega \delta(x) v(x)\, dx \right| = |v(0)| \ll \|v\|_{W^r_{2,\alpha}(\Omega)}$$

for $v \in \mathscr{D}$, and this shows that

$$\|\delta\|_{H_{r,0}} = \|\delta\|_{\overset{*}{V}{}^{r}_{2,\alpha}(\Omega)} = \sup_{\|v\|_{W^r_{2,\alpha}(\Omega)} \leqslant 1} \left| \int_\Omega \delta(x) v(x)\, dx \right| \leqslant c < \infty,$$

i.e., the $H_{r,0}$-norm of the δ-function is finite.

8.2. Proof of Theorem 5 when $0 \leqslant \gamma \leqslant r$. Let $U \in W^{r+\gamma}_{2,\alpha-\gamma}(\Omega)$. The function $F(x)$ obtained from it by means of L is determined as the function $F \in D'$ such that (see §2.6) for all $v \in \mathscr{D}$

$$\begin{aligned}
\int_\Omega F(x) v(x)\, dx &= \int_\Omega \sum_{|k|,\,|l| \leqslant r} a_{kl} U^{(k)} v^{(l)}\, dx \\
&= \sum_{\substack{|k|,\,|l| \leqslant r \\ |l| \leqslant r-\gamma}} \int_\Omega a_{kl} U^{(k)} v^{(l)}\, dx + \sum_{\substack{|k|,\,|l| \leqslant r \\ |l| > r-\gamma}} \int_\Omega a_{kl} U^{(k)} v^{(l)}\, dx \\
&= J_1 + J_2.
\end{aligned} \tag{9}$$

The terms of the sum J_1 can be estimated on the basis of (2) as follows (explanation below):

$$\begin{aligned}
\int_\Omega |a_{kl} U^{(k)} v^{(l)}|\, dx &\ll \int_\Omega \left| \frac{U^{(k)}}{\rho^\alpha} \frac{v^{(l)}}{\rho^\alpha} \right| dx \ll \left\| \frac{U^{(k)}}{\rho^\alpha} \right\|_{L_2(\Omega)} \left\| \frac{v^{(l)}}{\rho^\alpha} \right\|_{L_2(\Omega)} \\
&\ll \|U\|_{W^r_{2,\alpha}(\Omega)} \left\| \frac{v^{(l)}}{\rho^{r+\alpha-|l|}} \right\|_{L_2(\Omega)} \ll \|U\|_{W^{r+\gamma}_{2,\alpha-\gamma}(\Omega)} \|v\|_{V^{r-\gamma}_{2,\alpha+\gamma}(\Omega)}.
\end{aligned} \tag{10}$$

In the third relation in (10) we have made use of inequality (6) in §2 in estimating U, and the fact that $r - |l| > 0$ in estimating v. In the last relation in (10) it is necessary to take into account (7) and (8) in §2.

Inequality (10) says that the linear functional $\Sigma_\Omega a_{kl} U^{(k)} v^{(l)}\, dx$ of $v \in \mathscr{D}$ is bounded on the space of functions $v \in \mathscr{D}$, with the norm $\|v\|_{V^{r-\gamma}_{2,\alpha+\gamma}(\Omega)}$, and the norm of this functional does not exceed $c\|U\|_{W^{r+\gamma}_{\alpha-\gamma}(\Omega)}$, where c does not depend on U. Thus, the functional $J_1 = J_1(v)$ also has this property.

Let us proceed to estimate $J_2 = J_2(v)$. Assuming that $0 \leqslant l^0 < l$, $\|l^0\| = r - \gamma$, and

$$s = l - l^0, \qquad |s| \leqslant r - (r - \gamma) = \gamma$$

we have for any number of terms in J_2 that (explanation below)

$$\left|\int_\Omega a_{kl}U^{(k)}v^{(l)}dx\right| = \left|\int_\Omega \left(a_{kl}U^{(k)}\right)^{(s)} v^{(l^0)}dx\right| \leqslant \sum_{0\leqslant\lambda\leqslant s} C_s^\lambda \int_\Omega \left|a_{kl}^{(\lambda)}U^{(k+s-\lambda)}v^{(l^0)}\right|dx$$

$$\ll \sum_{0\leqslant\lambda\leqslant s}\int_\Omega \left|\frac{U^{(k+s-\lambda)}}{\rho^{\alpha-\gamma+|\lambda|}}\frac{v(l^0)}{\rho^{r+\alpha-(r-\gamma)}}\right|dx$$

$$\leqslant \sum_{0\leqslant\lambda\leqslant s}\left\|\frac{U^{(k+s-\lambda)}}{\rho^{\alpha-(\gamma-|\lambda|)}}\right\|_{L_2(\Omega)}\left\|\frac{v^{(l^0)}}{\rho^{r+\alpha-(r-\gamma)}}\right\|_{L_2(\Omega)}$$

$$\ll \|U\|_{W_{\alpha-\gamma}^{r+\gamma}(\Omega)}\|v\|_{V_{2,\alpha}^{r-\gamma}(\Omega)}. \tag{11}$$

In the first relation (11) we transferred s derivatives from $v^{(l)}$ to the adjacent factor; this is possible because a_{kl} is by assumption γ times continuously differentiable ($|s| \leqslant \gamma$!). The Leibniz formula was used in the second relation in (11), inequality (2) was used in the third, and, finally, inequality (6) in §2 ($\alpha - (\gamma - |\lambda|) \leqslant \alpha < \frac{1}{2}$!) and equality (8) in §2 were used for U and v, respectively, in the last.

We see anew that the integral on the left-hand side in (11) is a bounded linear functional of $v \in \mathscr{D}$ with norm $\|v\|_{V_{2,\alpha}^{r-\gamma}(\Omega)}$, and the norm of this functional is at most $c\|U\|_{W_{\alpha-\gamma}^{r+\gamma}(\Omega)}$, where $c > 0$ does not depend on U. Thus, the linear functional $J_2 = J_2(v)$ has the same property.

At this stage of the arguments we have proved that each function $U \in W_{\alpha-\gamma}^{r+\gamma}(\Omega)$ is mapped by L into the function $F(x) = LU \in H_{r,\gamma}(\Omega)$, and that inequality (4) is valid.

We now take a function $\Phi(x) \in W_{\alpha-\gamma}^{r+\gamma}(\Omega)$ and an $F(x) \in H_{r,\gamma}(\Omega)$. Subtraction of $L\Phi$ from the left-hand and right-hand sides of (1) gives us

$$L(U - \Phi) = F - L\Phi \tag{12}$$

or the equivalent equation

$$L(\Psi) = F - L\Phi, \qquad \Psi = U - \Phi \tag{12'}$$

in the desired function Ψ, which obviously belongs to $\overset{0}{W}{}_{2,\alpha-\gamma}^{r+\gamma}(\Omega)$. Moreover, the right-hand side of (12′) has a finite norm in $H_{r,\gamma}(\Omega)$, because

$$\|F\|_{H_{r,\gamma}(\Omega)} < \infty$$

by assumption, and

$$\|L\Phi\|_{H_{r,\gamma}(\Omega)} \subset \|\Phi\|_{W_{2,\alpha-\gamma}^{r+\gamma}(\Omega)}$$

on the basis of inequality (4), already proved. But then Theorem 1 is applicable to the boundary value problem (12′): there exists a unique function Ψ in $\overset{0}{W}{}_{2,\alpha-\gamma}^{r+\gamma}(\Omega)$ solving (12′), and

$$\|\Psi\|_{\overset{0}{W}{}_{2,\alpha-\gamma}^{r+\gamma}(\Omega)} \leqslant c_1\|F - L\Phi\|_{H_{r,\gamma}(\Omega)} \leqslant c_2\left(\|F\|_{H_{r,\gamma}(\Omega)} + \|\Phi\|_{W_{2,\alpha-\gamma}^{r+\gamma}(\Omega)}\right),$$

where c_1 and c_2 do not depend on the indicated F and Φ.

But then (5) is obviously also true. The uniqueness of the solution $U \in \mathfrak{M}_\Phi$ of (1) follows trivially from the uniqueness of the soltuion $\Psi \in \overset{0}{W}{}^{r+\gamma}_{2,\alpha-\gamma}(\Omega)$ of (12′).

Theorem 5 is proved.

REMARK 10. The converse part of Theorem 2 can also be given the following formulation (see the remark after Theorem 7 in [5] for an explanation):

Suppose that the conditions of Theorem 5 hold, and that functions

$$\varphi_s \in B_2^{(r+\gamma-1/2-s)}(\Gamma) \qquad (s = 0, \dots, s_0 - 1)$$

are given on the boundary $\Gamma \in C^{(r+\gamma+1)}$, *where* $B_2^{(\nu)}(\Gamma)$ *is the corresponding Besov class of functions on* Γ. *Then for each given function* $F(x) \in H_{r,\gamma}(\Omega)$ *there exists a unique function* $U \in W^{r+\gamma}_{2,\alpha-\gamma}(\Omega)$ *satisfying the boundary conditions*

$$\left.\frac{\partial^s U}{\partial \nu^2}\right|_\Gamma = \varphi_s \qquad (s = 0, \dots, s_0 - 1)$$

and solving the equation

$$LU = F(x) \qquad (x \in \Omega).$$

In [5] this assertion was prove both for $\gamma \geqslant r$ and for $0 \leqslant \gamma < r$. However, in the last case it was assumed that $\|F\|_{H_{r,\gamma}(\Omega)} = \|F\rho^{r+\alpha}\|_{L_2(\Omega)}$. The assertion has now been proved under the interpretation of the $H_{r,\gamma}$-metric considered here for $0 \leqslant \gamma < r$ (see (14) in §2), which is a natural interpretation, as is clear from Theorems 1 and 3.

Appendix

For completeness of the exposition we prove the assertion used in §2.2.

LEMMA (on denseness). *Suppose that* Ω *is a bounded domain in* R^n, μ *is a nonnegative integer*, α *is a real number, and* $1 \leqslant p < \infty$. *Then the set* $C_0^\infty(\Omega)$ *of infinitely differentiable functions with compact support in* Ω *is dense in the space* $V^\mu_{p,\alpha}(\Omega)$ *with the norm*

$$\|f\|_{V^{\mu}_{2,\cdot}(\Omega)} = \left\{ \sum_{|k| \leqslant \mu} \left\| \frac{f^{(k)}}{\rho^{\mu+\alpha-|k|}} \right\|^p_{L_p(\Omega)} \right\}^{1/p}.$$

PROOF. We construct a "cutoff" function. Let $\Omega_{2\delta}$ be the collection of points $x \in \Omega$ whose distance from $\partial\Omega$ is greater than 2δ, where $\delta > 0$:

$$\Omega_{2\delta} = \{x;\, x \in \Omega,\, \rho(x, \partial\Omega) > 2\delta\}.$$

Let $\chi_{2\delta}(x)$ be the characteristic function of $\Omega_{2\delta}$, and form its Sobolev average

$$\eta_\delta(x) = \frac{1}{\delta^n} \int \omega\left(\frac{x-y}{\delta/2}\right) \chi_{2\delta}(y)\, dy. \tag{1}$$

Clearly, $\eta_\delta(x) \in C_0^\infty(\Omega)$, and

$$\eta_\delta(x) = \begin{cases} 0, & x \in \Omega \setminus \Omega_\delta, \\ 1, & x \in \Omega_{3\delta}. \end{cases} \tag{2}$$

From (1) and (2) we easily get that

$$|D^k \eta_\delta| \leqslant C\delta^{-|k|} \leqslant C_1 \rho^{-|k|}.$$

It follows from this that

$$|D^k(1-\eta_\delta)| \leqslant C\rho^{-|k|}. \tag{3}$$

Suppose now that $f \in V^{\mu}_{p,\alpha}(\Omega)$. We prove that as $\delta \to 0$

$$\|f - f\eta_\delta\|_{V^{\mu}_{p,\alpha}(\Omega)} \to 0.$$

Since $f = f\eta_\delta$ for $x \in \Omega_{3\delta}$, it suffices to see that for each k with $|k| \leqslant \mu$ the quantity

$$J^k_\delta(f) = \left(\int_{\Omega-\Omega_{3\delta}} \left|\frac{(f-f\eta_\delta)^{(k)}}{\rho^{\mu+\alpha-|k|}}\right|^p dx\right)^{1/p}$$

tends to zero as $\delta \to 0$.

Using the Leibniz formula, Minkowski's inequality, and (4), we get that

$$J^k_\delta(f) = \left(\int_{\Omega-\Omega_{3\delta}} \left|\rho^{-(\mu+\alpha-|k|)}\left(\sum_{j\leqslant k} C^k_j(1-\eta_\delta)^{(k-j)} f^{(j)}\right)\right|^p dx\right)^{1/p}$$
$$\leqslant C\sum_{j\leqslant k}\left(\int_{\Omega-\Omega_{3\delta}} \left|\rho^{-(\mu+\alpha-|j|)} f^{(j)}\right|^p dx\right)^{1/p}.$$

Each term under the summation sign tends to zero as $\delta \to 0$, because f belongs to $V^{\mu}_{p,\alpha}(\Omega)$.

It remains to note that for fixed δ the function $f\eta_\delta$ can be approximated with any degree of accuracy by $C^\infty_0(\Omega)$-functions in the norm of $V^{\mu}_{p,\alpha}(\Omega)$. This fact follows in an obvious way from the theory of unweighted Sobolev spaces.

The lemma is proved.

The denseness lemma was used in §2.2 for $p = 2$ and in the case of a domain Ω with boundary $\Gamma \in C^{s_0+1}$. We remark that for a domain with Lipschitz boundary it is essentially contained in the results of [2] for $-\alpha + 1/p \neq 1,\dots,r$.

BIBLIOGRAPHY

1. O. V. Besov, *On the denseness of compactly supported functions in weighted Sobolev spaces*, Trudy Inst. Mat. Steklov. **161** (1983), 29–47; English transl. in this issue.

2. O. V. Besov, Ya. Kadlets [J. Kadlec] and A. Kufner, *Some properties of weighted classes*, Dokl. Akad. Nauk SSSR **171** (1966), 514–516; English transl. in Soviet Math. Dokl. 7 (1966).

3. G. G. Kazaryan, *On the denseness of the smooth compactly supported functions in $W^r_p(\Omega)$*, Mat. Zametki **2** (1967), 42–45; English transl. in Math. Notes **2** (1967).

4. P. I. Lizorkin and S. M. Nikol'skiĭ, *On some inequalities for functions in weighted classes and boundary value problems with strong degeneracy on the boundary*, Dokl. Akad. Nauk SSSR **159** (1964), 512–515; English transl. in Soviet Math. Dokl. **5** (1964).

5. ______, *Coercive properties of elliptic equations with degeneracy. The variational method*, Trudy Mat. Inst. Steklov. **157** (1981), 90–118; English transl. in Proc. Steklov Inst. Math. **1983**, no. 3(157).

6. S. M. Nikol'skiĭ, *Approximation of functions of several variables and embedding theorems*, 2nd rev. aug. ed. "Nauka", Moscow, 1977; English transl., Springer-Verlag, 1974.

7. ______, *A variational problem for an equation of elliptic type with degeneracy on the boundary*, Trudy Mat. Inst. Steklov. **150** (1979), 212–238; English transl. in Proc. Steklov Inst. Math. **1981**, no. 4 (150).

8. Mario Troisi, *Teoremi di inclusione negli spazi de Sobolev conpeso*, Ricerche Mat. **18** (1969), 49–74.

9. Hans Triebel, *Interpolation theory, function spaces, differential operators*, VEB Deutscher Verlag Wiss., Berlin, 1977, and North-Holland, Amsterdam, 1978.

Translated by H. H. McFADEN

Труды Матем. Инст. Стеклов.
161(1983)

Proceedings of the Steklov
Institute of Mathematics
1984, Issue 3

INEQUALITIES FOR DERIVATIVES AND DUALITY

UDC 517

G. G. MAGARIL-IL'YAEV

ABSTRACT. Duality relations are derived between best constants and extremal functions in inequalities for derivatives of the form

$$\|x^{(k)}\|_{L_\infty(I)} \leqslant K\|x\|^{\alpha}_{L_p(I)}\|x^{(n)}\|^{1-\alpha}_{L_r(I)}, \tag{1}$$

where $I = R$ or $R_+ = [0, \infty)$, $n \geqslant 1$ and $0 \leqslant k < n$ are integers, $1 \leqslant p, r \leqslant \infty$, and $\alpha \geqslant 0$, and the analogous objects in the inequalities dual (in a certain sense) to (1). A number of new sharp constants are obtained as consequences, along with duality relations between the values of best approximations of linear functionals on classes of functions.

Bibliography: 12 titles.

This article deals with the derivation of duality relations between best constants and extremal functions in inequalities for derivatives of the form

$$\|x^{(k)}\|_{L_\infty(I)} \leqslant K\|x\|^{\alpha}_{L_p(I)}\|x^{(n)}\|^{1-\alpha}_{L_r(I)}, \tag{1}$$

where $I = R$ or $R_+ = [0, \infty)$, $n \geqslant 1$ and $0 \leqslant k < n$ are integers, $1 \leqslant p, r \leqslant \infty$ and $\alpha \geqslant 0$, and the analogous objects in the inequalities dual (in a certain sense) to (1). As consequences we obtain some new sharp constants, along with duality relations between the values of best approximations of linear functionals on classes of functions.

The proofs of our main assertions are based on methods in duality theory for convex functions. The scheme for the arguments is as follows. With inequality (1) we associate a certain convex programming problem such that computation of the best constant and determination of the extremal functions in this inequality (i.e., the functions at which it becomes an equality) is equivalent to computation of the infimum and determination of the exact solution of the given problem. Then we choose a "perturbation" of the indicated problem such that the problem dual to it is associated in the same way as the original problem with another inequality of the type (1). The known relations between the dual problems lead to corresponding assertions with respect to the inequalities for derivatives.

The article has two sections. In the first we introduce the concept of duality of two inequalities, give a general scheme for constructing the problem dual to a given problem, and formulate the main results (Theorems 1 and 2) and their corollaries. The second section contains the proofs.

1980 *Mathematics Subject Classification*. Primary 26D10, 41A44, 41A50; Secondary 46E35.

0081-5438/84 $1.00 + $.25 per page

1. Let $\boldsymbol{I} = \boldsymbol{R}$ or $\boldsymbol{R}_+ = [0, \infty)$, or $[0, T]$ $(T > 0)$. As usual, $L_p(\boldsymbol{I})$ denotes the collection of all measurable functions on $\boldsymbol{I}$ whose pth powers are integrable for $1 \leqslant p < \infty$, and essentially bounded for $p = \infty$. Denote the norm of an $x \in L_p(\boldsymbol{I})$ by $\|x\|_{L_p(\boldsymbol{I})}$, or simply $\|x\|_p$ if the domain of x is clear or unimportant.

Suppose, further, that $n \geqslant 1$ is an integer and $1 \leqslant p, r \leqslant \infty$. Denote by $W^n_{p,r}(\boldsymbol{I})$ the set of functions $x \in L_p(\boldsymbol{I})$ whose $(n-1)$st derivatives $x^{(n-1)}$ are locally absolutely continuous on $\boldsymbol{I}$ (i.e., absolutely continuous on each closed interval in $\boldsymbol{I}$) and whose nth derivatives $x^{(n)}$ are in $L_r(\boldsymbol{I})$. This is a normed space with the norm $\|x\|_p + \|x^{(n)}\|_r$.

It is known (see [1]) that inequality (1) holds with

$$\alpha = (n - k - 1/r)(n + 1/p - 1/r)$$

$(\alpha = 1$ if $n + 1/p - 1/r = 0)$ for all $x \in W^n_{p,r}(\boldsymbol{I})(\boldsymbol{I} = \boldsymbol{R}$ or $\boldsymbol{R}_+)$ with some constant $K > 0$. The smallest such constant is called the *best constant* and denoted by $K_n(k; p, r; \boldsymbol{I})$.

Let $J = \{0, 1, \ldots, n-1\}$ and $J_m = J \setminus \{m\}$ for $m \in J$. The (possibly empty) subsets of J_m are denoted by $Q_m, P_m, \ldots$. If $Q \subset J$, then let

$$W^n_{p,r}(\boldsymbol{I}; Q) = \left\{x \in W^n_{p,r}(\boldsymbol{I}) \,|\, x^{(i)}(0) = 0,\ i \in Q\right\} \quad \left(W^n_{p,r}(\boldsymbol{I}; \varnothing) = W^n_{p,r}(\boldsymbol{I})\right).$$

Along with (1) we also consider inequalities of the form

$$\left|x^{(k)}(0)\right| \leqslant K\|x\|^{\alpha}_{L_p(\boldsymbol{I})}|x^{(n)}|^{1-\alpha}_{L_r(\boldsymbol{I})}, \qquad x \in W^n_{p,r}(\boldsymbol{I}; Q_k), \tag{2}$$

where α is the same as above.

Denote by $K_n(k; p, r; \boldsymbol{I}; Q_k)$ the best constant in (2). It is easy to see that if $Q_k = \varnothing$, then the inequalities (1) and (2) (with the same parameters) are or are not valid simultaneously, and $K_n(k; p, r; \boldsymbol{I}; \varnothing) = K_n(k, p, r; \boldsymbol{I})$. Inequalities of type (2) thus form a larger class than the inequalities (1).

An extremal function in (2) is defined to be any nontrivial function in $W^n_{p,r}(\boldsymbol{I}; Q_k)$ at which this inequality with $K = K_n(k; p, r; \boldsymbol{I}; Q_k)$ becomes an equality. It is clear that if $Q_k = \varnothing$, then an extremal function in (2) is also extremal in (1).

For each fixed $n \geqslant 1$ inequality (2) is obviously determined completely by specifying a quintuple $\xi = (k; p, r; \boldsymbol{I}; Q_k)$, and the best constant in it can be regarded as a function on the set M_n of all such quintuples.

Our main goal is to prove that on some (sufficiently large) subset $L_n \subset M_n$ it is possible to define an involutive mapping $\Phi_n\colon L_n \to L_n$ such that the best constants and the extremal functions (if they exist) in the inequalities corresponding to the quintuples ξ and $\xi' = \Phi_n(\xi)$ are connected by definite relations.

We give the precise definitions. Let $k \in J$, and let

$$\mathfrak{A}_k = \{k + 2, |m = -[k/2], \ldots, -1, 1, \ldots, [(n-k-1)/2]\}.$$

Then

$$L_n = \{\xi = (k\colon p, \boldsymbol{I}; Q_k) \in M_n | Q_k \subset \mathfrak{A}_k, 1 + 1/p - 1/r > 0\}.$$

The second condition here means that only the single point $(\infty, 1)$ is excluded from the set of all pairs (p, r) with $1 \leqslant p, r \leqslant \infty$.

Suppose next that $\mathfrak{G}\colon J \to J$ is a permutation such that $\mathfrak{G}(m) = n - m - 1$, $m \in J$, and $\mathfrak{G}(\varnothing) = \varnothing$. For each $k \in J$ let a mapping $2^{\mathfrak{A}_k} \to J$ be defined by the rule

$$\mathfrak{A}_k \supset Q_k \to Q'_k = \mathfrak{G}(\mathfrak{A}_k \setminus Q_k).$$

It is easy to see that $Q'_k \subset \mathfrak{A}_{n-k-1}$ and $Q''_k = (Q'_k)' = Q_k$, i.e., the mapping is involutive.

If $1 \leqslant s \leqslant \infty$, then s' will always stand for the exponent conjugate to s, i.e., $1/s + 1/s' = 1$, with $s' = \infty$ when $s = 1$ and vice versa.

Suppose now that $\xi = (k; p, r; \boldsymbol{I}; Q_k) \in L_n$. By definition, we set

$$\xi' = \Phi_n(\xi) = (n - k - 1; r', p'; \boldsymbol{I}'; Q'_k),$$

where r' and p' are the exponents conjugate to r and p, respectively, and $\boldsymbol{I}' = \boldsymbol{R}_+$ if $\boldsymbol{I} = \boldsymbol{R}$, and $\boldsymbol{I}' = \boldsymbol{R}$ if $\boldsymbol{I} = \boldsymbol{R}_+$. It is obvious that Φ_n is an involution.

It should be noted that it is also possible to determine other subsets and involutive mappings on them, where the best constants and extremal functions for the corresponding inequalities are connected by certain relations. However, it is only in this case that the mapping turns out to be invariant (for some $n > 1$ and $k \in J$) with respect to the set of $\xi \in L_n$ such that $Q_k = \varnothing$ (i.e., the set of ξ corresponding to inequalities of the type (1)), and thus in this case it is possible to obtain as a consequence new sharp constants in (1) (see Corollary 1).

The inequalities corresponding to quintuples $\xi \in L_n$ and $\xi' = \Phi_n(\xi)$ (or simply the inequalities ξ and ξ') are said to be *dual* to each other.

THEOREM 1. *Let* $\xi = (k; p, r; \boldsymbol{R}; Q_k)$ *and* $\xi' = (n - k - 1; r', p'; \boldsymbol{R}_+; Q'_k)$ *be two inequalities in duality.*

a) *The best constants in* ξ *and* ξ' *are connected by the relation*

$$K_n(\xi)K_n(\xi') = \alpha^{-\alpha}(1 - \alpha)^{-(1-\alpha)}2^{-\gamma},$$

where

$$\alpha = \frac{n - k - 1/r}{n + 1/p - 1/r}, \qquad \gamma = \frac{(n - k)/p + k/r}{n + 1/p - 1/r}.$$

b) *If* $p < \infty$ *and* $r > 1$, *then there exist extremal functions* $\hat{x}$ *and* $\hat{y}$ *in* ξ *and* ξ', *respectively, such that* $\hat{x}$ *is even (odd) if* k *is even (odd), and a.e. on* $\boldsymbol{R}_+$

$$\begin{aligned} \hat{x}^{(n)}(t) &= (-1)^{n-k}\|\hat{x}^{(n)}\|_{L_r(\boldsymbol{R})}|\hat{y}(t)|^{r'-1}\operatorname{sgn}\hat{y}(t),\\ \hat{y}^{(n)}(t) &= (-1)^{k+1}\|\hat{x}\|^{1-p}_{L_p(\boldsymbol{R})}|\hat{x}(t)|^{p-1}\operatorname{sgn}\hat{x}(t). \end{aligned} \tag{3}$$

COROLLARY 1. *The following relations are valid:*

$$K_2(0; p, 1; \boldsymbol{R}) = ((p + 1)/8)^{1/(p+1)}, \qquad 1 \leqslant p < \infty; \tag{4}$$

$$K_2(0; p, 1; \boldsymbol{R}_+) = (p + 1)^{1/(p+1)}, \qquad 1 \leqslant p < \infty; \tag{5}$$

$$K_3(1; p, 1; \boldsymbol{R}) = \begin{cases} \kappa(p)K_3^{-1}(1; \infty, p'; \boldsymbol{R}_+), & 1 < p < \infty,\\ 1/\sqrt[3]{36}, & p = 1; \end{cases} \tag{6}$$

$$K_3(1; p, 1; \boldsymbol{R}_+) = \begin{cases} 8^{1/(2p+1)}\kappa(p)K_3^{-1}(1; \infty, p'; \boldsymbol{R}), & 1 < p < \infty,\\ \sqrt[3]{6}, & p = 1, \end{cases} \tag{7}$$

where

$$\kappa(p) = 2^{-(p+2)/(2p+1)}\left(\frac{2p+1}{p}\right)^{p/(2p+1)}\left(\frac{2p+1}{p+1}\right)^{(p+1)/(2p+1)},$$

and $K_3(1;\infty,p';\boldsymbol{R}_+)$ *and* $K_3(1;\infty,p';\boldsymbol{R})$, $1 < p < \infty$, *are known quantities.*

Before stating the second corollary we give a formulation of a particular problem of Stechkin on approxination of linear operators [2]. Suppose, as above, that $n \geqslant 1$ and $0 \leqslant k < n$ are integers, $1 \leqslant p, r \leqslant \infty$, $\boldsymbol{I} = \boldsymbol{R}$ or $\boldsymbol{R}_+$, and $Q_k \subset J_k$. Further, let

$$B^n_{p,r}(\boldsymbol{I}; Q_k) = \left\{ x \in W^n_{p,r}(\boldsymbol{I}; Q_k) \,\middle|\, \|x^{(n)}\|_r \leqslant 1 \right\},$$

let $L_p^*(\boldsymbol{I})$ be the dual of $L_p(\boldsymbol{I})$ (the norm of an element $l \in L_p^*(\boldsymbol{I})$ is denoted by $\|l\|$), and let $N > 0$. The quantity

$$E_n(N; k; p, r; \boldsymbol{I}; Q_k) = \inf_{\substack{\|l\| \leqslant n \\ l \in L_p^*(\boldsymbol{I})}} \sup_{x \in B^n_{p,r}(\boldsymbol{I}: Q_k)} \left(x^{(k)}(0) - \langle l, x\rangle\right) \tag{8}$$

characterizes the best approximation of the linear functional $x \to x^{(k)}(0)$ by the continuous linear functionals on $L_p(\boldsymbol{I})$ with norm at most N on the given class $B^n_{p,r}(\boldsymbol{I}; Q_k)$.

It is obvious that problem (8) is completely determined by the specification of the sextuple $\eta = (N; \xi) = (N; k; p, r; \boldsymbol{I}; Q_k)$ (for each fixed $n \geqslant 1$).

COROLLARY 2. *Suppose that the sextuples* $\eta = (N; \xi) = (N; k; p, r; \boldsymbol{R}; Q_k)$ *and* $\eta' = (N'; \xi') = (N^{-\alpha/(1-\alpha)}; n-k-1; r', p'; \boldsymbol{R}_+; Q'_k)$ *are such that* $\xi \in L_n$. *Then*

$$E_n^{1-\alpha}(\eta) E_n^{\alpha}(\eta') = \alpha^\alpha (1-\alpha)^{1-\alpha} 2^{-\gamma},$$

where α *and* γ *are the same as in Theorem* 1.

The assertions of Theorem 1 are a consequence of the relations between the extremal problems corresponding to the dual inequalities. These relations are indicated in Theorem 2; for its statement we need additional definitions.

First of all we give a general scheme for constructing the problem dual to a given one (see [3] or [4] for details). Let X and X^* be real vector spaces in duality with respect to the bilinear form $\langle\cdot,\cdot\rangle_1$: $X^* \times X \to \boldsymbol{R}$, and let f: $X \to \overline{\boldsymbol{R}} = \boldsymbol{R} \cup (-\infty) \cup (+\infty)$ be a given function.

We consider the problem

$$f(x) \to \inf, \qquad x \in X, \tag{9}$$

which consists in minimizing f on the whole space X. Further, let Y and Y^* be another pair of spaces in duality with respect to the bilinear form $\langle\cdot,\cdot\rangle_2$: $Y^* \times Y \to \boldsymbol{R}$ and a function F: $X \times Y \to \overline{\boldsymbol{R}}$ such that $F(x,0) = f(x)$ for all $x \in X$.

The problem dual to (9) (with respect to the triple (Y, Y^*, F)) is defined to be the problem

$$-F^*(0, y^*) \to \sup, \qquad y^* \in Y^*, \tag{10}$$

where F^*: $X^* \times Y^* \to \overline{\boldsymbol{R}}$ is the function conjugate to F, i.e.,

$$F^*(x^*, y^*) = \sup_{x \in X,\, y \in Y} \left\{\langle x^*, x\rangle_1 + \langle y^*, y\rangle_2 - F(x,y)\right\}.$$

Note that if X is a normed space, X^* is its dual space, and $\langle x^*, x\rangle$ is the value of a linear functional $x^* \in X^*$ at an element $x \in X$, then (as is well known) X and X^* are in duality with respect to the bilinear form $(x^*, x) \to \langle x^*, x\rangle$.

The values S and S^* of problems (9) and (10) are defined to be the corresponding values of the infumum and supremum in these problems, and the solutions are defined to be any elements of X and Y^* at which the infimum or supremum is attained.

It follows easily from the definitions that $S^* \leqslant S$ always holds (in the sense of the natural ordering on $\overline{\boldsymbol{R}}$). We give conditions for these quantities to coincide in a form convenient for our purposes (see [4]).

PROPOSITION. *Let Y be a normed space and Y^* its dual space. If S is finite, the function $F\colon X \times Y \to \overline{\boldsymbol{R}}$ is convex, and there exists an $x_0 \in X$ such that the mapping $F(x_0, \cdot)\colon Y \to \overline{\boldsymbol{R}}$ is continuous at zero, then $S = S^*$ and there is a solution problem* (10).

Suppose that $n \geqslant 1$ and $0 \leqslant k, l < n$ are integers, $1 \leqslant p, r, s, t \leqslant \infty$, $\boldsymbol{I} = \boldsymbol{R}$ or $\boldsymbol{R}_+$, $Q_k \subset J_k$, $P_l \subset J_l$, and a, b, c, and d are positive numbers. Consider the problems:

$$a\|x\|_{L_p(\boldsymbol{I})} + b\|x^{(n)}\|_{L_r(\boldsymbol{I})} \to \inf, \quad x^{(k)}(0) = 1, \qquad x \in W^r_{p,r}(\boldsymbol{I}; Q_k), \tag{11}$$

$$y^{(l)}(0) \to \sup, \quad \|y\|_{L_s(\boldsymbol{I})} \leqslant c, \quad \|y^{(n)}\|_{L_t(\boldsymbol{I})} \leqslant d, \qquad y \in W^r_{s,t}(\boldsymbol{I}; P_t). \tag{12}$$

For each particular $n \geqslant 1$ problems (11) and (12) are completely determined by the respective septuples $\zeta = (k; p, r; \boldsymbol{I}; Q_k; a, b)$ and $\zeta^* = (l; s, t; \boldsymbol{I}; P_l; c, d)$. In what follows, these septuples themselves will sometimes be called problems, with ζ understood to be problem (11) and ζ^* problem (12).

Let $S_n = S_n(\zeta)$ and $S^*_n = S^*_n(\zeta^*)$ be the values of problems (11) and (12). For each $k \in J$ we define a mapping $2^{J_k} \to J$ by the rule

$$J_k \supset Q_k \to \tilde{Q}_k = \mathfrak{S}(J_k \setminus Q_k),$$

where $\mathfrak{S}\colon J \to J$ is the same permutation as before, i.e., $\mathfrak{S}(m) = n - m - 1$, $m \in J$ and $\mathfrak{S}(\varnothing) = \varnothing$.

THEOREM 2. *The following assertions are valid*:

a) *If $p < \infty$, then the problem dual to $\zeta = (k; p, r; \boldsymbol{R}_+; Q_k; a, b)$ is the problem $\zeta^* = (n - k - 1; r'; \boldsymbol{R}_+; \tilde{Q}_k; b, a)$, and $0 < S_n(\zeta) = S^*_n(\zeta^*) < \infty$.*

b) *If $r, t > 1$, then problems* (11) *and* (12) *have solutions.*

c) *If $p < \infty$, $r > 1$, and $\hat{x}$ and $\hat{y}$ are solutions of the problem $\zeta = (k; p, r; \boldsymbol{R}_+; Q_k; a, b)$ and the (dual) problem $\zeta^* = (n - k - 1; r', p'; \boldsymbol{R}_+; \tilde{Q}_k; b, a)$, respectively, then a.e. on $\boldsymbol{R}_+$*

$$\begin{aligned} \hat{x}^{(n)}(t) &= (-1)^{n-k} b^{1-r'} \|\hat{x}^{(n)}\|_r |\hat{y}(t)|^{r'-1} \operatorname{sgn} \hat{y}(t), \\ \hat{y}^{(n)}(t) &= (-1)^{k+1} a \|\hat{x}\|_p^{1-p} |\hat{x}(t)|^{p-1} \operatorname{sgn} \hat{x}(t) \end{aligned} \tag{13}$$

and, moreover, $\|\hat{y}\|_{r'} = b$ and $\|\hat{y}^{(n)}\|_{p'} = a$.

2. In this section we also consider the following spaces of functions. Let $\boldsymbol{I} = \boldsymbol{R}$ or $\boldsymbol{R}_+$ or $[0, T]$ $(T > 0)$. Denote by $C(\boldsymbol{I})$ the collection of all continuous real functions

on I, which tend to zero at infinity if $I = R$ or R_+, with the norm $\|x\|_{C(I)} = \max\{|x(t)| \mid t \in I\}$ (or we simply write $\|x\|_C$). Let $V(I)$ be the collection of all finite real Borel measures μ on I, with norm equal to the total variation of μ (we write $\|\mu\|_{V(I)}$ or $\|\mu\|_V$). If $\mu \in V(I)$ is absolutely continuous with respect to Lebesque measure on I, then its density is denoted by $\dot{\mu}$. The collection of all n times continuously differentiable functions x on $[0, T]$ such that $x^{(i)}(0) = x^{(i)}(T) = 0$, $i = 0, 1, \ldots, n-1$, is denoted by $C_0^n([0, T])$, $n \geqslant 1$. We precede the proof of Theorems 1 and 2 by four lemmas.

LEMMA 1. *Suppose that* $n \geqslant 1$, $T > 0$, $y \in L_1([0, T])$, *and* $\mu \in V([0, T])$ *are such that*

$$\int_0^T xy\,dt + \int_0^T x^{(n)}\,d\mu = 0$$

for all $x \in C_0^n([0, T])$. *Then μ is absolutely continuous*, $\dot{\mu} \in W_{1,1}^n([0, T])$, *and* $(\dot{\mu})^{(n)}(t) = (-1)^{n+1}y(t)$ *a.e. on* $([0, T])$.

The proof is analogous to that of the generalized Du Bois-Reymond lemma (see [5], p. 306).

LEMMA 2. *Suppose that* $n \geqslant 1$, $1 \leqslant p < \infty$, $1 \leqslant r \leqslant \infty$, *and* $I = R$ *or* R_+. *If* $x \in W_{p,r}^n(I)$ *and* $0 \leqslant i \leqslant n-1$, *then* $x^{(i)}(t) \to 0$ *as* $|t| \to \infty$.

PROOF. Without loss of generality it can be assumed that $I = R_+$. In this case the lemma was proved in [6] for $n = 2$. The proof is completely analogous for $n = 1$. Suppose that $n > 1$ and assume that the lemma is valid for all functions in $W_{p,r}^{n-1}(I)$. If now $x \in W_{p,r}^n(I)$, then a theorem of Gabushin [1] gives us that, in particular, $x^{(n-1)} \in L_q(I)$, where $q = n/(1/p + (n-1)/r)$. Consequently, $x^{(i)}(t) \to 0$ as $\|t\| \to \infty$ for $i = 0, 1, \ldots, n-2$ by assumption; but since $x^{(n-1)} \in W_{q,r}^{(1)}(I)$ and, obviously, $1 \leqslant q < \infty$, it also follows that $x^{(n-1)}(t) \to 0$ as $\|t\| \to \infty$. The lemma is proved.

LEMMA 3. *Suppose that the values S_n and S_n^* of problems* (11) *and* (12) *are finite and positive. Then*

$$K_n(k; p, r; I; Q_k) = a^\alpha b^{1-\alpha}\left[\alpha^\alpha(1-\alpha)^{1-\alpha}S_n(k; p, r; I; Q_k; a, b)\right]^{-1}, \quad (14)$$

$$K_n(l; s, t; I; P_l) = c^{-\beta}d^{-(1-\beta)}S_n^*(l; s, t; I; P_l; c, d), \quad (15)$$

where

$$\alpha = \frac{n-k-1/r}{n+1/p-1/r}, \qquad \beta = \frac{n-l-1/t}{n+1/s-1/t},$$

and any solutions of (11) *and* (12) *are extremal functions in the corresponding inequalities.*

PROOF. It is assumed below that $0 < \alpha, \beta < 1$, since (14) and (15) can easily be established in the extreme cases. Let us prove the first one.

To do this we show first that for all $x \in W_{p,r}^n(I; Q_k)$ and each $\rho > 0$

$$S_n|x^{(k)}(0)| \leqslant a\|x\|_p\rho^{-k-1/p} + b\|x^{(n)}\|_r\rho^{n-k-1/r}. \quad (16)$$

Indeed, let $x \in W^n_{p,r}(\boldsymbol{I}; Q_k)$. If $x^{(k)}(0) = 0$, then (16) obviously holds. But if $x^{(k)}(0) \neq 0$, then consider the function $\tilde{x}(t) = (x^{(k)}(0))^{-1}\rho^{-k}x(\rho t)$. It is clear that $\tilde{x} \in W^n_{p,r}(\boldsymbol{I}; Q_k)$ and $\tilde{x}^{(k)}(0) = 1$. Consequently,

$$S_n \leqslant a\|\tilde{x}\|_p + b\|\tilde{x}^{(n)}\|_r. \tag{17}$$

From this we get (16) at once by taking into account the equality

$$\|y^{(m)}\|_s = |\gamma|\xi^{m-1/s}\|x^{(m)}\|_s,$$

which is valid if $y(t) = \gamma x(\xi t)$, $\gamma \in \boldsymbol{R}$, $\xi > 0$, and $x^{(m)} \in L_s(\boldsymbol{I})$, where $m \geqslant 0$, $1 \leqslant s \leqslant \infty$, and $\boldsymbol{I} = \boldsymbol{R}$ or $\boldsymbol{R}_+$.

Suppose now that $x \in W^n_{p,r}(\boldsymbol{I}; Q_k)$ and $x \neq 0$. In (16) let

$$\rho = \left[ab^{-1}\left(\frac{k+1/p}{n-k-1/r}\right)\|x\|_p\|x^{(n)}\|_r^{-1}\right]^{1/(n+1/p-1/r)}, \tag{18}$$

then after simple transformations we get inequality (2) (corresponding to the quintuple $\xi = (k; p, r; \boldsymbol{I}; Q_k)$), valid for all $x \in W^n_{p,r}(\boldsymbol{I}; Q_k)$ and with a constant equal to the right-hand side of (14). This constant (denote it by K_1) is the smallest possible constant. Indeed, let $0 < \varepsilon < K_1$ and $\varepsilon_1 = S_n\varepsilon/(K_1 - \varepsilon)$. According to the definition of S_n, there is a function $x_\varepsilon \in W^n_{p,r}(\boldsymbol{I}; Q_k)$, $x_\varepsilon^{(k)}(0) = 1$, such that

$$S_n + \varepsilon_1 > a\|x_\varepsilon\|_p + b\|x_\varepsilon^{(n)}\|_r.$$

Consequently, the function $\tilde{x}_\varepsilon(t) = x_\varepsilon^{(k)}(0)\rho^k x_\varepsilon(\rho^{-1}t)$ satisfies the inequality opposite to (16) with $S_n + \varepsilon_1$ instead of S_n. Let us now substitute in it the ρ in (18), where $x = \tilde{x}_\varepsilon$. Then we get that $\tilde{x}_\varepsilon$ satisfies the inequality opposite to (2) with the constant $K_1 - \varepsilon$. Since ε is arbitrary, this proves that K_1 cannot be improved and thus (14) holds.

If x_0 is a solution of (11), then (17) becomes an equality at this function. Passage from (17) to the multiplicative inequality obviously preserves the equality of the left-hand and right-hand sides; hence, x_0 is an extremal function in $\xi = (k; p, r; \boldsymbol{I}; Q_k)$.

The relation (15) can be proved by similar arguments.

LEMMA 4. *Suppose that $Q_k \subset \mathfrak{A}_k$ and $P_l \subset \mathfrak{A}_l$ in problems* (11) *and* (12) (*the sets $\mathfrak{A}_k$ were defined above for all $k \in J$*). *Suppose, further, that $\mathfrak{B}_k = J_k \setminus \mathfrak{A}_k$ and $\mathfrak{B}_l = J_l \setminus \mathfrak{A}_l$. Then*

$$S_n(k; p, r; \boldsymbol{R}; Q_k; a, b) = S_n(k; p, r; \boldsymbol{R}_+; Q_k \cup \mathfrak{B}_k; a2^{1/p}, b2^{1/r}), \tag{19}$$

$$S_n^*(l; s, t; \boldsymbol{R}; P_l; c, d) = S_n^*(l; s, t: \boldsymbol{R}_+; P_l \cup \mathfrak{B}_l; c2^{-1/s}, d2^{-1/t}). \tag{20}$$

The equalities are understood in the sense that the finiteness of the left-hand sides implies that of the right-hand sides and conversely.

PROOF. Suppose that $\boldsymbol{I} = \boldsymbol{R}$ in problem (11) (which is obviously always finite) and for definiteness that k is even. We show that the value of this problem does not change if we confine ourselves to the even functions in $W^n_{p,r}(\boldsymbol{R}; Q_k)$. Indeed, if this were not so and f is a functional to be minimized in (11), while S_n' is the value of this problem on the even functions, then there is a function $x \in W^n_{p,r}(\boldsymbol{R}; Q_k)$, $x^{(k)}(0) = 1$, such that $f(x) < S_n'$. We associate with x the even function $\tilde{x}(t) = \frac{1}{2}(x(t) + x(-t))$.

Clearly, $\tilde{x} \in W^n_{p,r}(\boldsymbol{R}; Q_k)$ and $\tilde{x}^{(k)}(0) = 1$. On the other hand, $S'_n \leqslant f(\tilde{x}) < \frac{1}{2}f(x) + \frac{1}{2}f(x) = f(x)$ because of the convexity of f and the properties of the norm. The contradiction shows that in fact $S_n = S'_n$.

If k is odd, we can prove in a completely analogous way that it suffices to consider only the odd functions in the problem (11).

A simple check shows that for the corresponding even or odd functions $x \in W^n_{p,r}(\boldsymbol{R}; Q_k)$ it is necessary that $x^{(i)}(0) = 0$ for $i \in \mathfrak{B}_k$. Moreover, it follows from the evenness or oddness of x that

$$\|x^{(m)}\|_{L_s(\boldsymbol{R})} = 2^{1/s}\|x^{(m)}\|_{L_s(\boldsymbol{R})_+}$$

if $x^{(m)} \in L_s(\boldsymbol{R})$, $m \geqslant 0$, $1 \leqslant s \leqslant \infty$.

Thus, with each admissible even (or odd) function in the problem $\zeta_1 = (k; p, r; \boldsymbol{R}; Q_k; a, b)$ we have associated its restriction to $\boldsymbol{R}_+$, which is an admissible function in the problem $\zeta_2 = (k; p, r; \boldsymbol{R}_+; Q_k \cup \mathfrak{B}_k; a2^{1/p}, b2^{1/r})$, and the functionals being minimized in ζ_1 and ζ_2 are equal at a function and at its restriction, respectively. From this it obviously follows that $S_n(\zeta_1) \geqslant S_n(\zeta_2)$.

Conversely, if we consider the problem ζ_2, then by using the fact that $x^{(i)}(0) = 0$, $i \in \mathfrak{B}_k$, we can extend the functions to $\boldsymbol{R}$ in an even (or odd) way. The same arguments as before (recall that $Q_k = \mathfrak{B}_k = \varnothing$) now lead to the inequality $S_n(\zeta_2) \geqslant S_n(\zeta_1)$, which proves (19). The equality (20) can be established in a completely analogous way. The lemma is proved.

Proof of Theorem 2. a) Suppose that $X = W^n_{p,r}(\boldsymbol{R}_+)$, X^* is the space dual to X, $A = \{x \in X | x^{(k)}(0) = 1,\ x^{(i)}(0) = 0,\ i \in Q_k\}$, and $\delta(\cdot|A)$ is the indicator function of A, i.e., the function equal to zero on A and $+\infty$ outside A. Then (11) can be written in the form

$$a\|x\|_p + b\|x^{(n)}\|_r + \delta(x|A) \to \inf, \qquad x \in X. \tag{21}$$

Using the general scheme in §1, we write the problem dual to (21) with respect to $Y = L_p(\boldsymbol{R}_+) \times L_r(\boldsymbol{R}_+)$ with dual space $Y^* = L_{p'}(\boldsymbol{R}_+) \times L_{r'}(\boldsymbol{R}_+)$ (the elements in Y and Y^* are denoted by $y = (y_1, y_2)$ and $y^* = (y_1^*, y_2^*)$, respectively) and with respect to the function

$$F(x, y) = a\|x - y_1\|_p + b\|x^{(n)} - y_2\|_r + \delta(x|A).$$

By definition

$$\begin{aligned} F^*(0, y^*) &= \sup_{x \in X,\, y \in Y} \{\langle y_1^*, y_1\rangle + \langle y_2^*, y_2\rangle - F(x, y)\} \\ &= \sup_{x \in X}\sup_{y \in Y}\left\{\langle y_1^*, y_1\rangle + \langle y_2^*, y_2\rangle - a\|x - y_1\|_p - b\|x^{(n)} - y_2\|_r\right\}. \end{aligned}$$

Let $z_1 = x - y$ and $z_2 = x^{(n)} - y_2$; then

$$\begin{aligned} F^*(0, y^*) = \sup_{x \in A}\ \sup_{z_1 \in L_p}\ \sup_{z_2 \in L_r} \Big\{&\langle y_1^*, x\rangle + \langle y_2^*, x^{(n)}\rangle + \langle -y_1^*, z_1\rangle \\ &- a\|z_1\|_p + \langle -y_2^*, z_2\rangle - b\|z_2\|_r\Big\}. \end{aligned}$$

The successive computation of the suprema with respect to z_2 and z_1 now leads to the following relation:

$$F^*(0, y^*) = \begin{cases} \sup\limits_{x\in A} \{\langle y_1^*, x\rangle + \langle y_2^*, x^{(n)}\rangle\} & \text{if } \|y_1^*\|_{p'} \leqslant a, \|y_2^*\|_{r'} \leqslant b, \\ +\infty & \text{otherwise.} \end{cases}$$

The problem dual to (21) thus takes the form

$$-\sup_{x\in A} \{\langle y_1^*, x\rangle + \langle y_2^*, x^{(n)}\rangle\} \to \sup, \qquad \|y_1^*\|_{p'} \leqslant a, \|y_2^*\|_{r'} \leqslant b. \tag{22}$$

Let $\varphi\colon Y^* \to \overline{\boldsymbol{R}}$ be the functional to be maximized in (22), and let

$$\operatorname{dom}\varphi = \{y^* = (y_1^*, y_2^*) \in Y^* | \varphi(y^*) > -\infty\}.$$

Our immediate goal is to prove that

$$\operatorname{dom}\varphi = \{y^* = (y_1^*, y_2^*) \in Y^* | y_2^* \in W^n_{r',p'}(\boldsymbol{R}_+; \tilde{Q}_k), \\ (y_2^*)^{(n)}(t) = (-1)^{n+1} y_1^*(t) \text{ a.e. on } \boldsymbol{R}_+\}, \tag{23}$$

$$\varphi(y^*) = (-1)^{n-k-1}(y_2^*)^{(n-k-1)}(0), \qquad y^* \in \operatorname{dom}\varphi. \tag{24}$$

Denote the set on the right-hand side in (23) by G, and let $y^* = (y_1^*, y_2^*) \in \operatorname{dom}\varphi (\operatorname{dom}\varphi \neq \varnothing$, because $0 \in \operatorname{dom}\varphi)$. We show that $y^* \in G$. The quantity $-\varphi(y^*)$ is the supremum of the linear functional $x \to \langle y_1^*, x\rangle + \langle y_2^*, x^{(n)}\rangle$ on the linear manifold A, and $-\varphi(y^*) < +\infty$ by assumption. Consequently, this functional is actually constant on A and is thereby equal to zero on the subspace $A_0 = \{x \in W^n_{p,r}(\boldsymbol{R}_+) | x^{(i)}(0) = 0, i \in Q_k \cup \{k\}\}$, i.e.,

$$\langle y_1^*, x\rangle + \langle y_2^*, x^{(n)}\rangle = \int_0^\infty (y_1^* x + y_2^* x^{(n)})\, dt = 0 \tag{25}$$

for all $x \in A_0$.

We now fix some $T > 0$ and denote by A_0^T the collection of functions in A_0 whose supports are concentrated on $[0, T]$. It is clear that $C_0^n([0, T]) \subset A_0^T$, and Lemma 1 (where μ is absolutely continuous and $\mu = y_2^*$) gives us at once that $(y_2^*)^{(n-1)}$ is absolutely continuous on $[0, T]$ and $(y_2^*)^{(n-1)}(t) = (-1)^{(n+1)} y_1^*(t)$ a.e. on this interval. Since this is true for any $T > 0$, $(y_2^*)^{(n-1)}$ is locally absolutely continuous on $\boldsymbol{R}_+$ and $(y_2^*)^{(n)}(t) = (-1)^{n+1} y_1^*(t)$ a.e. on $\boldsymbol{R}_+$. In particular, $y_2^* \in W^n_{r',p'}(\boldsymbol{R}_+)$.

To prove the inclusion $\operatorname{dom}\varphi \subset G$ it remains to verify that $(y_2^*)^{(j)}(0) = 0, j \in \tilde{Q}_k$. Using the properties already proved for y^*, we compute $\varphi(y^*)$. Since $y_2^* \in W^n_{r',p'}(\boldsymbol{R}_+)$, it follows from Gabushin's theorem [1] that the functions $(y_2^*)^{(i)}$, $i = 0, \dots, n-1$, are bounded on $\boldsymbol{R}_+$. Further, by Lemma 2, $x^{(i)}(t) \to 0$ as $t \to \infty$ for $i = 0, \dots, n-1$, provided that $x \in W^n_{p,r}(\boldsymbol{R}_+)$. Consequently, for each $x \in W^n_{p,r}(\boldsymbol{R}_+)$ we have after n-fold integration by parts of the second term in the integrand in (25) that

$$\langle y_1^*, x\rangle + \langle y_2^*, x^{(n)}\rangle = (-1)^{n+1}\int_0^\infty (y_2^*)^{(n)} x\, dt + (-1)^{n-k}(y_2^*)^{(n-k-1)}(0) \\ + \sum_{i \notin Q_k \cup \{k\}} (-1)^{n-i}(y_2^*)^{(n-i-1)}(0) x^{(i)}(0) \\ + (-1)^n \int_0^\infty (y_2^*)^{(n)} t\, dx.$$

By assumption, the supremum of this expression over all $x \in A$ is finite, and thus it is necessary that $(y_2^*)^{(n-i-1)}(0) = 0$, $i \in J \setminus (Q_k \cup \{k\}) = J_k \setminus Q_k$, or, what is the same, $(y_2^*)^{(j)}(0) = 0, j \in \tilde{Q}_k$. Moreover, we get that

$$\varphi(y^*) = (-1)^{n-k-1}(y_2^*)^{(n-k-1)}(0).$$

This proves (24) and the fact that $\operatorname{dom}\varphi \subset G$.

Conversely, if $y^* = (y_1^*, y_2^*) \in G$, then after integration by parts we obtain $\varphi(y^*) = (-1)^{(n-k-1)}(y_2^*)^{(n-k-1)}(0)$, i.e., $y^* \in \operatorname{dom}\varphi$.

Clearly, it suffices to consider problem (22) only on $\operatorname{dom}\varphi$. But then with the notation $y = (-1)^{n-k-1}y_2^*$ this is exactly problem (12), and it is thus dual to problem (11) under the conditions of a).

Let us now show that $S_n = S_n^*$. We consider two separate cases.

1) $1 \leqslant p, r < \infty$. In this situation all the conditions of the proposition in §1 are obviously satisfied (any element of A can be taken as x_0), and thus $S_n = S_n^*$.

2) $1 \leqslant p < \infty$, $r = \infty$. Here it is no longer possible to use the proposition in §1, because Y^* is not the dual of Y. We proceed as follows. Consider the auxiliary problem

$$a\|x\|_p + b\|x^{(n)}\|_C \to \inf, \quad x^{(k)}(0) = 1, \qquad x \in W^n_{p,C}(\boldsymbol{R}_+; Q_k), \tag{26}$$

where $W^n_{p,C}(\boldsymbol{R}_+)$ is defined in a way analogous to that for the space $W^n_{p,r}(\boldsymbol{R}_+)$ ($L_r(\boldsymbol{R}_+)$ is replaced by $C(\boldsymbol{R}_+)$). The problem dual to it with respect to $\tilde{Y} = L_p(\boldsymbol{R}_+) \times C(\boldsymbol{R}_+)$ and $\tilde{Y}^* = L_{p'}(\boldsymbol{R}_+) \times V(\boldsymbol{R}_+)$ (with elements $\tilde{y} = (\tilde{y}_1, \tilde{y}_2)$ and $\tilde{y}^* = (\tilde{y}_1^*, \tilde{y}_2^*)$), and the function $\tilde{F}$ differing from F by the replacement of $L_r(\boldsymbol{R}_+)$ by $C(\boldsymbol{R}_+)$ can be written out just as above. As a result we get problem (22) with $V(\boldsymbol{R}_+)$ instead of $L_{r'}(\boldsymbol{R}_+)$.

Further, if $\tilde{\varphi}\colon \tilde{Y}^* \to \overline{\boldsymbol{R}}$ is the functional to be maximized in this problem, then Lemma 1 can be used in a way completely analogous to the preceding to verify that the measure $\tilde{y}_2^*$ is absolutely continuous if $\tilde{\varphi}(\tilde{y}^*) > -\infty$. This means that it suffices to consider problem (22) on $Y^* = L_{p'}(\boldsymbol{R}_+) \times L_1(\boldsymbol{R}_+) \subset \tilde{Y}^*$ (i.e., its value $\tilde{S}_n^*$ does not then change). But in this case (22) is dual to the original problem with $1 \leqslant p < \infty$ and $r = \infty$, and thus $S_n^* = \tilde{S}_n^*$.

Since $\tilde{Y}^*$ is the space dual to $\tilde{Y}$, the proposition in §1 gives us at once that $\tilde{S}_n = \tilde{S}_n^*$, and since, obviously, $S_n \leqslant \tilde{S}_n$, it follows that $S_n \leqslant \tilde{S}_n = \tilde{S}_n^* = S_n^*$. The reverse inequality $S_n^* \leqslant S_n$ is always true (see §1), and a) is thereby proved.

b) We prove the existence of a solution of (11). Let $\{x_m\}$ be a minimizing sequence in (11). Then, clearly, $\|x_m\|_p \leqslant c_1$ and $\|x_m^{(n)}\|_r \leqslant c_2$ for some $c_i > 0$ ($i = 1, 2$). Since $r > 1$, it can be assumed that $\{x_m^{(n)}\}$ converges weakly to some $z \in L_r(\boldsymbol{I})$.

All the functions $x_m^{(n-1)}$ ($m = 1, 2, \ldots$) are locally absolutely continuous on $\boldsymbol{I}$, and, consequently, for each $t \in \boldsymbol{I}$

$$x_m(t) = \frac{t^k}{k!} + \sum_{i \neq Q_k \cup \{k\}} \frac{t^i}{i!}x_m^{(i)}(0) + \frac{1}{(n-1)!}\int_0^t (t-\tau)^{n-1}x_m^{(n)}(\tau)\,d\tau. \tag{27}$$

By Gabushin's theorem [1], the sequences $\{x_m^{(i)}(0)\}$, $i = 0, 1, \ldots, n-1$, are uniformly bounded with respect to m, and it can thus be assumed that $x_m^{(i)}(0) \to \xi_i$ as $m \to \infty$. Passing to the limit as $m \to \infty$ in (27), we then get a function

$$x(t) = \lim_{m \to \infty} x_m(t)$$

with the following properties: $x^{(i)}(0) = 0$, $i \in Q_k$; $x^{(k)}(0) = 1$; $x^{(n-1)}$ is locally absolutely continuous, and $x^{(n)}(t) = z(t)$ a.e. on $\boldsymbol{I}$ (that is, $x^{(n)} \in L_r(\boldsymbol{I})$). Let us verify that $x \in L_p(\boldsymbol{I})$. If $p = \infty$, this is obvious, If $p < \infty$, then by Fatou's lemma,

$$\int_I |x|^p \, dt = \int_I \lim_{m\to\infty} |x_m|^p \, dt \leqslant \varliminf_{m\to\infty} \int_I |x_m|^p \, dt \leqslant C_1^p. \tag{28}$$

Thus, it has been established that $x \in W^n_{p,r}(\boldsymbol{I}; Q_k)$ and $x^{(k)}(0) = 1$. We show that x is a solution of (11). Indeed, since $x_m^{(n)}$ converges weakly to $x^{(n)}$, it follows that

$$\|x^{(n)}\|_r \leqslant \lim_{m\to\infty} \|x_m^{(n)}\|_r,$$

and thus, by (28),

$$S_n = \lim_{m\to\infty} \left(a\|x_m\|_p + b\|x_m^{(n)}\|_r\right) \geqslant \varliminf_{m\to\infty} a\|x_m\|_p + \varliminf_{m\to\infty} b\|x_m^{(n)}\|_r$$
$$\geqslant a\|x\|_p + b\|x^{(n)}\|_r.$$

The proof of the existence of a solution is similar in problem (12). But it is also possible to use the proposition in §1 by regarding (12) as the problem dual to (11) with $p = t'$ and $r = s'$.

c) By the preceding there exist solutions $\hat{x}$ and $\hat{y}$ of the respective problems (11) and (12), and the values of these problems are equal. From the reduction of problem (22) to (12) it follows at once that the pair $\hat{y}^* = (\hat{y}_1^*, \hat{y}_2^*)$ with $\hat{y}_1^* = (-1)^k \hat{y}^{(n)}$ and $\hat{y}_2^* = (-1)^{n-k-1}\hat{y}$ is a solution of (22). Further, since $\varphi(\hat{y}^*) > -\infty$, it has been shown that

$$\varphi(\hat{y}^*) = -(\langle y_1^*, x\rangle + \langle \hat{y}_2^*, x^{(n)}\rangle)$$

for any $x \in A$. If we now set $x - \hat{x}$, we have

$$a\|\hat{x}\|_p + b\|\hat{x}^{(n)}\|_r = (-1)^{k+1}\int_0^\infty \hat{x}\hat{y}^{(n)}\, dt + (-1)^{n-k}\int_0^\infty \hat{x}^{(n)}\hat{y}\, dt. \tag{29}$$

By Hölder's inequality,

$$(-1)^{k+1}\int_0^\infty \hat{x}\hat{y}^{(n)}\, dt \leqslant \int_0^\infty |\hat{x}|\,|\hat{y}^{(n)}|\, dt \leqslant \|\hat{x}\|_p \|\hat{y}^{(n)}\|_{p'} \leqslant a\|\hat{x}\|_p, \tag{30}$$

$$(-1)^{n-k}\int_0^\infty \hat{x}^{(n)}\hat{y}\, dt \leqslant \int_0^\infty |\hat{x}^{(n)}|\,|\hat{y}|\, dt \leqslant \|\hat{x}^{(n)}\|_r \|\hat{y}\|_{r'} \leqslant b\|\hat{x}^{(n)}\|_r. \tag{31}$$

Because of (29) all the inequalities in (30) and (31) are equalities. In particular, $\|\hat{y}^{(n)}\|_{p'}\hat{y} = a$ and $\|\hat{y}\|_{r'} = b$.

Suppose first that $p > 1$. Then it is known that the second equality in (30) means that $|\hat{y}^{(n)}|^{p'} = \lambda|\hat{x}|^p$ for some $\lambda > 0$. Integrating, we get that $\lambda = a^{p'}\|\hat{x}\|_p^{-p}$. Further, since

$$(-1)^{k+1}\hat{x}\hat{y}^{(n)} \leqslant |\hat{x}|\,|\hat{y}^{(n)}| = \lambda^{1/p'}|\hat{x}|^{p/p'}|\hat{x}| = \lambda^{1/p'}|\hat{x}|^p,$$

it follows that

$$f = \lambda^{1/p'}|\hat{x}|^p - (-1)^{k+1}\hat{x}\hat{y}^{(n)} \geqslant 0$$

a.e. on $\boldsymbol{R}_+$. On the other hand, the first equality in (30) implies that the integral of f is equal to zero and, consequently,

$$(-1)^{k+1}\hat{x}\hat{y}^{(n)} = \lambda^{1/p'}|\hat{x}|^p = \lambda^{1/p'}|\hat{x}|^{p-1}\hat{x}\,\mathrm{sgn}\,\hat{x}$$

a.e. on $\boldsymbol{R}_+$. If we now divide this equality by $(-1)^{k+1}\hat{x}(t)$ when $\hat{x}(t) \neq 0$ and take into account the expression for λ, we get the second relation in (13). This relation is true also when $\hat{x}(t) = 0$, because $|\hat{x}|^p$ and $|\hat{y}^{(n)}|^{p'}$ are proportional.

Let $p = 1$. Then

$$(-1)^{k+1}\hat{x}\hat{y}^{(n)} \leqslant |\hat{x}|\,|\hat{y}^{(n)}| \leqslant |\hat{x}|\,\|\hat{y}^{(n)}\|_\infty \leqslant a|\hat{x}|,$$

and, consequently,

$$g = a|\hat{x}| - (-1)^{k+1}\hat{x}\hat{y}^{(n)} \geqslant 0$$

a.e. on $\boldsymbol{R}_+$. The integral of g is equal to zero because the first and last terms in (30) are equal, and it thus follows in exactly the same way as above that the second equality in (13) holds also when $p = 1$.

The proof of the first relation in (13) is completely analogous, with use of the equalities in (31). The theorem is proved.

PROOF OF THEOREM 1. a). Suppose first that $p < \infty$. Since $Q_k \subset \mathfrak{A}_k$ and $\mathfrak{B}_k = J_k \setminus \mathfrak{A}_k$, it follows that

$$\widetilde{Q_k \cup \mathfrak{B}_k} = \mathfrak{G}[J_k \setminus (Q_k \cup \mathfrak{B}_k)] = \mathfrak{G}(\mathfrak{A}_k \setminus Q_k) = Q_k'.$$

If we now apply successively (14), (19), assertion a) in Theorem 2, and (15), we get that

$$\begin{aligned} K_n(k;\, p, r;\, \boldsymbol{R};\, Q_k) &= \left[\alpha^\alpha(1-\alpha)^{1-\alpha}S_n(k;\, p, r;\, \boldsymbol{R};\, Q_k;\, 1, 1)\right]^{-1} \\ &= \left[\alpha^\alpha(1-\alpha)^{1-\alpha}S_n(k;\, p, r;\, \boldsymbol{R}_+;\, Q_k \cup \mathfrak{B}_k;\, 2^{1/p}, 2^{1/r})\right]^{-1} \\ &= \left[\alpha^\alpha(1-\alpha)^{1-\alpha}S_n^*(n-k-1;\, r', p';\, \boldsymbol{R}_+;\, Q_k';\, 2^{1/r}, 2^{1/p})\right]^{-1} \\ &= \left[\alpha^\alpha(1-\alpha)^{1-\alpha}2^\gamma K_n(n-k-1;\, r', p';\, \boldsymbol{R}_+;\, \tilde{Q}_k')\right]^{-1}. \end{aligned}$$

Let $r > 1$. Then $r' < \infty$, and, by the simple relation $\tilde{Q}_k' = Q_k \cup \mathfrak{B}_k$, successive application of (14), Theorem 2, (20) and (15) gives us that

$$\begin{aligned} &K_n(n-k-1;\, r', p';\, \boldsymbol{R}_+;\, Q_k') \\ &\quad= \left[(1-\alpha)^{1-\alpha}\alpha^\alpha S_n(n-k-1;\, r', p';\, \boldsymbol{R}_+;\, Q_k';\, 1, 1)\right]^{-1} \\ &\quad= \left[\alpha^\alpha(1-\alpha)^{1-\alpha}S_n^*(k;\, p, r;\, \boldsymbol{R}_+;\, Q_k \cup \mathfrak{B}_k;\, 1, 1)\right]^{-1} \\ &\quad= \left[\alpha^\alpha(1-\alpha)^{1-\alpha}S_n^*(k;\, p, r;\, \boldsymbol{R};\, Q_k;\, 2^{1/p}, 2^{1/r})\right]^{-1} \\ &\quad= \left[\alpha^\alpha(1-\alpha)^{1-\alpha}2^\gamma K_n(k;\, p, r;\, \boldsymbol{R};\, Q_k)\right]^{-1}. \end{aligned}$$

Thus proves a).

b) Since $r > 1$, there exists a solution $\hat{x}$ of the problem $\zeta_1 = (k;\, p, r;\, \boldsymbol{R};\, Q_k;\, 1/2, 1/2)$ by Theorem 2. By Lemma 4 (see its proof), it can be assumed that $\hat{x}$ is an even or odd function according to whether k is even or odd, and its restriction to $\boldsymbol{R}_+$ is a solution of the problem

$$\zeta_2 = (k;\, p, r;\, \boldsymbol{R}_+;\, Q_k \cup \mathfrak{B}_k;\, 2^{-1/p'}, 2^{-1/r'}).$$

Further, since $p < \infty$, there exists a solution $\hat{y}$ of the dual problem

$$\zeta_2^* = (n-k-1;\, r', p';\, \boldsymbol{R}_+;\, Q_k';\, 2^{-1/r'}, 2^{-1/p'})$$

that is connected with $\hat{x}$ by the relations (13), which in this case are precisely (3). It now remains only to observe that, by Lemma 3, the solutions of ζ_1 and ζ_2^* are extremal functions in the respective inequalities $\xi = (k; p, r; \boldsymbol{R}; Q_k)$ and $\xi' = (n-k-1; r', p'; \boldsymbol{R}_+; Q'_k)$. The theorem is proved.

PROOF OF COROLLARY 1. The set $\mathfrak{A}_k$ is empty if and only if $n \leqslant 2$ or $n = 3$ and $k = 1$. Consequently, only in these cases are $K_n(\xi)$ and $K_n(\xi')$ (simultaneously) the best constants in inequalities of the type (1). The quantities $K_2(1; \infty, p'; \boldsymbol{I})$ for $1 \leqslant p' \leqslant \infty$ and $K_3(1; \infty, p'; \boldsymbol{I})$ for $1 \leqslant p' < \infty$ ($\boldsymbol{I} = \boldsymbol{R}$ or $\boldsymbol{I} = \boldsymbol{R}_+$) were found by Arestov [7], with the first two in explicit form (the case $p' = \infty$ corresponds to the well-known inequalities of Landau ($\boldsymbol{I} = \boldsymbol{R}_+$) and Hadamard ($\boldsymbol{I} = \boldsymbol{R}$)). The last two constants were computed by Kolmogorov [8] ($\boldsymbol{I} = \boldsymbol{R}$) and Matorin [9] ($\boldsymbol{I} = \boldsymbol{R}_+$). The relations (4)–(7) now follow at once from these results and part a) of Theorem 1. Corollary 1 is proved.

PROOF OF COROLLARY 2. By a minimax theorem of Ky Fan [10], the inf and sup in (8) can be interchanged. Then it is easy to see that E_n is the value of the following problem:

$$x^{(k)}(0) - N\|x\|_p \to \sup, \quad \|x^{(n)}\|_r \leqslant 1, \qquad x \in W^n_{p,r}(\boldsymbol{I}; Q_k).$$

It can next be established just as in Lemma 3 that

$$K_n(k; p, r; \boldsymbol{I}; Q_k) = \alpha^{-\alpha}(1-\alpha)^{-(1-\alpha)} N^\alpha E_n^{1-\alpha}(N; k; p, r; \boldsymbol{I}; Q_k). \qquad (32)$$

This relation was proved earlier by Gabushin for $Q_k = \varnothing$ [11].

The required stability is now a simple consequence of (32) and the first assertion in Theorem 1. Corollary 2 is proved.

We make some remarks.

1) Theorem 2 remains valid for problems (11) and (12), considered on a finite closed interval (for example, for periodic functions). Hence, certain duality relations between the suprema of the intermediate derivatives (at a point, in general) are obtained also in this case under restrictions on the functions themselves and their highest derivatives.

2) It can be shown that Theorem 1 is true also for pairs of inequalities of the form $\xi = (k; \infty, 1; \boldsymbol{I}; Q_k)$ and $\xi' = (n-k-1; \infty, v; \boldsymbol{I}'; Q'_k)$, where $Q_k \subset \mathfrak{A}_k$, and v symbolizes the space $V(\boldsymbol{I}')$.

3) The relations (3) can be obtained in a number of cases also from other considerations; for example, application of the Pontryagin maximum principle to the problem (11) (on this see the author's article [12]), where sharp constants are computed and extremal functions are found in the pair of dual inequalities $\xi = (k; p, \infty; \boldsymbol{I}; \varnothing)$ and $\xi' = (2-k-1; 1; p'; \boldsymbol{I}'; \varnothing)$, $1 \leqslant p < \infty$, $k = 0, 1$.

In conclusion the author thanks V. M. Tikhomirov for his attention to this work.

BIBLIOGRAPHY

1. V. N. Gabushin, *Inequalities for norms of a function and its derivatives in the L_p-metrics*, Mat. Zametki **1** (1967), 291–298; English transl. in Math. Notes **1** (1967).

2. S. B. Stechkin, *Best approximation of linear operators*, Mat. Zametki **1** (1967), 137–148; English transl. in Math. Notes **1** (1967).

3. Pierre-Jean Laurent, *Approximation et optimisation*, Hermann, Paris, 1972.

4. Ivar Ekeland and Roger Teman, *Analyse convexe et problèmes variationnels*, Dunod and Gauthier-Villars, 1974; English transl., North-Holland, 1976.

5. V. M. Alekseev, V. M. Tikhomirov, and S. V. Fomin, *Optimal control*, "Nauka", Moscow, 1979. (Russian)

6. Leonard D. Berkovitz and Harry Pollard, *A non-classical variational problem arising from an optimal filter problem*, Arch. Rational Mech. Anal. **26** (1967), 281–304; corrections, ibid. **28** (1967/68), 399.

7. V. V. Arestov, *On sharp inequalities between the norms of functions and their derivatives*, Acta. Sci. Math. (Szeged) **33** (1972), 243–267. (Russian)

8. A. N. Kolmogorov, *On inequalities between the suprema of successive derivatives of an arbitrary function on an infinite interval*, Uchen. Zap. Moskov. Gos. Univ. Matematika **30** (1939), no. 3, 3–16; English transl. in Amer. Math. Soc. Transl. (1) **2** (1962).

9. A. P. Matorin, *On inequalities between the largest values of the moduli of a function and its derivatives on the half-lane*, Ukrain. Mat. Zh. **7** (1955), 262–266; English transl. in Amer. Math. Soc. Transl. (2) **8** (1958).

10. Ky Fan, *Minimax theorems*, Proc. Nat. Acad. Sci. U.S.A. **39** (1953), 42–47.

11. V. N. Gabushin, *Best approximation of functionals on certain sets*, Mat. Zametki **8** (1970), 551–562; English transl. in Math. Notes **8** (1970).

12. G. G. Magaril-Il'yaev, *On Kolmogorov's inequalities on a half-line*, Vestnik Moskov. Univ. Ser. 1 Mat. Mekh. **1976**, no. 5, 33–41; English transl. in Moscow Univ. Math. Bull. **1976**.

Translated by H. H. McFADEN

Труды Матем. Инст. Стеклов.
161(1983)

Proceedings of the Steklov
Institute of Mathematics
1984, Issue 3

COERCIVE ESTIMATES AND SEPARATION THEOREMS FOR ELLIPTIC EQUATIONS IN R^n

UDC 517.43

M. OTELBAEV

ABSTRACT. The differential operator

$$Lu = \rho(x)(-\Delta)^l \rho(x)u(x) + q(x)u(x)$$

is studied in the space $L_2(R^n)$. For a rather large class of coefficients $\rho(x)$, $q(x)$ and weight function $r(x)$ conditions are given which ensure:

a) the existence of a coerciveness estimate (separation)

$$\left\|\rho(x)(-\Delta)^l\rho(x)u\right\|_2 + \|q(x)u\|_2 \leqslant C(\|Lu\|_2 + \|u\|_2),$$

where C does not depend on $u \in D(L)$ ($D(\cdot)$ is the domain), and

b) the existence of an estimate

$$\|r(x)D^\alpha\rho(x)u\|_\theta \leqslant C(\|Lu\|_2 + \|u\|_2).$$

Here $\|\cdot\|_\theta$ is the norm in $L_\theta(R^n)$, and

$$D^\alpha = \partial^{\alpha_1+\cdots+\alpha_n}/\partial x_1^{\alpha_1}\cdots\partial x_n^{\alpha_n}, \qquad \alpha = (\alpha_1,\dots,\alpha_n).$$

Bibliography: 22 titles.

In a rather large class of boundary value problems for elliptic equations $Lu = f$ it is possible to estimate (in some norm) the derivatives of order at most m (m is equal to the order of L) of the function u in terms of the norm of Lu and suitable norms of the boundary functions. Such problems are called coercive problems. They have now been well studied.

For elliptic equations with nonconstant coefficients in the case of degeneracy or in the case of unbounded domains this circumstance, generally speaking, does not hold. For example, there exists a function $y(x) \in L_2(J)$ ($J = (-\infty, \infty)$) such that $-y'' - x^2 \in L_2(J)$ but $y'' \notin L_2(J)$ (see [1] and [2]).

In this paper we study the operator

$$Lu = \rho(x)(-\Delta)^l\rho(x)u(x) + q(x)u(x) \tag{0.1}$$

in $L_2(R^n)$. Here $l > 0$ is an integer, Δ is the Laplace operator, and $x = (x_1,\dots,x_n) \in R^n$.

We shall be interested in the existence of estimates

$$\left\|\rho(-\Delta)^l\rho u\right\|_2 + \|qu\|_2 \leqslant c(\|Lu\|_2 + \|u\|_2), \tag{0.2}$$

$$\|r(x)D^\alpha\rho u\|_\theta \leqslant c(\|Lu\|_2 + \|u\|_2), \tag{0.3}$$

1980 *Mathematics Subject Classification*. Primary 35J30, 47F05; 35B45.

where $\|\cdot\|$ and $\|\cdot\|_\theta$ are norms in the respective spaces $L_2(R^n)$ and $L_\theta(R^n)$, and $r(x)$ is a continuous function in R^n.

If the coercive estimate (0.2) holds for all u in $D(L)$, the domain of the operator L, then we say that the operator L is *separated*. The term "separation" was first used by Everitt and Giertz [2]–[6], who studied the separation of the Sturm-Liouville operator in $L_2(J)$. Their results were later strengthened in [7]–[13].

The present paper consists of seven sections. In §§1–4 some auxiliary estimates are obtained. The main results, namely, estimates of the type (0.2) and (0.3), are proved in §§5 and 6. In the seventh and last section examples are presented, and a method is indicated for generalizing the main results to the case of operators of the general form

$$Lu = \rho(-\Delta)^l \rho u + \sum_{|\alpha|\leqslant 2l} r_\alpha D^\alpha \rho u + qu. \tag{0.4}$$

In almost all the main theorems sufficient conditions are given which coincide with the necessary conditions or are close to them. The main results are formulated in terms of certain auxiliary functions which can be constructed from the coefficients of the equation in an efficient manner.

In the case where the coefficients of the equation behave "well" at infinity our results admit formulation in terms of the coefficients themselves. For example, we have the following result.

THEOREM 0.1. *Let L be the closure in $L_2(R^n)$ of the operator $lu = \rho(-\Delta)^l\rho u + qu$ defined on $C_0^\infty(R^n)$, where $\rho(x)$ and $q(x)$ are sufficiently smooth functions in R^n such that*

$$\rho(x) > 0, \quad \operatorname{Im} q(x) > -c > -\infty, \quad \operatorname{Re} q(x) > -c > -\infty,$$
$$\inf\{\rho^{-2}(y)(|q(y)|+1) : y \in R\} \neq 0,$$
$$c_1^{-1} \leqslant \rho(x)\rho^{-1}(y) \leqslant c_1, \qquad c^{-1} \leqslant (|q(x)|+1)(|q(y)|+1)^{-1} \leqslant c$$

for $|x-y| \leqslant [\rho^{-2}(y)(|q(y)|+1)]^{-1/2l}k(y)$, where $k(y) \geqslant 4$ is a continuous function which tends to $+\infty$ as $|y| \to +\infty$. Then the following assertions are true:

a) *The operator L is maximal dissipative and separated.*

b) *If $r(x)$ is a continuous function in R^n and $|\alpha| = k \leqslant 2l-1$, $2 < \theta < \infty$, then the operator $r(x)D^\alpha\rho(x)(L+\lambda E)^{-1}$ is bounded (compact) from $L_2(R^n)$ to $L_\theta(R^n)$ if and only if*

$$B = \sup_{x\in R^n} B(x) < \infty \quad \left(B < \infty, \ \lim_{|x|\to\infty} B(x) = 0\right),$$

where λ is any resolvent point of the operator L and $B(x)$ is the function defined by

$$B(x) = \sup_{0<\eta\leqslant m^{-1}(x)} \left[\eta^{-p}\int_{|t-x|\leqslant\eta}\left(\frac{k|r(t)|}{\rho(t)}\right)^\theta dt\right]^{1/\theta}$$
$$\left(p = \theta(n/2 - 2l + k),\ m(x) = \left[\rho^{-2}(x)(|q(x)|+1)\right]^{1/2l}\right).$$

c) $$c_\lambda^{-1}B \leqslant \left\|r(x)D^\alpha\rho(x)(L+\lambda E)^{-1}\right\|_{L_2\to L_\theta} \leqslant c_\lambda B,$$

where c_λ does not depend on $r(x)$.

This theorem is a simple corollary of the stronger Theorems 5.1 and 6.1; in it no conditions (except the natural conditions) are imposed on $r(x)$, while power functions, for example, satisfy the conditions imposed on $q(x)$ and $\rho(x)$.

As we shall see below in §§5 and 6, estimates of the type (0.2) and (0.3) hold for a very large class of operators including "interesting operators" with coefficients which oscillate rapidly at infinity. An example is the operator

$$Lu = -\rho(x)(u(x))'' + e^{|x|}(\sin^2 e^{x^2})u, \tag{0.5}$$

which is separated in $L_2(J)$ (see §7, Example 1). Here, $\rho(x)$ is an infinitely smooth function equal to $e^{-|x|/2}$ for $|x| \geqslant 1$. Moreover, if $r(x)$ is a continuous function in J, then the operator

$$r(x)\frac{d}{dx}e^{-|x|/2}L^{-1}$$

is bounded from $L_2(J)$ to $L_\theta(J)$ $(2 \leqslant \theta < \infty)$ if and only if

$$\sup_{t\in J}\left(\int_{|t-x|\leqslant e^{-|t|}}\left|r(t)e^{|t|/2}\right|^\theta dt\right)^{1/\theta} < \infty.$$

Particular parts of this paper (including the formulation of the problems) were discussed at different times with Sh. A. Alimov, A. G. Kostyuchenko, B. M. Levitan, and P. I. Lizorkin. I therefore use this opportunity to express to them by gratitude.

NOTATION. $D(\cdot)$ is the domain, and $R(\cdot)$ is the range.

$\|\cdot\|_{X\to Y}$ is the norm of an operator from X to Y. For $X = L_p$ and $Y = L_\theta$ we often denote $\|\cdot\|_{X\to Y}$ by $\|\cdot\|_{p\to\theta}$.

$\|\cdot\|_{p,\Omega}$ is the norm of an element in $L_p(\Omega)$. If it leads to no confusion we often omit Ω in the notation $L_p(\Omega)$. For $\Omega = R^n$ in place of $\|\cdot\|_{p,\Omega}$ we write $\|\cdot\|_p$; if $p = 2$ in the notation $\|\cdot\|_{p,\Omega}$ or $\|\cdot\|_p$, we omit it.

The expression

$$D^\alpha u = \partial^{|\alpha|}u/\partial x_1^{\alpha_1}\cdots\partial_{x_n}^{\alpha_n}, \qquad \alpha = (\alpha_1,\ldots,\alpha_n)$$

is a multi-index, and $|\alpha| = \alpha_1 + \cdots + \alpha_n$.

$c, c_1, c_2, \ldots$ are various constants whose exact values do not interest us.

§1. Some known assertions and localization

In this section we present a number of elementary assertions on separation and a lemma on localization which reduces questions of separation for an operator with nonregular coefficients to the case of an operator with smooth periodic coefficients.

Let $\rho(x)$ be a real, positive function in R^n which is continuously differentiable $2l$ times, and let $q(x)$ be a bounded function on each compact set with $\operatorname{Im} q(x)$ and $\operatorname{Re} q(x)$ lower semibounded. The operator L defined on $C_0^\infty(R^n)$ by the equality

$$Lu = \rho(-\Delta)^l\rho u + \operatorname{sign} qu, \tag{1.1}$$

where $l > 0$ is an integer, admits closure, which we also denote by the letter L.

DEFINITION 1.1. Following Everitt and Giertz [2]–[6], we say that L is *separated* if

$$\|qu\|_2 + \left\|\rho(-\Delta)^l\rho u\right\|_2 \leqslant c(\|Lu\|_2 + \|u\|_2)$$

where c does not depend on $u \in D(L)$.

The following result is easily deduced from the definition and the triangle inequality.

LEMMA 1.1. *The following assertions* 1)–4) *are pairwise equivalent*:

1) *The operator L is separated.*

2) *The operator $L - \lambda E$ is separated for at least one complex λ.*

3) $\sup\{\|qu\|_2(\|Lu\|_2 + \|u\|_2)^{-1}: u \in D(L)\} < \infty$;

4) $\sup\{\|\rho(-\Delta)^l\rho u\|_2(\|Lu\|_2 + \|u\|_2)^{-1}: u \in D(L)\} > \infty$.

Because of this lemma, we may henceforth assume with no loss of generality that

$$\operatorname{Im} q(x) \geqslant 1, \qquad \operatorname{Re} q(x) \geqslant 1. \tag{1.2}$$

We recall ([14], Chapter IV, §4) that an operator A in a Hilbert space H is called *dissipative* if

$$\operatorname{Im}\langle Au, u\rangle_H \geqslant 0, \qquad u \in D(A).$$

A dissipative operator A is called *maximal dissipative* ([14], loc. cit.) if the set $\{Au + \lambda u: u \in D(A)\}$ ($\operatorname{Im}\lambda > 0$) is dense in H.

We note that if A is maximal dissipative, then the upper half-plane of the complex plane is contained in the resolvent set of A.

We choose a collection $\{\varphi_j\}_{j\geqslant 1}$ of nonnegative functions in $C_0^\infty(R^n)$ such that

$$\sum_{j\geqslant 1} \varphi_j^2 = 1, \qquad \operatorname{supp}\varphi_j \subset U_j, \qquad \bigcup_{j\geqslant 1} U_j = R^n, \tag{1.3}$$

where the U_j are open sets whose multiplicity of intersection is no greater than some number $\xi = \xi(n) < \infty$.(1)

We extend ρ and q from U_j to all of R^n so that their extensions $\rho_j(x)$ and $q_j(x)$ are bounded, periodic functions of the same period and the following conditions are satisfied: $\operatorname{Im} q_j(x) > 0.5$, $\operatorname{Re} q_j(x) > 0.5$, $\inf\{\rho_j(x): x \in R^n\} > 0$, and $\sup\{|D^\alpha\rho_j(x)|: x \in R^n\} < \infty$ for any $|\alpha| < 2l$. Such an extension is possible because of the conditions imposed on $q(x)$ and $\rho(x)$. We denote by L_j the closure of the operator $\rho_j(-\Delta)^l\rho_j u + q_j u$ defined on $C_0^\infty(R^n)$.

LEMMA 1.2. *Suppose that L is maximal dissipative, $f \in C_0^\infty(R^n)$, and* $\operatorname{Im}\lambda > 0$. *Then*

$$(L + \lambda E)^{-1} f = (L + \lambda E)^{-1} B_\lambda f + M_\lambda f, \tag{1.4}$$

where

$$\begin{aligned} B_\lambda f &= f - \sum_{j\geqslant 1} (L_j + \lambda E)\varphi_j(L_j + \lambda E)^{-1}\varphi_j f, \\ M_\lambda f &= \sum_{j\geqslant 1} \varphi_j(L_j + \lambda E)^{-1}\varphi_j f. \end{aligned} \tag{1.5}$$

PROOF. Since $F \in C_0^\infty(R^n)$, the sums defining $B_\lambda f$ and $M_\lambda f$ contain only finitely many terms. From general theorems for elliptic operators it follows that the operator L_j is maximal dissipative and separated, and, in addition,

$$\|D^\alpha u\|_2 \leqslant c_j\left(\|L_j u\|_2 + \|u\|_2\right)$$

(1) We say that a covering $\{U_j\}_{j\geqslant 1}$ has multiplicity ξ if no set $U_j \in \{U_j\}_{j\geqslant 1}$ intersects more than ξ sets of the family $\{U_j\}_{j\geqslant 1}$.

for $|\alpha| \leqslant 2l$. Because of what has been said, all the following computations are legitimate:

$$(L+\lambda E)^{-1}f = (L+\lambda E)^{-1}\left[f - \sum_{j\geqslant 1}(L_j+\lambda E)\varphi_j(L_j+\lambda E)^{-1}\varphi_j f\right]$$
$$+(L+\lambda E)^{-1}\sum_{j\geqslant 1}(L_j+\lambda E)\varphi_j(L_j+\lambda E)^{-1}\varphi_j f.$$

Since on the support of the function φ_j the coefficients of the operators L and L_j coincide, we have

$$(L+\lambda E)^{-1}(L_j+\lambda E)\varphi_j(L_j+\lambda E)^{-1}\varphi_j f = \varphi_j(L_j+\lambda E)^{-1}\varphi_j f.$$

From these equalities it follows that $(L+\lambda E)^{-1}f = (L+\lambda E)^{-1}B_\lambda f + M_\lambda f$. The lemma is proved.

LEMMA 1.3. *Suppose that* $\|B_\lambda\| < 1$ *for some* λ $(\operatorname{Im}\lambda > 0)$. *Then the operator L is maximal dissipative. If in addition the inequality* $\sup\{\|q_j(L_j+\lambda E)^{-1}\|: j = 1,2,\dots\} < \infty$ *holds, then L is separated.*

PROOF. If $\|B_\lambda\| < 1$, then the operator equation $X = XB_\lambda + M_\lambda$ has a unique solution X_λ. It is not hard to verify that the operator inverse to X_λ coincides with the closure in L_2 of the operator $L + \lambda E$ defined on $C_0^\infty(R^n)$. From what has been said it follows that L is maximal dissipative. From (1.4) we therefore find that $(L+\lambda E)^{-1} = M_\lambda(E - B_\lambda)^{-1}$. Since $(E-B_\lambda)^{-1}$ is bounded, we have that $\|q(L+\lambda E)^{-1}\| \leqslant c(\lambda)\|qM_\lambda\|$. Taking into account the finite multiplicity of the covering $\{U_j\}_{j\geqslant 1}$, from this inequality and the definition of M_λ we obtain

$$\left\|q(L+\lambda E)^{-1}\right\| \leqslant c(\lambda)\sup\left\{\left\|q_j(L_j+\lambda E)^{-1}\right\|: j = 1,2,\dots\right\},$$

which by Lemma 1.1 completely proves Lemma 1.3.

LEMMA 1.4. *Suppose that A is a maximal dissipative operator and* $\operatorname{Re}\langle Ax, x\rangle \geqslant 0$ *for* $x \in D(A)$. *If the operator* $B(A+iE)^{-1}$ *is compact, then* $\|B(A+\lambda E)^{-1}\| \to 0$ *as* $\operatorname{Re}\lambda \to +\infty$.

This lemma is well known. It can be proved by contradiction.

LEMMA 1.5 (on localization). *If*

$$\lim_{j\to\infty}\left\|\rho_j\left(D^\beta\varphi_j\right)\left(D^\alpha\rho_j L_j^{-1}\right)\right\| = 0 \tag{1.6}$$

for $|\alpha| \leqslant 2l - 1$, $|\alpha| + |\beta| \leqslant 2l$ *and*

$$\overline{\lim_{j\to\infty}}\left\|q_j L_j^{-1}\right\| < \infty, \tag{1.7}$$

then the following assertions are true:

a) *The operator L is maximal dissipative and separated.*

b) *For* λ $(\operatorname{Im}\lambda \geqslant 0, \operatorname{Re}\lambda \geqslant 0)$ *of sufficiently large modulus the equality* $(L+\lambda E)^{-1} = M_\lambda(E-B_\lambda)^{-1}$ *holds, in which* $\|B_\lambda\| \leqslant 0.5$.

PROOF. For $\operatorname{Im}\lambda \geqslant 0$ and $\operatorname{Re}\lambda \geqslant 0$ we have $\|L_j(L_j + \lambda E)^{-1}\| \leqslant c$, where c does not depend on λ or j. Therefore,

$$\left\|\rho_j\left(D^\beta\varphi_j\right)D^\alpha\rho_j\left(L_j + \lambda E\right)^{-1}\right\| \leqslant c\left\|\rho_j D^\beta\varphi_j D^\alpha\rho_j L_j^{-1}\right\|, \tag{1.8}$$

$$\sup\left\{\left\|q_i\left(L_j + \lambda E\right)^{-1}\right\| : j = 1,2,\ldots\right\} \leqslant \sup\left\{\left\|q_j L_j^{-1}\right\| : j = 1,2,\ldots\right\} < \infty. \tag{1.9}$$

By (1.6) and (1.8), for any $\delta > 0$ it is possible to find j_0 such that for $j \geqslant j_0$

$$\left\|\rho_j\left(D^\beta\varphi_j\right)D^\alpha\rho_j\left(L_j + \lambda E\right)^{-1}\right\| \leqslant \delta \tag{1.10}$$

for all λ, $\operatorname{Im}\lambda \geqslant 0$, $\operatorname{Re}\lambda \geqslant 0$.

Since $|\alpha| \leqslant 2l - 1$ and the φ_j have compact support, the operators $\rho_j(D^\beta\varphi_j)D^\alpha\rho_j(L_j + \lambda E)^{-1}$ are compact. Therefore, according to Lemma 1.4, it is possible to find $\lambda_0 = \lambda_0(j_0)$ depending on j_0 such that for $\operatorname{Im}\lambda \geqslant \operatorname{Im}\lambda_0$ and $\operatorname{Re}\lambda \geqslant \operatorname{Re}\lambda_0$ (1.10) holds for all $j \leqslant j_0$. By choosing $\delta > 0$ sufficiently small, from (1.10) we obtain $\|B_\lambda\| \leqslant 0.5$. The present lemma therefore follows from Lemma 1.3 and (1.9).

LEMMA 1.6. *Suppose that* $|\alpha| \leqslant 2l - 1$ *and* $r(x)$ *is a continuous function in* R^n. *Assume that conditions* (1.6) *and* (1.7) *of Lemma* 1.5 *are satisfied. If* $2l - |\alpha| \geqslant n(1/2 - 1/\theta)$ *and* $2 \leqslant \theta < \infty$, *then*

$$\|r(x)D^\alpha\rho L^{-1}\|_{L_2\to L_\theta} \leqslant c\sup\left\{\left\|r(x)D^\alpha\rho\varphi_j L_j^{-1}\varphi_j\right\|_{L_2\to L_\theta} : j = 1,2,\ldots\right\}.$$

PROOF. Let $f \in L_2$. Choosing λ appropriately and using Lemma 1.5b), we have

$$\|r(x)D^\alpha\rho L^{-1}f\|_\theta = \left\|r(x)D^\alpha\rho\left(L + \lambda E\right)^{-1}g\right\|_\theta = \left\|r(x)D^\alpha\rho M_\lambda\left(E - B_\lambda\right)^{-1}g\right\|,$$

where $q = (L + \lambda E)L^{-1}f$.

Using the finite multiplicity of the covering $\{U_j\}_{j\geqslant 1}$ and the definition of M_λ, we find further that

$$\begin{aligned}\|r(x)D^\alpha\rho L^{-1}f\|_\theta &\leqslant c_1\left\|r(x)D^\alpha\rho M_\lambda\left(E - B_\lambda\right)^{-1}g\right\|_\theta \\ &\leqslant c_2\left(\sum_{j\geqslant 1}\left\|r(x)D^\alpha\rho\varphi_j\left(L_j + \lambda E\right)^{-1}\varphi_j\left(E - B_\lambda\right)^{-1}g\right\|_\theta^\theta\right)^{1/\theta}\end{aligned}$$

Using the known inequality $(\Sigma|a_j|^\theta)^{1/\theta} \leqslant (\Sigma|a_j|^2)^{1/2}$, valid for $\theta \geqslant 2$, we thus arrive at

$$\begin{aligned}\|r(x)D^\alpha\rho L^{-1}f\|_\theta &\leqslant c_2\left(\sum_{j\geqslant 1}\left\|r(x)D^\alpha\rho\varphi_j\left(L_j + \lambda E\right)^{-1}\varphi_j\left(E - B_\lambda\right)^{-1}g\right\|_\theta^2\right)^{1/2} \\ &\leqslant c_3\left(\sum_{j\geqslant 1}\left\|r(x)D^\alpha\rho\varphi_j\left(L_j + \lambda E\right)^{-1}\varphi_j\right\|_{2\to\theta}^2\left\|\chi_j\left(E - B_\lambda\right)^{-1}g\right\|_2^2\right)^{1/2},\end{aligned}$$

where χ_j is the characteristic function of the set $\operatorname{supp}\varphi_j$.

Since the multiplicity of the covering $\{U_j\}_{j\geqslant 1}$ is no greater than $\xi = \xi(n) < \infty$, from this we obtain

$$\|r(x)D^\alpha\rho L^{-1}f\|_\theta \leqslant c\sup\left\{\left\|r(x)D^\alpha\rho\varphi_j(L_j+\lambda E)^{-1}\varphi_j\right\|_{2\to\theta}: j=1,2,\dots\right\}$$
$$\times\left\|(E-B_\lambda)^{-1}g\right\|_2,$$

which because of the inequalities

$$\left\|(E-B_\lambda)_g^{-1}\right\| \leqslant \left\|(E-B_\lambda)^{-1}\right\|\|g\| \leqslant 2\|g\| = 2\left\|(L+\lambda E)^{-1}f\right\| \leqslant 4(\|\lambda\|+1)\|f\|$$

proves Lemma 1.6.

LEMMA 1.7. *Suppose that* $|\alpha| \leqslant 2l-1$, $2l-|\alpha| > n(1/2-1/\theta)$, $2\leqslant\theta<\infty$, *and* $r(x)$ *is a continuous function. If*

$$\lim_{j\to\infty}\left\|r(x)D^\alpha\rho\varphi_j(L_j+\lambda E)^{-1}\varphi_j\right\|_{L_2\to L_\theta} = 0,$$

then the operator $r(x)D^\alpha\rho(L+\lambda E)^{-1}$ *is compact from* L_2 *to* L_θ.

PROOF. Because of Lemma 1.5, for λ of sufficiently large modulus with $\operatorname{Re}\lambda \geqslant 0$ and $\operatorname{Im}\lambda \geqslant 0$ we have

$$r(x)D^\alpha\rho(L+\lambda E)^{-1} = r(x)D^\alpha\rho M_\lambda(E-B_\lambda)^{-1}$$
$$= \left[\sum_{j\geqslant 1} r(x)D^\alpha\rho\varphi_j(L_j+\lambda E)^{-1}\varphi_j\right](E-B_\lambda)^{-1}, \quad \|B_\lambda\| \leqslant 0.5.$$

Each term of the first factor on the right side is a compact operator from L_2 to L_θ because of the conditions $|\alpha| \leqslant 2l-1$, $2l-|\alpha| > n(1/2-1/\theta)$. This follows from the fact that the L_j are elliptic. Therefore,

$$r(x)D^\alpha\rho(L+\lambda E)^{-1} = T_{j_0} + \sum_{j>j_0} r(x)D^\alpha\rho\varphi_j(L_j+\lambda E)^{-1}\varphi_j(E-B_\lambda)^{-1},$$

where T_{j_0} is a compact operator. The second term on the right side of this inequality tends to zero in norm as $j_0 \to +\infty$. This can be proved by means of the relations used in the proof of Lemma 1.6. This implies the compactness of the operator $r(x)D^\alpha\rho(L+\lambda E)^{-1}$ from L_2 to L_θ.

§2. Auxiliary estimates and assertions

From the lemmas proved in §1 it is evident that the choice of the functions $\{\varphi_j\}$ and the covering $\{U_j\}$ of (1.3) must play an important role in solving the problems we have posed. For this reason we need the following lemma.

LEMMA 2.1. *Let* $m(x) \geqslant \varepsilon > 0$ *be a function satisfying the condition*

$$c^{-1} \leqslant m(x)m^{-1}(y) \leqslant c \quad \text{for } |x-y| \leqslant k(y)m^{-1}(y), \tag{2.0}$$

where $k(y) \geqslant 4$ *is a continuous function tending to* $+\infty$ *as* $|y| \to +\infty$.

Then there exists a covering $\{U_j\}$ *of the space* R^n *and a collection of functions* $\{\varphi_j\}$ *for which the following conditions hold*:

a) $\sum_{j\geqslant 1}\varphi_j^2 = 1$, $\varphi_j \in C_0^\infty(U_j)$, $\bigcup_{j\geqslant 1} U_j = R^n$;
b) $c_1^{-1} \leqslant m(x)m^{-1}(y) \leqslant c_1$ for $x, y \in U_j$;
c) $\lim_{j\to\infty}\sup\{|D^\beta\varphi_j(x)|m^{-|\alpha|}(x): x \in R^n\} = 0$ for $|\beta| \leqslant |\alpha|$;

d) *U_j is contained in the cube Q^j with edge length equal to $\frac{1}{10}\sqrt{k(y_j)}\, m^{-1}(y_j)$ and center at the point y_j, where $y_j \in U_j$.*

e) *Each set U_j can intersect no more than $\xi = \xi(n)$ sets of the family $\{U_j\}_{j\geqslant 1}$.*

A covering lemma of closely related content was established (apparently for the first time) by Boĭmatov [15], and again somewhat later by Feigin [16]. We shall not prove Lemma 2.1, since it can be proved in the same way as the analogous lemma in [16].

Let $m(y) \geqslant \varepsilon_0$ and $k(y)$ be functions satisfying (2.0), and let $Q^{(y)}$ and $\tilde{Q}^{(y)}$ be cubes with centers at y and edge lengths equal to $\sqrt{k(y)}\, m^{-1}(y)$ and $\frac{1}{10}\sqrt{k(y)}\, m^{-1}(y)$ respectively. We assume that the function $\rho(y)$ satisfies the condition

$$c^{-1} < \rho(x)\rho^{-1}(y) < c \quad \text{for } |x-y| \leqslant k(y)m^{-1}(y), \tag{2.1}$$

where c does not depend on $y \in R^n$.

We extend the function $q(x)\rho^{-2}(x)$ from the cube $Q^{(y)}$ periodically with period equal to $\sqrt{k(y)}\, m^{-1}(y)$ in each variable. We denote this extension by $(q\rho^{-2})_y(x)$; we extend $\rho(x)$ from the cube $\tilde{Q}^{(y)}$ in some fashion so that this extension has period equal to $\sqrt{k(y)}\, m^{-1}(y)$,

$$\begin{aligned} &\left|D^\alpha \rho_y(x)\right| \leqslant b_\alpha < \infty \quad \text{for } |\alpha| \leqslant 2l \quad \text{and} \\ &c_2^{-1} \leqslant \rho(y)\rho_y^{-1}(x) \leqslant c_2 \quad \text{for all } x \in R^n, \end{aligned} \tag{2.2}$$

where ρ_y is the extension of $\rho(x)$ from the cube $\tilde{Q}^{(y)}$, and c_2 does not depend on $y \in R^n$. Of the constants b_α it is required only that they be finite.

We denote by $q_y(x)$ the function $\rho_y^2(x)(q\rho^{-2})_y(x)$. We then have $(q\rho^{-2})_y(x) = \rho_y^{-2}(x)q_y(x)$. It is obvious that the function $q_y(x)$ has period $\sqrt{k(y)}\, m^{-1}(y)$ and coincides with $q(x)$ in the cube $\tilde{Q}^{(y)}$. We further subject q_y to the condition $\operatorname{Re} q_y(x) \geqslant 0.5$, $\operatorname{Im} q_y(x) \geqslant 0.5$. Such extensions are possible because of (2.1) and (1.2). The notation used here we shall use throughout the remainder of the paper. We do not indicate the dependence of ρ_y and q_y on the function m if no confusion is possible.

LEMMA 2.2. *Suppose that $m(x) \geqslant \varepsilon_0 > 0$ satisfies condition* (2.0), *and $\rho(x)$ satisfies condition* (2.1); *let ρ_y and q_y be the functions constructed above. Suppose that for any $y \in R^n$*(2)

$$\begin{aligned} &\left\|(-\Delta)^{l/2}u\right\|^2 + \int \left|\left(q_y\rho_y^{-2l}\right)(x)\right| \cdot |u(x)|^2\, dx \\ &\qquad \geqslant c_2^{-1}\left(\left\|(-\Delta)^{l/2}u\right\|^2 + m^{2l}(y)\|u\|^2\right), \end{aligned} \tag{2.3}$$

$$\|q_y u\|^2 \leqslant c_2\rho^4(y)m^{2l}(y)\left(\left\|(-\Delta)^{l/2}u\right\|^2 + m^{2l}(y)\|u\|^2\right), \tag{2.4}$$

where c_2 does not depend on $y \in R^n$ or $u(x) \in C_0^\infty(R^n)$. Then

a) *the operator L is maximal dissipative and separated; and*

b) $\sup\{\|q_y(L_y + \lambda E)^{-1}\|: y \in R^n\} < \infty$ *where* $\operatorname{Im}\lambda \geqslant 0$, $\operatorname{Re}\lambda \geqslant 0$, *and L_y is the operator analogous to L in which $\rho(x)$ and $q(x)$ are replaced by $\rho_y(x)$ and $q_y(x)$.*

---

(2) For fractional $l/2$, by definition, $((-\Delta)^{l/2}u)(x) = F^{-1}_{\xi\to x}|\xi|^l F_{y\to\xi}u(y)$, where F and F^{-1} are the direct and inverse Fourier transforms.

PROOF. It may be assumed that $\lambda = 0$. We shall estimate the norm of the operator $q_y L_y^{-1}$. We set $A_y = \|q_y L_y^{-1}\|$. We have

$$A_y = \left\{\left\|q_y L_y^{-1} u\right\| \cdot \|u\|^{-1} : u \in L_2(R^n)\right\}.$$

The operator L_y is maximal dissipative. Therefore, it is easy to show that

$$A_y = \sup\left\{\|q_y u\| \|L_y u\|^{-1} : u \in C_0^\infty(R^n)\right\}.$$

From this, by virtue of the inequalities

$$\|L_y u\|^2 \geqslant |\langle L_y u, u\rangle|^2 \|u\|^{-2} = \|u\|^{-2} \left|\left(\left\|(-\Delta)^{l/2} \rho_y u\right\|^2 + \int q_y(x) |u(x)|^2 \, dx\right)\right|^2$$

$$\geqslant 2^{-2} \|u\|^{-2} \left(\left\|(-\Delta)^{l/2} \rho_y u\right\|^2 + \int |q_y(x)| |u(x)|^2 \, dx\right)^2,$$

which are consequences of the Cauchy-Schwarz-Bunyakovskiĭ inequality and (1.2), we find that

$$A_y^2 \leqslant 2 \sup\left\{\|q_y u\|^2 \|u\|^2 \left(\left\|(-\Delta)^{l/2} \rho_y u\right\|^2 + \left\|\sqrt{q_y}\, u\right\|^2\right)^{-2} : u \in C_0^\infty(R^n)\right\}$$

$$= 2 \sup\left\{\left\|q_y \rho_y^{-1} u\right\|^2 \left\|\rho_y^{-1} u\right\|^2 \left(\left\|(-\Delta)^{l/2} u\right\|^2 + \left\|\sqrt{q_y}\, \rho_y^{-1} u\right\|^2\right)^{-2} : u \in C_0^\infty(R^n)\right\}.$$

We now use (2.2) and (2.3):

$$A_y^2 \leqslant c_3 \rho^{-4}(y) \sup\left\{\|q_y u\|^2 \|u\|^2 \left(\left\|(-\Delta)^{l/2} u\right\|^2 + m^{2l}(y) \|u\|^2\right)^{-2} : u \in C_0^\infty(R^n)\right\}$$

$$\leqslant c_4 \rho^{-4}(y) m^{-2l}(y) \sup\left\{\|q_y u\|^2 \left(\left\|(-\Delta)^{l/2} u\right\|^2 + m^{2l}(y) \|u\|^2\right)^{-1} : u \in C_0^\infty(R^n)\right\},$$

where c_4 does not depend on $y \in R^n$. Part b) of the lemma follows from this and (2.4).

We now estimate the norm of the operator $r(x) D^\alpha \rho_y L_y^{-1}$, $|\alpha| \leqslant 2l$, where $r(x)$ is a continuous function. We set $S_y = \|r(x) D^\alpha \rho_y L_y^{-1}\|$. For $u \in D(L_y)$ we find that

$$\left\|\rho_y (-\Delta)^l \rho_y u\right\| = \|L_y u - q_y u\| \leqslant \|L_y u\| + \|q_y u\|$$

$$= \|L_y u\| + \left\|q_y L_y^{-1} L_y u\right\| \leqslant (1 + A_y) \|L_y u\|.$$

From this it follows that

$$\|q_y u\|^2 + \left\|\rho_y (-\Delta)^l \rho_y u\right\|^2 \leqslant 2(A_y + 1)^2 \|L_y u\|^2.$$

Using this inequality, since $A_y \leqslant c_5 < \infty$ by the part of the lemma already proved, we find that

$$\begin{aligned}
&\left\|r(x) D^{\alpha}\rho_y L_y^{-1}\right\|^2 \\
&\quad = \sup\left\{\left\|r(x) D^{\alpha}\rho_y u\right\|^2 \left\|L_y u\right\|^{-2} : u \in D(L_y)\right\} \\
&\quad \leqslant c_6 \sup\left\{\left\|r(x) D^{\alpha}\rho_y u\right\|^2 \left(\left\|q_y u\right\|^2 + \left\|\rho_y(-\Delta)^l \rho_y u\right\|^2\right)^{-1} : u \in D(L_y)\right\} \\
&\quad = c_6 \sup\left\{\left\|r(x) D^{\alpha} v\right\|^2 \left(\left\|q_y \rho_y^{-1} v\right\|^2 + \left\|\rho_y(-\Delta)^l v\right\|^2\right)^{-1} : v\rho_y^{-1} \in D(L_y)\right\}.
\end{aligned}$$

These inequalities give, by (2.2),

$$\begin{aligned}
&\left\|r(x) D^{\alpha}\rho_y L_y^{-1}\right\|^2 \\
&\quad \leqslant c_7 \rho^{-2}(y) \sup_{v\rho_y^{-1} \in \mathscr{D}(L_y)} \left\{\left\|r(x) D^{\alpha} v\right\|^2 \left(\left\|q_y(x)\rho_y^{-2}(x) v\right\|^2 + \left\|(-\Delta)^l v\right\|^2\right)^{-1}\right\}. \qquad (2.5)
\end{aligned}$$

From the triangle and Cauchy-Schwarz-Bunyakovskiĭ inequalities it follows that

$$\begin{aligned}
\left\|(-\Delta)^l_v\right\|^2 + \left\|q_y \rho_y^{-2} v\right\|^2 &\geqslant \tfrac{1}{4}\left(\left\|(-\Delta)^l v\right\|^2 + \left\|(-\Delta)^l v + |q_y| \rho_y^{-2} v\right\|^2\right) \\
&\geqslant \tfrac{1}{4}\left\|(-\Delta)^l v\right\|^2 + \tfrac{1}{4}\|v\|^{-2}\left(\left\|(-\Delta)^{l/2} v\right\|^2 + \left\|\sqrt{q_y}\,\rho_y^{-1} v\right\|^2\right)^2.
\end{aligned}$$

From this and (2.3) we find

$$\left\|(-\Delta)^l v\right\|^2 + \left\|q_y \rho_y^{-2} v\right\|^2 \geqslant \tfrac{1}{4}\left(\left\|(-\Delta)^l v\right\|^2 + m^{4l}(y)(y)\|v\|^2\right).$$

We substitute this inequality into (2.5):

$$\begin{aligned}
&\left\|r(x) D^{\alpha}\rho_y L_y^{-1}\right\|^2 \\
&\quad \leqslant c_8 \rho^{-2}(y) \sup_{v \in C_0^{\infty}(R^n)} \left\{\left\|r(x) D^{\alpha} v\right\|^2 \left(\left\|(-\Delta)^l v\right\|^2 + m^{4l}(y)\|v\|^2\right)^{-1}\right\}. \qquad (2.6)
\end{aligned}$$

Using a suitable similarity transformation, from (2.6) with $r(x) \equiv 1$ we obtain

$$\begin{aligned}
\left\|D^{\alpha}\rho_y L_y^{-1}\right\|^2 &\leqslant c_8 \rho^{-2}(y) m^{2l(-2+|\alpha|/l)}(y) \\
&\quad \times \sup_{u \in C_0^{\infty}(R^n)} \left\{\left\|D^{\alpha} u\right\|^2 \left(\left\|(-\Delta)^l u\right\|^2 + \|u\|^2\right)^{-1}\right\} \\
&\leqslant c_9 \rho^{-2}(y) m^{-4l+2|\alpha|}(y). \qquad (2.7)
\end{aligned}$$

Let $\{\varphi_j\}_{j \geqslant 1}$ and $\{y_j\}_{j \geqslant 1}$ be the collections of functions and points of Lemma 2.1 corresponding to the function $m(x)$. By the part of Lemma 2.2. already proved and Lemma 1.5 it follows that Lemma 2.2 will be completely proved if we show that

$$\lim_{j \to +\infty} \left\|\rho_{y_j}\left(D^{\beta}\varphi_j\right)\left(D^{\alpha}\rho_{y_j} L_{y_j}^{-1}\right)\right\| = 0 \qquad (2.8)$$

for $|\beta| \neq 0$, $|\beta| + |\alpha| \leqslant 2l$, $|\alpha| \leqslant 2l - 1$.

We have

$$\left\|\rho_{y_j}\left(D^{\beta}\varphi_j\right) D^{\alpha}\rho_{y_j} L_{y_j}^{-1}\right\| \leqslant \left\|\rho_{y_j} D^{\beta}\varphi_j\right\|_{C(R^n)} \left\|D^{\alpha}\rho_{y_j} L_{y_j}^{-1}\right\|.$$

Using (2.7) and (2.4), we now get

$$\left\|\rho_{y_j}\left(D^{\beta}\varphi_j\right)D^{\alpha}\rho_{y_j}L_{y_j}^{-1}\right\| \leqslant c_{10}\left\|D^{\beta}\varphi_j\right\|_{C(R^n)}m^{-2l+|\alpha|}(y_j).$$

According to the choice of the collection $\{y_j\}_{j\geqslant 1}$, the right side of this expression tends to zero as $j \to +\infty$. Equality (2.8) is thus proved.

The notation of Lemma 2.2 is used in the next lemma.

LEMMA 2.3. *Let $m(x)$ be a function satisfying condition (2.1), and let $r(x)$ be a continuous function. Assume that the conditions of Lemma 2.2 are satisfied. If $2 \leqslant \theta < \infty$, $\gamma = -4l + 2n(1/2 - 1/\theta) + 2|\alpha| \leqslant 0$ and $|\alpha| < 2l$, then $R(y) \leqslant c\lambda(y)$ for any $y \in R^n$, where*

$$R(y) = \left\|r_y(x)D^{\alpha}\rho_y L_y^{-1}\right\|_{L_1\to L_\theta}^2, \qquad r_y(x) = r(x)\chi_y(x),$$

$$\lambda(y) = \rho^{-2}(y)m^{\gamma}(y)\left\|r_y(tm^{-1}(y))(-\Delta+E)^{-l+|\alpha|/2}\right\|_{L_2\to L_\theta}^2,$$

$\chi_y(x)$ is the characteristic function of the cube $\tilde{Q}^{(y)}$ with center at y and edge length equal to $\frac{1}{10}\sqrt{k(y)}\,m^{-1}(y)$, and $(-\Delta+E)^{-l+|\alpha|/2}$ is the closure of the operator defined on $C_0^\infty(R^n)$ by the equality

$$(-\Delta+E)^{-l+|\alpha|/2}u = F_{\xi\to x}\left(|\xi|^2+1\right)^{\theta}F_{y\to\xi}u(y).$$

Here F and F^{-1} are the direct and inverse Fourier transforms.

PROOF. Just as in (2.6), we obtain

$$R(y) \leqslant c_1\rho^{-2}(y)\sup_{v\in C_0^\infty(R^n)}\left\{\|r(x)D^{\alpha}v\|_\theta^2\left(\left\|(-\Delta)^l v\right\|^2 + m^{4l}(y)\|v\|^2\right)^{-1}\right\}.$$

Using the similarity transformation $x \to t/m(y)$, we now deduce that

$$R(y) \leqslant c_1\rho^{-2}m^{\gamma}(y)\sup_{v(x)\in C_0^\infty(R^n)}\left\{\|s(t)D^{\alpha}v(x)\|_\theta^2\left(\left\|(-\Delta)^l v\right\|^2 + \|v\|^2\right)^{-1}\right\},$$

where $s(t) = r_y(tm^{-1}(y))$. However

$$\left\|(-\Delta)^l v\right\|^2 + \|v\|^2 \leqslant c_2\left\|(-\Delta+E)^l v\right\|^2.$$

Therefore,

$$\begin{aligned}
R(y) &\leqslant c_3\rho^{-2}(y)m^{\gamma}(y)\sup_{\|v\|=1}\left\|s(t)D^{\alpha}(-\Delta+E)^{-l}v(t)\right\|_\theta^2 \\
&= c_3\rho^{-2}(y)m^{\gamma}(y)\left\|s(t)D^{\alpha}(-\Delta+E)^{-l}\right\|_{L_2\to L_\theta}^2 \\
&= c_3\rho^{-2}(y)m^{\gamma}(y)\left\|s(t)(-\Delta+E)^{-l+|\alpha|/2}\left(D^{\alpha}(-\Delta+E)^{-|\alpha|/2}\right)\right\|_{L_2\to L_\theta}^2 \\
&\leqslant c_3\rho^{-2}(y)m^{\gamma}(y)\left\|s(t)(-\Delta+E)^{-l+|\alpha|/2}\right\|_{L_2\to L_\theta}^2\left\|D^{\alpha}(-\Delta+E)^{-|\alpha|/2}\right\|^2 \\
&\leqslant c_4\lambda(y).
\end{aligned}$$

The lemma is proved.

LEMMA 2.4. *Let $m(x) \geqslant \varepsilon_0 > 0$ be a function satisfying* (2.0), *and let $r(x)$ be a continuous function. Assume that the conditions of Lemma 2.2 are satisfied. If $2 \leqslant \theta < \infty$, $\gamma = -4l + 2n(1/2 - 1/\theta) + 2|\alpha| \leqslant 0$ and $|\alpha| < 2l$, then*

$$\|r(x)D^\alpha\rho(x)L^{-1}\|^2_{L_2\to L_\theta} \leqslant c\sup\{\lambda(y): y \in R^n\},$$

where $\lambda(y)$ is the same as in Lemma 2.3.

PROOF. Let $\{\varphi_j\}_{j\geqslant 1}$ and $\{y_j\}_{j\geqslant 1}$ be the collections of functions and corresponding points of Lemma 2.1 connected with the function $m(x)$. According to Lemma 1.6, it suffices to prove that

$$\left\|r(x)\chi_{y_j}(x)D^\alpha\rho_{y_j}\varphi_j L^{-1}_{y_j}\varphi_j\right\|^2_{L_2\to L_\theta} \leqslant c_0\sup\{\lambda(y): y \in R^n\}. \tag{2.9}$$

We have

$$\left\|r(x)\chi_{y_j}D^\alpha\rho_{y_j}\varphi_j L^{-1}_{y_j}\varphi_j\right\|^2_{L_2\to L_\theta} \leqslant c_1 \sum_{|\beta|+|\alpha'|=|\alpha|}\left\|r(x)\chi_{y_j}(x)\left(D^\beta\varphi_j\right)D^{\alpha'}\rho_{y_j}L^{-1}_{y_j}\right\|^2_{L_2\to L_\theta}.$$

Using the choice of $\{\varphi_j\}_{j\geqslant 1}$ and Lemma 2.3, from this we obtain

$$\begin{aligned}
&\left\|r(x)\chi_{y_j}(x)D^\alpha\rho_{y_j}\varphi_j L^{-1}_{y_j}\varphi_j\right\|^2_{L_2\to L_\theta} \\
&\quad\leqslant c_2\rho^{-2}(y_j)\sum_{|\beta|+|\alpha'|=|\alpha|} m^{2|\beta|-4l+2n(1/2-1/\theta)+2|\alpha'|} \\
&\qquad\qquad\times\left\|r\left(tm^{-1}(y_j)\right)\chi_{y_j}\left(tm^{-1}(y_j)\right)(-\Delta+E)^{-l+|\alpha'|/2}\right\|^2_{L_2\to L_\theta} \\
&\quad= c_2\rho^{-2}(y_j)\sum_{|\beta|+|\alpha'|=|\alpha|} m^\gamma(y_j)\left\|r\left(tm^{-1}(y_j)\right)\chi_{y_j}\left(tm^{-1}(y_j)\right)\right. \\
&\qquad\qquad\left.\times(-\Delta+E)^{-l+|\alpha'|/2}\right\|^2_{L_2\to L_\theta} \\
&\quad\leqslant c_2\rho^{-2}(y_j)m^\gamma(y_j) \\
&\qquad\times\sum_{|\beta|+|\alpha'|=|\alpha|}\left\|r\left(tm^{-1}(y_j)\right)\chi_{y_j}\left(tm^{-1}(y_j)\right)(-\Delta+E)^{-l+|\alpha|/2}\right\|_{L_2\to L_\theta} \\
&\qquad\times\left\|(-\Delta+E)^{-|\alpha'|+|\alpha|/2}\right\|^2_{L_2\to L_\theta} \\
&\quad\leqslant c_3\rho^{-2}(y_j)m^\gamma(y_j)\left\|r\left(tm^{-1}(y_j)\right)\chi_{y_j}\left(tm^{-1}(y_j)\right)(-\Delta+E)^{-l+|\alpha|/2}\right\|^2_{L_2\to L_\theta}.
\end{aligned}$$

From these estimates and the definition of $\lambda(y)$ we obtain (2.9). The lemma is proved.

LEMMA 2.5. *Let $m(x) \geqslant \varepsilon_0 > 0$ be a function satisfying* (2.0), *andlet $r(x)$ be a continuous function. Assume that the conditions of Lemma 2.2 are satisfied. If $2 \leqslant \theta < \infty$, $\gamma = -4l + 2n(1/2 - 1/\theta) + 2|\alpha| < 0$, $|\alpha| < 2l$, and $\lim_{|y|\to\infty}\lambda(y) = 0$, then the operator $r(x)D^\alpha\rho(x)L^{-1}$ is compact from L_2 to L_θ. Here $\lambda(y)$ is the same function as in Lemma* 2.4.

This lemma is proved in the same way as Lemma 2.4; where Lemma 1.6 was used it is necessary to use Lemma 1.7.

Below Lemmas 2.2, 2.4, and 2.5 play a basic role. On the basis of them, we can already present some results. To illustrate what has been said, we shall prove the following result.

THEOREM 2.1. *Let $r(x)$ be a continuous function, $r(x) \neq 0$, and suppose that*

$$\begin{gathered}\inf\{\rho^{-2}(y)(|q(y)|+1), y \in R^n\} \neq 0, \\ c^{-1} \leqslant \rho(x)\rho^{-1}(y) \leqslant c, \qquad c^{-1} \leqslant r(x)r^{-1}(y) \leqslant c, \\ c^{-1} \leqslant |q(x)q^{-1}(y)| \leqslant c\end{gathered} \tag{2.10}$$

for $|x-y| \leqslant k(y)(\rho^{-2}(y)|q(y)|)^{-1/2l}$, where $k(y) \geqslant 4$ is a continuous function tending to $+\infty$ as $|y| \to +\infty$. Suppose that $\gamma < 0$ and $2 \leqslant \theta < \infty$, where

$$\gamma = -4l + 2|\alpha| + 2n(1/2 - 1/\theta) < 0.$$

Then the following assertions are true:

a) *The operator L is maximal dissipative and separated.*

b) *The operator $r(x)D^\alpha \rho L^{-1}$ is bounded if*

$$\sup\{B(y): y \in R^n\} < \infty,$$

where $B(y) = r(y)\rho^{-1}(y)m^{\gamma/2}(y)$, $m(y) = (\rho^{-2}(y)|q(y)|)^{1/2l}$.

c) *The operator $r(x)D^\alpha \rho L^{-1}$ is compact if $\lim_{|y| \to +\infty} B(y) = 0$.*

d)

$$\|rD^\alpha \rho L^{-1}\|_{L_2 \to L_\theta} \leqslant c_1 \sup\{B(y): y \in R^n\}.$$

PROOF. In Lemmas 2.2, 2.4, and 2.5 we take $m(x) = (\rho^{-2}(x)|q(x)|)^{1/2l}$. Using (2.10) and the known inequality $\|(-\Delta + E)^{-\alpha}\|_{L_2 \to L_\theta} < \infty$ valid for $\alpha > n(1/2 - 1/\theta)$, $2 \leqslant \theta < \infty$, one can then verify directly that the conditions of Lemma 2.2 are satisfied and $\lambda(y) \leqslant cB^2(y)$. The theorem is thus a corollary of Lemmas 2.4 and 2.5.

REMARK 2.1. As we shall see in §6, the conditions of parts b) and c) of Theorem 2.1 are also necessary, and, in addition to the inequality $\|rD^\alpha \rho L^{-1}\|_{2 \to \theta} \leqslant c_1 B$, we also have the inequality $B \leqslant c_2 \|rD^\alpha \rho L^{-1}\|_{2 \to \theta}$ where $c_2 < \infty$ depends on the constants of (2.10) and the function $k(y)$.

§3. Some known inequalities for Sobolev spaces

In order to prove theorems more precise than Theorem 2.1 we need to choose the function $m(x)$ in Lemmas 2.2, 2.4, and 2.5 in a suitable way and also be able to estimate the norms of operators of the form $\mathrm{i}r(x)(-\Delta + E)^{-\alpha}$. For this we need some auxiliary materials which will be presented in this and the next sections.

We choose a collection of functions $\{\mu_j^{(\varepsilon)}\}_{j \geqslant 1}$, $c_\varepsilon^{-1} \leqslant \sum_{j \geqslant 1} \mu_j^{(\varepsilon)}(x) \leqslant c_\varepsilon$ in R^n, where the $\mu_j^{(\varepsilon)}$ are obtained by translating a small fixed nonnegative function $\mu^{(\varepsilon)} \in C_0^\infty(R^n)$ with support contained in a cube of edge length ε.

The next theorem follows from results of Strichartz [17].

THEOREM 3.1. *For $1 < p < \infty$ and $\alpha > 0$*

$$c_p^{-1} \sum_{j \geqslant 1} \|(-\Delta)^\alpha \mu_j u\|_p^p \leqslant \|(-\Delta)^\alpha u\|_p^p + \|u\|_p^p \leqslant c_p \sum_{j \geqslant 1} \|(-\Delta)^\alpha \mu_j u\|_p^p,$$

where c_p does not depend on $u \in C_0^\infty(R^n)$.

The next result is well known for spaces of Sobolev type and is easily deduced from the results of Strichartz [17].

THEOREM 3.2. *If* $1 < p \leqslant \theta < \infty$ *and* $\alpha \geqslant 0$, *then*

$$\|(-\Delta)^{\alpha} u\|_p \leqslant c_{p,\theta}\|(-\Delta)^{\alpha} u\|_{\theta},$$

where $c_{p,\theta}$ *does not depend on* $u \in C_0^\infty(Q_1)$.

In [18] and [19] Adams proved the following theorem.

THEOREM 3.3. *Let* $1 < p < \theta < \infty$, $pl < n$, $\beta = (n/p - l)$, *and* $(R_l u)(x) = \int |x-y|^{l-n} u(y)\,dy$. *Then the inequality*

$$\|v(x) R_l u(x)\|_\theta \leqslant A_1 \|u\|_p, \tag{3.1}$$

where A_1 *does not depend on* $u \in L_p$, *holds if and only if*

$$A_2 \overset{\text{def}}{=} \left(\sup_{x \in R^n} M_p(v)(x)\right)^{1/\theta} < \infty, \tag{3.2}$$

where

$$M_p(v)(x) = \sup_{r>0} r^{-\beta}\left(\int_{|x-t|\leqslant r} |v(t)|^\theta\,dt\right)^{1/\theta}.$$

Moreover, for the least constant A_1 *in* (3.1) *there are the estimates*

$$c^{-1}A_2 \leqslant A_1 \leqslant cA_2, \tag{3.3}$$

where c *depends only on* n, l, p, *and* θ.

If $u \in C_0^\infty(R^n)$, then the function $(-\Delta)^{l/2}u$ belongs to L_p and $R_l(-\Delta)^{l/2}u = u$. Hence, from Theorem 3.3 it is not hard to obtain the following assertion.

THEOREM 3.3′.(³) *Let* $pl < n$ *and* $1 < p < \theta < \infty$. *Then the inequality*

$$\|v(x)u(x)\|_\theta \leqslant A_1\|(-\Delta)^{l/2}u\|_p,$$

where A_1 *does not depend on* $u \in C_0^\infty(R^n)$, *holds if* (3.2) *is satisfied. Moreover, inequality* (3.3) *hold for the least constant* A_1.

§4. Some integral estimates

We shall introduce the concept of $(2,l)$-capacity which provides us with the possibility of using the estimates given in [21].

DEFINITION [20], [21]. The $(2, l)$-*capacity* of a closed set e relative to an open set Ω containing e is the number

$$(2,l)\text{-cap}(e,\Omega) = \inf\|(-\Delta)^{l/2}u\|^2,$$

where the infimum is taken over all $u \in C_0^\infty(\Omega)$ equal to one in a neighborhood of e.

We denote by $N_{\varepsilon,a}(Q_d)$, $\varepsilon > 0$, $a > 1$, the collection of all closed subsets e of the cube $\overline{Q}_d$ such that

$$(2,l)\text{-cap}(e, Q_{ad}) \leqslant \varepsilon d^{n-2l}.$$

(³) For integers l Theorem 3.3′ was proved by V. G. Maz'ya before the appearance of Adams' work.

LEMMA 4.1. *If* $e \in N_{\varepsilon,a}(Q_d)$, $\varepsilon > 0$, $a > 1$, *then* meas $e \leqslant \varepsilon a^{2l} A d^n$, *where*

$$A = \sup\left\{\|u\|^2 \left\|(-\Delta)^{l/2} u\right\|^{-2} : u \in C_0^\infty(Q_1)\right\}$$

and meas *is Lebesgue measure.*

PROOF. By the definition of $N_{\varepsilon,a}(Q_d)$, for any $\delta > 0$ there exists a function $\varphi \in C_0^\infty(Q_{ad})$, equal to one in a neighborhod of e, such that

$$\left\|(-\Delta)^{l/2}\varphi\right\|^2 \leqslant (1+\delta)\varepsilon d^{n-2l}.$$

However,

$$\begin{aligned} \operatorname{meas} e &\leqslant \int \varphi^2(x)\,dx \leqslant \|\varphi\|^2 \\ &\leqslant \left\|(-\Delta)^{l/2}\varphi\right\|^2 \sup\left\{\|u\|^2\left\|(-\Delta)^{l/2}u\right\|^{-2} : u \in C_0^\infty(Q_{ad})\right\} \\ &= a^{2l}d^{2l}\left\|(-\Delta)^{l/2}\varphi\right\|^2 \sup\left\{\|u\|^2\left\|(-\Delta)^{l/2}u\right\|^{-2} : u \in C_0^\infty(Q_1)\right\} \\ &= a^{2l}d^{2l}A\left\|(-\Delta)^{l/2}\varphi\right\|^2. \end{aligned}$$

Therefore,

$$\operatorname{meas} e \leqslant a^{2l}d^{2l}A(1+\delta)d^{n-2l} = A\varepsilon a^{2l}d^n(1+\delta).$$

The lemma follows from this and the fact that $\delta > 0$ is arbitrary.

With a measurable function $V(t)$ that is bounded in each compact set we associate the function

$$V_\varepsilon^*(t) = \inf\left\{d^{-1} : d^{n-2l} \geqslant \inf_{e \in F_d^{(\varepsilon)}(t)} \int_{Q_d(t)\setminus e} |V(x)|\,dx\right\}, \tag{4.1}$$

where $F_d^{(\varepsilon)}(t)$ is the empty set for $2l > n$, while for $2l \leqslant n$ it is the collection of all compact subsets e of the cube $Q_d(t)$ satisfying the inequality meas $e \leqslant \varepsilon d^n$, $\varepsilon > 0$. For $2l > n$ the function $V_\varepsilon^*(t)$ does not depend on ε and is defined by

$$V_\varepsilon^*(t) = V^*(t) = \inf\left\{d^{-1} : d^{n-2l} \geqslant \int_{Q_d(t)} |V(x)|\,dx\right\}. \tag{4.2}$$

Below we shall establish some properties of these functions which will be used in subsequent sections. We immediately note the obvious inequality

$$V_{\varepsilon_1}^*(t) \leqslant V_{\varepsilon_2}^*(t), \qquad 0 < \varepsilon_2 \leqslant \varepsilon_1 < 1. \tag{4.3}$$

LEMMA 4.2. *Let* $0 < \varepsilon < 1$ *and*

$$C^2 = \sup_{u \in C_0^\infty(R^n)} \left\{\|y\|^2\left[\left\|(-\Delta)^{l/2}y\right\|^2 + \int |V(t)|\cdot\|y(t)\|^2\,dt\right]^{-1}\right\}.$$

Then

$$\inf\left\{V_\varepsilon^{*l}(t) : t \in R^n\right\} \leqslant c(\varepsilon)C^{-1},$$

where $c(\varepsilon)$ *depends only on* ε, l, *and* n.

PROOF. By (4.3) it suffices to prove the lemma for small $\varepsilon > 0$. Accordiing to Theorem 4.1 of [21] the quantity C satisfies the estimate

$$c_\gamma^{-1}C^{-1} \leqslant D_\gamma^l \leqslant c_\gamma C^{-1},$$

where

$$D_\gamma = \inf_{Q_d \subset R^n} \left\{ d^{-1} : d^{n-2l} \geqslant \inf_{e \in N_{\gamma,2}(Q_d)} \int_{Q_d \setminus \varepsilon} |V(t)|\, dt \right\}.$$

Here γ is any sufficiently small positive number not exceeding some number γ_0 depending on n and l.

It follows from Lemma 4.1 that for $\varepsilon \in (0, 1)$ it is possible to find a γ such that $\gamma \in (0, \gamma_0)$ and $N_{\gamma,2}(Q_d(t)) \subseteq F_d^{(\varepsilon)}(t)$. Therefore, $D_\gamma \geqslant \inf\{V^*(t) : t \in R^n\}$. The lemma follows from this and the estimate for D_γ given above.

LEMMA 4.3. *Suppose that* $8^n\varepsilon \in (0,1)$, $s \geqslant 100A^{-1}$, *and in the cube* $Q_s(y)$ *the inequality* $(8^{-n}V)^*_{8^n\varepsilon}(x) > A$ *is satisfied. Denote by* $\tilde{V}(t)$ *the periodic function with period s which coincides with* $V(t)$ *in* $Q_s(y)$. *Then*

$$\inf\{\tilde{V}_\varepsilon^*(t) : t \in R^n\} \geqslant A.$$

PROOF. From the definition of $\tilde{V}_\varepsilon^*(t)$ it follows that there exists a cube Q_{d_0} such that

$$d_0^{-1} \leqslant 2\inf\{\tilde{V}_\varepsilon^*(t) : t \in R^n\}; \quad d_0^{-2l+n} \geqslant \inf_{\operatorname{meas} e \leqslant \varepsilon d_0^n} \left\{ \int_{Q_{d_0} \setminus e} |\tilde{V}(t)|\, dt \right\}. \qquad (4.4)$$

From the periodicity of $\tilde{V}(t)$ it follows that $\tilde{V}_\varepsilon^*(t)$ is also a periodic function, and their periods coincide. It may therefore be assumed that the center x_0 of the cube Q_{d_0} is contained in $Q_s(y)$. if $d_0^{-1} \geqslant 10^{-1}A$, then Lemma 4.3 follows from the first inequality of (4.4).

Suppose that $d_0^{-1} < 10^{-1}A$. Two cases are possible: $Q_{d_0} \supseteq Q_{4s}(y)$ or $Q_{4s}(y) \not\subseteq Q_{d_0}$. If $Q_{4s}(y) \not\subseteq Q_{d_0}$, then it is not hard to see that $Q_{4s}(y) \cap Q_{d_0}$ contains a cube $Q_{d_1}(x_1)$, where $d_1 = 8^{-1}d_0$ and $x_1 \in Q_{4s}(y) \cap Q_{d_0}$. Because of (4.4), for this cube we obtain

$$d_1^{-2l+n} \geqslant \inf\left\{ \int_{Q_{d_1}(x_2) \setminus e} 8^{-n}|V(t)|\, dt : e, \operatorname{meas} e \leqslant \varepsilon 8^n d_1^n \right\}.$$

From this and the definition of $(8^{-n}V)^*_{\varepsilon 8^n}(x)$ we find that $d_1^{-1} \geqslant (8^{-n}V)^*_{\varepsilon 8^n}(x_1)$. Therefore,

$$\inf\{\tilde{V}_\varepsilon^*(t) : t \in R^n\} \geqslant 2^{-1}d_0^{-1} \geqslant \tfrac{1}{16}(8^{-n}V)^*_{\varepsilon 2^n}(x_1).$$

By the hypothesis of the lemma this inequality proves the required assertion.

If $Q_{4s}(y) \subseteq Q_{d_0}$, then Q_{d_0} contains at least $k = ([d_0/s] - 2)^n \geqslant 2^n$ cubes $Q_s(y_i)$ $(i = 1, \ldots, k)$ obtained by translating $Q_s(y)$ by vectors whose components have ratios to the number s which are integers. For any cube $Q_\alpha(x) \subseteq Q_s(y_i)$, where $\alpha^{-1} \in (\frac{1}{2}A, A)$, since $(8^{-n}V)^*_{8^n\varepsilon}(x) > A$ by the definition of $(8^{-n}V)^*_{8^n\varepsilon}(x)$ (by an argument by contradiction), we find the inequality

$$\alpha^{-2l+n} < \inf\left\{ \int_{Q_\alpha(x) \setminus e} |8^{-n}V(t)|\, dt : e, \operatorname{meas} e \leqslant 8^n\varepsilon\alpha^n \right\}. \qquad (4.5)$$

We choose the number $\alpha^{-1} \in (\frac{1}{2}A, A)$ so that $s\alpha^{-1}$ is an integer. Such a choice is possible since $s > 100A^{-1}$. We then find that it is possible to construct at least $k(s/\alpha)^n$ cubes $Q_\alpha(x_j) \subset Q_{d_0}$ $(j = 1,\ldots,k(s/\alpha)^n)$ whose interiors do not intersect and which are such that (4.5) is satisfied. Obviously,

$$k(s/\alpha)^n \geqslant (d_0/s - 3)^n (s/\alpha)^n = (d_0 - 3s)^n \alpha^{-n} \geqslant 4^{-n} d_0^n \alpha^{-n}. \tag{4.6}$$

We choose a set $\tilde{e} \subseteq Q_{d_0}$ such that meas $\tilde{e} \leqslant \varepsilon d_0^n$. We consider the set $e_j = \tilde{e} \cap Q_\alpha(x_j)$ $(j = 1,\ldots,k(s/\alpha)^n)$. We shall show that among these sets there are at least $k' \geqslant 8^{-n} d_0^n \alpha^{-n}$ sets $\{e_{j_m}\}_{m=1}^{k'}$ such that meas $e_{j_m} \leqslant 8^n \varepsilon \alpha^n$. Indeed, if there are fewer than $8^{-n} d_0^n \alpha^{-n}$ such sets, then the number of those sets e_j for which meas $e_j > 8^n \varepsilon \alpha^n$ is not less than $8^{-n} d_0^n \alpha^{-n}$ by (4.6). We therefore obtain the following contradictory inequalities:

$$\text{meas } \tilde{e} \geqslant \text{meas } e_j 8^{-n} d_0^n \alpha^{-n} > 8^n \varepsilon \alpha^n 8^{-n} d_0^n \alpha^{-n} \geqslant \varepsilon d_0^n.$$

Hence, $k' \geqslant 8^{-n} d_0^n \alpha^{-n}$.

From (4.5) we find that

$$\alpha^{-2l+n} < \int_{Q_\alpha(x_j')\backslash e_{jm}} |8^{-n} V(t)|\, dt, \qquad m = 1,\ldots,k'.$$

We sum these inequalities over all $m = 1,\ldots,k'$:

$$k'\alpha^{-2l+n} < \sum_{m=1}^{k'} \int_{Q_\alpha(x'_{j_m})\backslash e_{j_m}} |8^{-n} V(t)|\, dt.$$

We strengthen these inequalities:

$$8^{-n} d_0^n \alpha^{-2l} = 8^{-n} d_0^n \alpha^{-n} \alpha^{-2l+n} \leqslant k'\alpha^{-2l+n} < \int_{Q_{d_0}\backslash e} |8^{-n} V(t)|\, dt.$$

Now $\alpha^{-2l} > 8^{2l} d_0^{-2l}$, and hence

$$8^{2l} d_0^{n-2l} < \int_{Q_{d_0}\backslash e} |V(t)|\, dt.$$

Since $\tilde{e}$ is any set satisfying the inequality meas $\tilde{e} \leqslant \varepsilon d_0^n$, this inequality contradicts (4.4). The lemma is proved.

§5. The main theorems on separation

In this section we shall prove the main theorem on separation. We recall that L denotes the closure in $L_2(R^n)$ of the differential operator defined on $C_0^\infty(R^n)$ by the equality

$$Ly = (-1)^l \rho(x)(-\Delta)^l \rho(x) y(x) + q(x) y(x), \tag{5.1}$$

where $\rho(x)$ is a sufficiently smooth positive function, and $q(x)$ is a complex-valued function bounded in each compact set such that Re $q(x)$ and Im $q(x)$ are lower semibounded. For the problems we are solving it may be assumed with no loss of generality that Re $q(x) \geqslant 1$ and Im $q(x) \geqslant 1$.

THEOREM 5.1. *Let* $V(xz) = q(x)\rho^{-2}(x)$ *and* $m(x) = m_\varepsilon(x) = V_\varepsilon^*(x)$, *where* $\varepsilon \in (0, 1)$ *and* $V_\varepsilon^*(x)$ *is the function defined by* (4.1). *Assume that the following conditions are satisfied*:

$$\inf\{m(x) \colon x \in R^n\} \neq 0 \quad \textit{and}$$

$$c^{-1} \leqslant m(x)m^{-1}(y) \leqslant c, \qquad c^{-1} \leqslant \rho(x)\rho^{-1}(y) \leqslant c$$
$$\textit{for } |x - y| \leqslant k(y)m^{-1}(y), \tag{5.2}$$

where $k(y)$ is a continuous function tending to $+\infty$ as $|y| \to +\infty$. Denote by $A_p(x)$ the function defined by

$$A_p(x) = \begin{cases} \rho^{-4}(x)m^{-4l+n}(x)\int_{|x-t|\leqslant m^{-1}(x)} |q(t)|^2\,dt & \text{for } 2l > n, \\ \rho^{-4}(x)m^{-4l+n-\beta}(x)\sup_{0<\eta<m^{-1}(x)} \eta^{-\beta}\int_{|y-t|\leqslant\eta} |q(t)|^2\,dt & \text{for } 2l \leqslant n, \end{cases}$$

where $\beta = 2(n/p - l)$ and p is any number in the interval $(1,2)$. If for some $p \in (1,2)$ the quantity $A_p = \sup\{A_p(x)\colon x \in R^n\}$ is finite, then the operator L is maximal dissipative and separated.

PROOF. By (5.2) the conditions (2.0) and (2.1) hold. Let y be any point of R^n, and let $\rho_y(x)$ and $q_y(x)$ be the functions of Lemma 2.2 constructed for the functions $\rho(x)$ and $q(x)$, where $\rho_y(x)$ satisfies (2.2). To prove the theorem it suffices to establish inequalities (2.3) and (2.4).

We prove (2.3). We choose $\delta > 0$ sufficiently small. Then by Lemma 4.2

$$\begin{aligned} &\left[\inf\left\{\left(8^n\rho_y^{-2}(t)q_y(t)\right)_\delta^{*l}\colon t \in R^n\right\}\right]^2\|u\|^2 \\ &\quad\leqslant c\left(\left\|(-\Delta)^{l/2}u\right\|^2 + \int 8^n\left|q_y(t)\rho_y^{-2}(t)\right|\left|u(t)\right|^2 dt\right) \\ &\quad\leqslant c_1\left(\left\|(-\Delta)^{l/2}u\right\|^2 + \int \left|q_y(t)\rho_y^{-2}(t)\right|\left|u(t)\right|^2 dt\right). \end{aligned}$$

We now use Lemma 4.3 and condition (5.2). Then

$$m^{2l}(y)\|u\|^2 \leqslant c\left(\left\|(-\Delta)^{l/2}u\right\|^2 + \int \left|q_y(t)\rho_y^{-2}(t)\right|\left|u(t)\right|^2 dt\right),$$

whence (2.3) follows.

We now prove (2.4). Using the similarity transformation $x = tm^{-1}(y)$, we find that (2.4) is equivalent to the inequality

$$\left\|\rho^{-2}(y)m^{-2l}(y)q_y\left(tm^{-1}(y)\right)u(t)\right\|^2 \leqslant c\left(\left\|(-\Delta)^{l/2}u\right\|^2 + \|u\|^2\right),$$

in which $c < \infty$ does not depend on $u(t) \in C_0^\infty(R^n)$ and $y \in R^n$. According to Strichartz' Theorem 3.1 the right side of this inequality is equivalent to the quantity

$$\sum_{j\geqslant 1} \left\|(-\Delta)^{l/2}\mu_j^{(\varepsilon)}u\right\|^2,$$

where $\{\mu_j^{(\varepsilon)}\}_{j\geqslant 1}$ is the collection of functions of Strichartz' theorem. Therefore, it is not hard to see that it suffices to prove the inequality

$$M = \sup_{y\in R^n}\ \sup_{j\geqslant 1}\ \sup_{u\in C_0^\infty(Q^{(j)})} \left\|r_y(t)u(t)\right\|^2 \cdot \left\|(-\Delta)^{l/2}u\right\|^2 \leqslant \infty, \tag{5.3}$$

where $r_y(t) = q_y(tm^{-1}(y))\rho^{-2}(y)m^{-2l}(y)$ and $Q^{(j)}$ is a cube with edge length ε containing the support of $\mu_j^{(\varepsilon)}(t)$.

If $2l > n$, then for $u \in C_0^\infty(Q^{(j)})$ we have $|u(t)|^2 \leqslant c_1\left\|(-\Delta)^{l/2}u\right\|^2$, where c_1 is a constant. This follows from the imbedding theorems for spaces of Sobolev type.

Hence,

$$M \leqslant c_2 \sup_{y\in R^n} \sup_{j\geqslant 1} \int_{Q_{(j)}} |r_y(t)|^2\,dt$$

$$\leqslant c_3 \sup_{y\in R^n} \rho^{-4}(y) m^{-4l}(y) \int_{|t-x|\leqslant\sqrt{n}\,\varepsilon} |q_y(tm^{-1}(y))|^2\,dt$$

$$= c_3 \sup_{y\in R^n} \rho^{-4}(y) M^{-4l+n}(y) \sup_{x\in R^n} \int_{|t-x|\leqslant m^{-1}(y)\sqrt{n}\,\varepsilon} |q_y(t)|^2\,dt. \tag{5.4}$$

Since $q_y(t)$ has period $\sqrt{k(y)}\,m^{-1}(y)$, using first (5.4) and then (5.2), we obtain

$$M \leqslant c_4 \sup_{y\in R^n} \rho^{-4}(y) m^{-4l+n}(y) \sup_{x\in Q^{(y)}} \int_{|t-x|\leqslant m^{-1}(y)\sqrt{n}\,\varepsilon} |q_y(t)|^2\,dt$$

$$\leqslant c_5 \sup_{y\in R^n} \sup_{x\in Q^{(y)}} \rho^{-4}(x) m^{-4l+n}(x) \int_{|t-x|\leqslant m^{-1}(y)\sqrt{n}\,\varepsilon} |q_y(t)|^2\,dt$$

$$\leqslant c_6 \sup_{y\in R^n} \sup_{x\in Q^{(y)}} \rho^{-4}(x) m^{-4l+n}(x) \int_{|t-x|\leqslant m^{-1}(x)c_6\varepsilon} |q_y(t)|^2\,dt. \tag{5.5}$$

But it is easy to see that for $x \in Q^{(y)}$

$$\int_{|t-x|\leqslant m^{-1}(x)c_6\varepsilon} |q_y(t)|^2\,dt \leqslant c_7 \sup_{\eta\in Q^{(y)}} \int_{|t-\eta|\leqslant m^{-1}(x)c_6\varepsilon} |q(t)|^2\,dt.$$

Substituting this into (5.5) and taking account of (5.2), we have

$$M \leqslant c_8 \sup_{y\in R^n} \sup_{x\in Q^{(y)}} \rho^{-4}(x) m^{-4l+n}(x) \int_{|t-x|\leqslant m^{-1}(x)c_6\varepsilon} |q(t)|^2\,dt$$

$$= c_8 \sup_{x\in R^n} \rho^{-4}(x) m^{-4l+n}(x) \int_{|t-x|\leqslant m^{-1}(x)c_6\varepsilon} |q(t)|^2\,dt.$$

Choosing ε from the condition $c_6\varepsilon < 1$, we get

$$M \leqslant c_9 \sup_{x\in R^n} \rho^{-4}(x) m^{-4l+n}(x) \int_{|x-t|\leqslant m^{-1}(x)} |q(t)|^2\,dt. \tag{5.6}$$

By the hypothesis of the theorem inequality (2.4) is proved for $2l > n$. Suppose now that $2l \leqslant n$. We choose any number $p \in (1,2)$. By Theorem 3.2 and (5.3) we find that it suffices to show that $M < \infty$, where

$$M = \sup_{y\in R^n} \sup_{j\geqslant 1} \sup_{u\in C_0^\infty(Q_{(j)})} \|r_y(t)u(t)\|^2 \cdot \|(-\Delta)^{l/2}u\|_p^{-2} < \infty. \tag{5.7}$$

But according to Theorem 3.3′

$$M \leqslant c_8 \sup_{y\in R^n} \sup_{j\geqslant 1} \left[\sup_{x\in R^n} \sup_{\eta>0} \eta^{-\beta} \int_{|x-t|\leqslant\eta} |r_y(t)\chi_j(t)|^2\,dt \right],$$

where $\chi_j(t)$ is the characteristic function of the cube $Q^{(j)}$ and $\beta = 2(n/p - l)$.

Hence, since the edge length of $Q^{(j)}$ is equal to ε,

$$M \leqslant c_9 \sup_{y\in R^n} \sup_{x\in R^n} \sup_{0<\eta\leqslant c\varepsilon} \eta^{-\beta}\int_{|x-t|\leqslant\eta} |r_y(t)|^2\,dt$$

$$= c_9 \sup_{y\in R^n} \rho^{-4}(y)m^{-4l}(y) \sup_{x\in R^n} \sup_{0<\eta\leqslant c\varepsilon} \eta^{-\beta}\int_{|x-t|\leqslant\eta} |q_y(tm^{-1}(y))|^2\,dt$$

$$= c_9 \sup_{y\in R^n} \rho^{-4}(y)m^{-4l+n-\beta}(y) \sup_{0<\eta\leqslant m^{-1}(y)c\varepsilon} \eta^{-\beta}\int_{|x-t|\leqslant\eta} |q_y(t)|^2\,dt.$$

From this we find in the same way as (5.6) was derived from (5.4) that

$$M \leqslant c_{10} \sup_{y\in R^n} \rho^{-4}(y)m^{-4l+\beta}(y) \sup_{0<\eta\leqslant m^{-1}(y)} \eta^{-\beta}\int_{|y-t|\leqslant\eta} |q(t)|^2\,dt.$$

From this inequality and the hypothesis of the theorem we obtain (2.4). The theorem is proved.

§6. Estimates of derivatives of solutions of the equation $\rho(-\Delta)^l\rho u + qu = f$

If $u(x)$ is a solution in L_2 of the equation $Lu = \rho(-\Delta)^l\rho u + qu = f \in L_2$, then to estimate $\|rD^\alpha\rho u\|_\theta$ it suffices to be able to estimate $\|rD^\alpha\rho(L-\lambda E)^{-1}\|_{L_2\to L_\theta}$ for some λ in the resolvent set R_L of the operator L. Indeed, from the equation $Lu = f$ $(u, f \in L_2)$ we obtain

$$(L+\lambda E)u = (f-\lambda u) \in L_2.$$

Hence,

$$rD^\alpha\rho u = rD^\alpha\rho(L+\lambda E)^{-1}(f-\lambda u).$$

Therefore,

$$\|rD^\alpha\rho u\|_\theta \leqslant \left\|rD^\alpha\rho(L+\lambda E)^{-1}\right\|_{L_2\to L_\theta}\|f-\lambda u\|.$$

In this section we shall concern ourselves with estimates of operators of the form $rD^\alpha\rho(L+\lambda E)^{-1}$, where $\lambda \in R_L$. Without mention, if there is no special need, we shall everywhere use the notation of §5 and assume that $\operatorname{Im} q(x) \geqslant 1$ and $\operatorname{Re} q(x) \geqslant 1$, since this does not limit the generality.

THEOREM 6.1. *Suppose that* $0 \leqslant |\alpha| < 2l$, $2l - |\alpha| > n(1/2 - 1/\theta)$, $2 < \theta < \infty$, $4l - 2|\alpha| \neq n$, *and* $A_p(x)$ *and* $m(x)$ *are the functions of Theorem* 5.1 $(1 < p < 2$, $0 < \varepsilon < 1)$. *Assume that conditions* (5.2) *of Theorem* 5.1 *are satisfied, and the quantity* $A_p = \sup\{A_p(t)\colon t \in R^n\}$ *for some* $p \in (1,2)$ *is finite. If* $r(x)$ *is a continuous function, then the following assertions are true*:

a) *The operator* $r(x)D^\alpha\rho(L+\lambda E)^{-1}$ $(\lambda \geqslant 0)$ *is bounded from* L_2 *to* L_θ *if and only if* $B = \sup\{B(x)\colon x \in R^n\} < \infty$, *where*

$$B(x) = \sup_{0\leqslant\eta\leqslant m^{-1}(x)} \eta^{-d}\left(\int_{|x-t|\leqslant\eta} |r(t)\rho^{-1}(t)|^\theta\,dt\right)^{1/\theta}, \quad d = -2l + |\alpha| + \frac{n}{2}.$$

b) *The operator* $r(x)D^\alpha\rho(L+\lambda E)^{-1}$ $(\lambda \geqslant 0)$ *is compact from* L_2 *to* L_θ *if and only if* $B < \infty$ *and* $\lim_{|x|\to\infty} B(x) = 0$.

c)

$$c^{-1}B \leqslant \left\|r(x)D^\alpha\rho(L+\lambda E)^{-1}\right\|_{L_2\to L_\theta} \leqslant cB, \tag{6.1}$$

where c *does not depend on* $r(x)$.

PROOF. It may be assumed that $\lambda = 0$. We shall prove the right inequality of (6.1) and the sufficiency of a) and b). Let $\lambda(y)$ be the function of Lemma 2.3. Using the inequality

$$\Big(\sum_{j\geqslant 1} |a_j|^2\Big)^{1/2} \geqslant \Big(\sum_{j\geqslant 1} |a_j|^\theta\Big)^{1/\theta},$$

valid for $\theta \geqslant 2$, and Strichartz' Theorem 3.1 as well as (5.3), we find that

$$\lambda(y) \leqslant c_1\rho^{-2}(y)m^\gamma(y) \times \sup_{u(t)\in C_0^\infty(Q^{(j)})} \Big\{\big\|r(tm^{-1}(y))\chi_y(tm^{-1}(y))u(t)\big\|_\theta^2 \big\|(-\Delta)^{l+|\alpha|/2}u\big\|^{-2}\Big\}, \tag{6.2}$$

where $\gamma = -4l + 2n(1/2 - 1/\theta) + 2|\alpha|$, $\{Q^{(j)}\}_{j\geqslant 1}$ is a family of cubes covering R^n consisting of cubes with edge lengths $\leqslant \varepsilon$ (ε is an arbitrary number), and $\chi_y(t)$ is the characteristic function of the cube $Q^{(y)}$ with center at y and edge length equal to $\sqrt{k(y)}\, m^{-1}(y)$.

If $4l - 2|\alpha| > n$, then for $u \in C_0^\infty(Q^{(j)})$ we have $|u(t)| \leqslant c_2\|(-\Delta)^{l+|\alpha|/2}u\|$. Therefore, from (6.2) we find that

$$\lambda(y) \leqslant c_2\rho^{-2}(y)m^\gamma(y)\sup_{\{j\}}\Big(\int \big|r(tm^{-1}(y))\chi_y(tm^{-1}(y))\big|^\theta \chi^{(j)}(t)\,dt\Big)^{2/\theta},$$

where $\chi^{(j)}$ is the characteristic function of the cube $Q^{(j)}$.

Choosing ε sufficiently small, we now obtain

$$\lambda(y) \leqslant c_3\rho^{-2}(y)m^\gamma \sup_{x\in R^n}\Big(\int_{|x-t|\leqslant\delta} \big|r(tm^{-1}(y))\big|^\theta \chi_y(tm^{-1}(y))\,dt\Big)^{2/\theta}$$
$$= c_3\rho^{-2}(y)m^\gamma(y)m^{2n/\theta}(y)\sup_{x\in R^n}\Big(\int_{|x-t|\leqslant m^{-1}(y)\delta} |r(t)|^\theta \chi_y(t)\,dt\Big)^{2/\theta},$$

where $\delta \in (0,1)$ is any fixed number.

Using the conditions imposed on $\rho(\cdot)$ and $m(\cdot)$, as in the proof of the analogous part of Theorem 5.1, we have

$$\lambda(y) \leqslant c_4 \sup_{x\in Q^{(y)}} m^{\gamma+2n/\theta}(x)\Big(\int_{|x-t|\leqslant m^{-1}(x)} |r(t)\rho^{-1}(t)|^\theta\,dt\Big)^{2/\theta}.$$

Since $\gamma + 2n/\theta = -4l + n + 2|\alpha|$, the right side of this expression is equal to

$$c_4\sup\{B^2(x) : x \in Q^{(y)}\}.$$

Hence,

$$\lambda(y) \leqslant c_4\sup\{B^2(x) : x\in Q^{(y)}\} \leqslant c_4\sup\{B^2(x): x\in R^n\}.$$

From the proof of Theorem 5.1 it follows that the conditions of Lemma 2.2 are satisfied. Therefore, for $4l - 2|\alpha| > n$ from the last inequality and Lemmas 2.4 and 2.5 we obtain the sufficiency of a) and b) and the right side of (6.1).

In the case $4l - 2|\alpha| < n$, from (6.2), using the Adams-Maz'ya Theorem 3.3', we obtain

$$\lambda(y) \leqslant c_5\rho^{-2}(y)m^{\gamma}(y)$$
$$\times \sup_{j\geqslant 1}\ \sup_{x\in R^n}\ \sup_{\eta>0}\eta^{-d}\left(\int_{|x-t|\leqslant\eta}\left|r\left(tm^{-1}(y)\right)\right|^{\theta}\chi_y\left(tm^{-1}(y)\right)\chi_j(t)\,dt\right)^{2/\theta},$$

where $d = 2\theta(n/2 - 2l + |\alpha|)$.

After easy transformations analogous to those made in the case $4l - 2|\alpha| > n$ (with account taken of the conditions imposed on $\rho(\cdot)$ and $m(\cdot)$), from this we obtain the inequality

$$\lambda(y) \leqslant c_6 \sup\{B^2(x): x \in Q^{(y)}\},$$

from which, as in the case $4l - 2|\alpha| > n$, we obtain the sufficiency of a) and b) and the right side of (6.1).

It remains to prove the necessity of a) and b) and the left side of (6.1). We choose an infinitely smooth function $w(t)$ equal to $e^{i(t_1+\cdots+t_n)}$ for $|t| \leqslant 0.5$ and to zero for $|t| \geqslant 0.6$. Let t_0 be any point in R^n, and let $\eta \in (0, 1]$. We set

$$w_{t_0,\eta}(t) = w\left(m(t_0)(t-{}_0)\eta^{-1}\right)\rho^{-1}(t).$$

Then

$$\left\|Lw_{t_0,\eta}\right\| \leqslant \left\|\rho(-\Delta)^l\rho w_{t_0,\eta}\right\| + \left\|qw_{t_0,\eta}\right\|$$
$$\leqslant \left\|\rho(-\Delta)^l w\left(m(t_0)(t-t_0)\eta^{-1}\right)\right\| + \left\|w\left(m(t_0)(t-t_0)\eta^{-1}\right)q\rho^{-1}(t)\right\|.$$

Using condition (5.2) imposed on $\rho(\cdot)$, we obtain

$$\left\|Lw_{t_0,\eta}\right\| \leqslant c_1\left[\rho(t_0)\left(m(t_0)\eta^{-1}\right)^{2l-n/2} + \rho^{-1}(t_0)\left(\int_{|t-t_0|\leqslant m^{-1}(t_0)\eta}|q(t)|^2\,dt\right)^{1/2}\right]. \tag{6.3}$$

By the hypothesis of the theorem the function $A_p(t)$ defined in Theorem 5.1 is bounded by a constant for some $p \in (1, 2)$. Thus

$$\rho^{-1}(t_0)\left(\int_{|t-t_0|\leqslant m^{-1}(t_0)\eta}|q(t)|^2\,dt\right)^{1/2} \leqslant c_2\rho(t_0)\left(m(t_0)\eta^{-1}\right)^{2l-n/2}. \tag{6.4}$$

Indeed, for $2l > n$ this inequality is obvious from the definition of $A_p(t)$. In the case $2l \leqslant n$, from the definition of $A_p(t)$ we obtain

$$\int_{|t_0-t|\leqslant\eta m^{-1}(t_0)}|q(t)|^2\,dt \leqslant c_3\left(\eta m^{-1}(t_0)\right)^{\beta}\rho^4(t_0)m^{4l-n+\beta}(t_0)A_p,$$

where $\beta = 2(n/p - l)$. This implies that

$$\rho^{-1}(t_0)\left(\int_{|t_0-t|\leqslant m^{-1}(t_0)\eta}\|q(t)\|^2\,dt\right)^{1/2} \leqslant c_4\rho(t_0)m^{2l-n/2}(t_0)\eta^{n/p-l}\sqrt{A_p}.$$

Since $1 < p < 2$ and $0 < \eta \leqslant 1$, the right side of this inequality does not exceed

$$c_5\rho(t_0)m^{2l-n/2}(t_0)\eta^{-2l+n/2}\sqrt{A_p}.$$

Therefore, (6.4) holds. From (6.3) and (6.4) we obtain

$$\left\|Lw_{t_0,\eta}\right\| \leqslant c_7\rho(t_0)\left(m(t_0)\eta^{-1}\right)^{2l-n/2}. \tag{6.5}$$

We have, furthermore,

$$\begin{aligned}\left\|r(t)D^\alpha \rho w_{t_0,\eta}\right\|_\theta &= \left\|r(t)D^\alpha w\left(m(t_0)(t-t_0)\eta^{-1}\right)\right\|_\theta \\ &\geqslant \left(\int_{|t-t_0|\leqslant 0.5\eta m^{-1}(t_0)} |r(t)|^\theta \left|D^\alpha w\left(m(t_0)(t-t_0)\eta^{-1}\right)\right|^\theta dt\right)^{1/\theta} \\ &= \left(\int_{|t-t_0|\leqslant 0.5\eta m^{-1}(t_0)} |r(t)|^\theta \left|m(t_0)\eta^{-1}\right|^{|\alpha|} dt\right)^{1/\theta}.\end{aligned}$$

From this and (6.5) it follows that

$$\begin{aligned}\left\|r(t)D^\alpha \rho w_{t_0,\eta}\right\|_\theta \left\|Lw_{t_0,\eta}\right\|^{-1} &\geqslant c_8^{-1}\left(\eta^{-1}m(t_0)\right)^{|\alpha|-2l-n/2} \\ &\quad\times \rho^{-1}(t_0)\left(\int_{|t-t_0|\leqslant 0.5m^{-1}(t_0)} |r(t)|^\theta dt\right)^{1/\theta}.\end{aligned} \tag{6.6}$$

Let $\|r(x)D^\alpha \rho L^{-1}\|_{L_2\to L_\theta} \leqslant k < \infty$. From (6.6) and (5.2) it then follows that

$$kc_{10} \geqslant \sup_{t\in R^n} \left(\eta^{-1}m(t_0)\right)^{|\alpha|-2l+n/2}\left(\int_{|t-t_0|\leqslant 0.5\eta m^{-1}(t_0)} |r(t)/\rho(t)|^\theta dt\right)^{1/\theta}.$$

But $\eta \in (0,1]$ is an arbitrary number; hence,

$$kc_{10} \geqslant \sup_{t\in R^n}\left[\sup_{0<\eta\leqslant 0.5m^{-1}(t_0)} \eta^{-|\alpha|+2l-n/2}\left(\int_{|t-t_0|\leqslant 0.5\eta m^{-1}(t_0)} |r(t)/\rho(t)|^\theta dt\right)^{1/\theta}\right].$$

Using condition (5.2) imposed on $m(\cdot)$ and arguing by contradiction, from this it is easy to obtain $kc_{11} \geqslant \sup\{B(x): x \in R^n\}$. Therefore, the necessity of a) and the left side of (6.1) have been proved.

If $\sup\{B(x): x \in R^n\} < \infty$, but $\lim_{|x|\to+\infty} B(x) \neq 0$, then with the help of (5.2) we easily find that

$$\overline{\lim_{|x|\to+\infty}} \sup_{0<\eta\leqslant 0.5m^{-1}(x)} \eta^{-|\alpha|+2l+n/2}\left(\int_{|t-x|\leqslant 0.5\eta} \left|r(t)\rho^{-1}(t)\right|^\theta\right)^{1/\theta} > 0.$$

Using (5.2), from this we again deduce that

$$\overline{\lim_{|x|\to+\infty}} \sup_{0<\eta\leqslant 0.5m^{-1}(x)} \rho^{-1}(x)\eta^{-|\alpha|+2l+n/2}\left(\int_{|t-x|\leqslant 0.5\eta} |r(t)|^\theta dt\right)^{1/\theta} > 0.$$

This inequality together with (6.5) implies that there exist sequences of points $\{t_k\}_{k\geqslant 0}$ and numbers $\{\eta_k\}_{k\geqslant 0}$ such that $\eta_k \in (0,1]$, $\lim|t_k| \to +\infty$, and

$$\left\|r(t)D^\alpha \rho w_{t_k\eta_k}\right\|_\theta \left\|Lw_{t_k\eta_k}\right\|^{-1} \geqslant \delta_0 > 0, \qquad k = 1,2,\ldots.$$

From this it follows that the operator $r(x)D^{\alpha}\rho L^{-1}$ is not compact. This is a contradiction.

REMARK 6.1. From the proof of this theoren. it is evident that if $4l - 2|\alpha| > n$, then its assertions hold also for $\theta = 2$.

THEOREM 6.2. *Suppose that* $0 \leqslant |\alpha| < 2l$, $2 \leqslant \theta < \infty$, $4l - 2|\alpha| \leqslant n$ *and* $A_p(\cdot)$ *and* $m(\cdot)$ *are the functions of Theorem* 5.1 ($1 < p < 2$, $0 < \varepsilon < 1$). *Assume that conditions* (5.2) *of Theorem* 5.1 *are satisfied, the quantity* $A_p = \sup\{A_p(x)\colon x \in R^n\}$ *for some* $p \in (1,2)$ *is finite, and* $r(x)$ *is a continuous function in* R^n. *Then the following assertions are true*:

a) *If* $B = \sup\{B(t)\colon t \in R^n\} < \infty$, *then the operator* $r(\cdot)D^{\alpha}\rho(L + \lambda E)^{-1}$ ($\lambda > 0$) *is bounded from* L_2 *to* L_θ, *where*

$$B(x) = m^{n(1/2-1/\theta)}(x) \sup_{0<\eta\leqslant m^{-1}(x)} \eta^{\beta}\left(\int_{|x-t|\leqslant\eta} |r(t)/\rho(t)|^{\theta}\,dt\right)^{1/\theta},$$

θ' *is any number in* $(1,2)$ *such that* $2l - |\alpha| > n(1/\theta' - 1/\theta)$, *and* $\beta = n/\theta' - 2l + |\alpha|$.

b) *If* $B < \infty$ *and* $\lim_{|x|\to\infty} B(x) = 0$, *then the operator* $r(x)D^{\alpha}\rho(L + \lambda E)^{-1}$ *is compact.*

c) $\|r(t)D^{\alpha}\rho(L + \lambda E)^{-1}\|_{L_2\to L_\theta} \leqslant cB$, *where* c *does not depend on* $r(t)$.

PROOF. In this case inequality (6.2) also holds. We choose any number $\theta' \in (1,2)$. From Theorem 3.2 and (6.2) we obtain

$$\lambda(y) \leqslant c_1\rho^{-2}(y)m^{\gamma}(y)$$
$$\times \sup_{u(t)\in C_0^{\infty}(Q^{(j)})} \left\{\left\|r(tm^{-1}(y))\chi_y(tm^{-1}(y))u(t)\right\|_{\theta}^{2}\left\|(-\Delta)^{l+|\alpha|/2}u\right\|_{\theta'}^{-2}\right\}.$$

Since $\theta(2l - |\alpha|) < n$ and $1 < \theta < 2$, using Theorem 3.3′, we have

$$\lambda(y) \leqslant c_2\rho^{-2}(y)m^{\gamma}(y)$$
$$\times \sup_{j\geqslant 1}\sup_{x\in R^n}\sup_{\eta>0}\eta^{d}\left(\int_{|x-t|\leqslant\eta} \left|r(tm^{-1}(y))\right|^{\theta}\chi_y(tm^{-1}(y))\chi^{(j)}(t)\,dt\right)^{2/\theta},$$

where $d = -2(n/\theta' - 2l + |\alpha|)$.

The rest of the proof of Theorem 6.2 does not differ from the proof of sufficiency of a) and b) and the right side of inequality (6.1) of Theorem 6.1.

PROOF OF THEOREM 0.1. Theorem 0.1 is a simple corollary of Theorems 5.1 and 6.1.

PROOF OF REMARK 2.1. Remark 2.1 is proved in the same way as the necessity of a) and b) of Theorem 6.1 and the left side of inequality (6.1).

§7. Generalizations. Examples

1. Let $Lu = (-\Delta)^{l+\beta}u + q(x)(-\Delta)^{\beta}u$, where $l > 0$, $\beta \geqslant 0$, and l is an integer. If $q(x)$ is such that the conditions of Theorem 5.1 are satisfied ($\rho(x) \equiv 1$), then from Theorem 5.1 we find that

$$\left\|(-\Delta)^{l+\beta}u\right\| + \left\|q(x)(-\Delta)^{\beta}u\right\| \leqslant c\left(\|Lu\| + \left\|(-\Delta)^{\beta}u\right\| + \|u\|\right).$$

But it is well known that

$$\|(-\Delta)^{\alpha}u\| \leqslant c_1\left(\varepsilon\|(-\Delta)^{l+\alpha}u\| + \|u\|\right),$$

where $\varepsilon \to 0$ as $\delta \to +\infty$, and c_1 does not depend on ε, δ, or $u \in C_0^\infty(R^n)$. Therefore, the preceding inequality gives

$$\|(-\Delta)^{l+\beta}u\| + \|q(x)(-\Delta)^{\beta}u\| \leqslant c_2(\|Lu\| + \|u\|).$$

We thus obtain the separation of the operator

$$Lu = (-\Delta)^{l+\beta}u + q(x)(-\Delta)^{\beta}u.$$

In a special case we shall formulate a corollary of the arguments presented above and of Theorem 0.1.

THEOREM 7.1. *Let $l_0 = y^{(\mathrm{IV})}(x) - q(x)y''(x)$, $x \in (-\infty, \infty)$. Assume that $q(x) \geqslant 1$ and $c^{-1} \leqslant q(x)q^{-1}(y) \leqslant c$ for $|x - y| \leqslant k(y)q^{-1/2}(y)$, where $k(y) \geqslant 4$ and $\overline{\lim}_{|y|\to+\infty} k(y) = +\infty$. Denote by L the closure in $L_2(J)$ of the operator l_0 defined on $C_0^\infty(R)$. Then the operator L is separated, i.e., if $y \in D(L)$, then $y^{(\mathrm{IV})} \in L_2(-\infty, \infty)$.*

2. Operators of the general form

$$Lu = \rho(-\Delta)^l\rho u + \sum_{|\alpha|\leqslant 2l} r_\alpha(x)D^\alpha\rho u + qu \tag{7.1}$$

in some cases can be studied by perturbation methods. For example, if $\rho(\cdot)$ and $q(\cdot)$ satisfy the conditions of Theorem 5.1, then with the help of Theorem 6.2 we obtain conditions on the coefficients $r_\alpha(x)$ ensuring that the intermediate terms are "subordinate" in a particular sense to a sum $\rho(-\Delta)^l\rho u + qu$. If "subordination" is present it is possible by perturbation methods to extend our previous results to the operators (7.1).

3. Boundary value problems in bounded domains for degenerate elliptic equations are of great interest. In some cases they reduce to an operator of the form (7.1) considered in the entire space R^n by means of a change of variables.

EXAMPLE 1. Let

$$Lu = -\rho(x)(u\rho(x))'' + e^{|x|}\sin^2 e^{x^2}u, \qquad x \in J, \tag{7.2}$$

where $\rho(x) > 0$ is an infinitely smooth function equal to $e^{-|x|/2}$ for $|x| \geqslant 1$. In this case the function $((q+1)\rho^{-2})^*(x)$ for large $|x|$ is defined by

$$((q+1)\rho^{-2})^*(x) \leqslant \inf\left\{d^{-1}\colon d^{-1} \geqslant \int_{x-d}^{x+d} e^{|2t|}\sin^2 e^{t^2}\,dt\right\}.$$

If $v(x) \leqslant u(x)$ for all $x \in J$, then $v^*(x) \geqslant u^*(x)$ for all $x \in J$. Therefore,

$$((q+1)\rho^{-2})^*(x) \leqslant \inf\left\{d^{-1}\colon d^{-1} \geqslant \int_{x-d}^{x+d} e^{|2t|}\,dt\right\} \leqslant ce^{|x|}.$$

We claim that

$$((q+1)\rho^{-2})^*(x) \geqslant c^{-1}e^{|x|}.$$

Suppose not. Then there exists a sequence x_n (it may be assumed that $x_n \to +\infty$) such that $d_n^{-1} = ((q+1)\rho^{-2})^*(x_n) \leqslant 1000^{-1}e^{x}n$. From the definition of d_n and $((q+1)\rho^{-2})^*(x_n)$ it follows that

$$d_n^{-1}l = \int_{x_n-d_n}^{x_n+d_n} e^{2t}\sin^2 e^{t^2}\,dt.$$

It is easy to see that $d_n \to 0$. Therefore, from the last equality we find that for sufficiently large h

$$d_n^{-1} \geqslant \frac{1}{2} e^{2x_n} \int_{x_n - d_n}^{x_n + d_n} \sin^2 e^{t^2}\, dt = \frac{1}{2} e^{2x_n} d_n - o\left(e^{2x_n} x_n^{-1} e^{-x_n^2}\right).$$

Since $d_n \geqslant 100 e^{-x_n}$, from this it follows that $d_n^{-1} \geqslant \frac{1}{10} e^{2x_n} d_n$ for sufficiently large n. Therefore, $d_n^{-1} \geqslant \frac{1}{10} e^{x_n}$, a contradiction.

Thus, $c^{-1} e^{|x|} \leqslant ((q+1)\rho^{-2})^*(x) \leqslant c e^{|x|}$.

From this inequality it follows easily that all the conditions of Theorem 5.1 are satisfied. Hence, the operator (7.2) with rapidly oscillating coefficients is separated.

EXAMPLE 2. The following operator is separated:

$$Lu = -\left(|x|^2 + 1\right)^{\alpha} \left(u\left(|x|^2 + 1\right)^{\alpha}\right)'' + \left(|x|^2 + 1\right)^{\gamma} e^{\eta |x|} \left|\sin e^{x^2}\right|^{\beta} u,$$
$$-\infty < \alpha < \infty,\ \beta \geqslant 0,\ -\infty < \gamma < \infty,\ 0 < \eta < \infty.$$

Separation of this operator can be established as in the preceding example.

EXAMPLE 3. Let L be the operator defined by (7.2), and let $r(x)$ be a continuous function. We assume that $2 \leqslant \theta < \infty$. From Theorem 6.1 and Example 1 we find that the operator $r(x)(d/dx)e^{-|x|/2}L^{-1}$ is bounded from $L_2(J)$ to $L_\theta(J)$ if and only if

$$\sup_{x \in R^n} e^{-|x|/2} \left(\int_{|x-t| \leqslant e^{-|x|}} \left| r(t) e^{|t|/2} \right|^{\theta} dt \right)^{1/\theta} < \infty.$$

Similarly, the operator $r(x) e^{-|x|/2} L^{-1}$ is bounded from $L_2(J)$ to $L_\theta(J)$ if and only if

$$\sup_{x \in R^n} e^{-3|x|/2} \left(\int_{|x-t| \leqslant e^{-|x|}} \left| r(t) e^{|t|/2} \right|^{\theta} dt \right)^{1/\theta} < \infty.$$

In particular, the operators

$$e^{|x|} L^{-1}, \qquad e^{|x|/2} \frac{d}{dx} e^{-|x|/2} L^{-1}, \qquad \frac{d}{dx} L^{-1}$$

are bounded from $L_2(J)$ to $L_2(J)$.

BIBLIOGRAPHY

1. A. L. Izmaĭlov and M. Otelbaev, *Summability with weight of the solution of a differential equation in an unbounded domain*, Izv. Akad. Nauk Kazakh. SSR Ser. Fiz.-Mat. **1977**, no. 1, 36–40. (Russian)

2. William N. Everitt and Magnus Giertz, *Some properties of the domains of certain differential operators*, Proc. London Math. Soc. (3) **23** (1971), 301–324.

3. ______, *Some inequalities associated with certain ordinary differential operators*, Math. Z. **126** (1972), 308–326.

4. ______, *On some properties of the powers of a formally self-adjoint differential expression*, Proc. London Math. Soc. (3) **24** (1972), 149–170.

5. ______, *On some properties of the domains of powers of certain differential operators*, Proc. London Math. Soc. (3) **24** (1972), 756–768.

6. M. Giertz, Report from the Conference on Ordinary and Partial Differential Equations (Dundee, 1976), Trita-Mat. 1976-7, Royal Inst. Technology, Stockholm, 1976.

7. M. Otelbaev, *The summability with weight of the solution of a Sturm-Liouville equation*, Mat. Zametki **16** (1974), 969–980; English transl. in Math. Notes **16** (1974).

8. K. Kh. Boĭmatov, *Separation theorems for the Sturm-Liouville operator*, Mat. Zametki **14** (1973), 349–359; English transl. in Math. Notes **14** (1973).

9. ______, *Theorems on the separation property*, Dokl. Akad. Nauk SSSR **213** (1973), 1109–1111; English transl. in Soviet Math. Dokl. **14** (1973).

10. M. Otelbaev, *On the smoothness of solutions of differential equations*, Izv. Akad. Nauk Kazakh. SSR Ser. Fiz.-Mat. **1977**, no. 5, 45–48. (Russian)

11. ______, *On the separation of elliptic operators*, Dokl. Akad. Nauk SSSR **234** (1977), 540–543; English transl. in Soviet Math. Dokl. **18** (1977).

12. A. L. Izmaĭlov, *Smoothness of solutions of differential operators and separation theorems*, Candidate's Dissertation, Alma-Ata, 1978. (Russian)

13. D. Zh. Raĭmbekov, *Smoothness of the solution in L_2 of a singular equation*, Izv. Akad. Nauk Kazakh. SSR Ser. Fiz.-Mat. **1974**, no. 3, 78–83. (Russian)

14. Béla Sz.-Nagy and Ciprian Foiaş, *Analyse harmonique des opérateurs de l'espace de Hilbert*, Masson, Paris, and Akad. Kiadó, Budapest, 1967; English transl., North-Holland, Amsterdam, and Akad. Kiadó, Budapest, 1970.

15. K. Kh. Boĭmatov, *L_2-estimates of generalized solutions of elliptic differential equations*, Dokl. Akad. Nauk SSSR **223** (1975), 521–524; English transl. in Soviet Math. Dokl. **16** (1975).

16. V. I. Feĭgin, *The Noether property of differential operators in R^n*, Differentsial'nye Uravneniya **11** (1975), 2231–2235; English transl. in Differential Equations **11** (1975).

17. Robert S. Stichartz, *Multipliers on fractional Sobolev spaces*, J. Math. Mech. **16** (1967), 1031–1060.

18. David R. Adams, *On the existence of capacitary strong type estimates in R^n*, Ark. Mat. **14** (1976), 125–140.

19. ______, *A trace inequality for generalized potentials*, Studia Math. **48** (1973), 99–105.

20. V. G. Maz'ya and M. Otelbaev, *Imbedding theorems and the spectrum of a certain pseudodifferential operator*, Sibirsk. Mat. Zh. **18** (1977), 1073–1087; English transl. in Siberian Math. J. **18** (1977).

21. V. G. Maz'ya, *On (p, l)-capacity, imbedding theorems, and the spectrum of a selfadjoint elliptic operator*, Izv. Akad. Nauk SSSR Ser. Mat. **37** (1973), 356–385; English transl. in Math. USSR Izv. **7** (1973).

22. M. Otelbaev, *Imbedding theorems for weighted spaces and their applications in the study of the spectrum of the Schrödinger operator*, Trudy Mat. Inst. Steklov. **150** (1979), 265–305; English transl. in Proc. Steklov Inst. Math. **1981**, no. 4 (150).

Translated by J. R. SCHULENBERGER

Труды Матем. Инст. Стеклов.
161(1983)

Proceedings of the Steklov
Institute of Mathematics
1984, Issue 4

CHARACTERISTIC PROPERTIES OF CONVOLUTION INTEGRAL OPERATORS

UDC 517.444

V. D. STEPANOV

ABSTRACT. Necessary and sufficient conditions are found for a measurable function to be the kernel of an integral convolution operation acting from $L^p(R^N)$ to the space of all measurable a.e. finite functions on R^N. Applications of the criterion are considered.

Bibliography: 12 titles.

If $k(x)$ and $f(x)$ are two measurable complex-valued functions on Euclidean space R^N, then the relation

$$T_R f(x) = \int_{R^N} k(x-y) f(y)\, dy \equiv (k * f)(x) \tag{1}$$

defines an integral convolution operator T_k with kernel $k(x)$ on some class of measurable functions whenever the integral in (1) converges in the Lebesgue sense for all $f(x)$ in this case and for almost all (a.a.) $x \in R^N$, i.e.,

$$\int_{R^N} |k(x-y) f(y)|\, dy < \infty \quad \text{for a.a. } x \in R^N. \tag{2}$$

In this article we consider integral convolution operators defined in the sense of (1) and (2) on the Lebesgue spaces $L^p \equiv L^p(R^N)$, $1 \leqslant p \leqslant \infty$. The following problem is well known for such operators.

PROBLEM A. *What are necessary and sufficient conditions for a measurable kernel $k(x)$ to generate by formula* (1) *an integral convolution operator T_k acting boundedly from L^p to L^q, where p and q are fixed and $1 \leqslant p \leqslant q \leqslant \infty$?*

The restriction $p \leqslant q$ is necessary by Hörmander's theorem [1], which implies that if T_k is bounded from L^p to L^q and $p > q$, $1 \leqslant p, q \leqslant \infty$, then $k(x) = 0$ for a.a. $x \in R^N$.

Problem A has apparently not yet been solved definitively. We mention [1]–[4], where further references can be found, as well as [5]–[8], where problems close to Problem A are considered. In the above formulation, Problem A actually splits into two independent problems. The first has to do with an estimate

$$\|T_k f\|_q \leqslant C \|f\|_p \tag{3}$$

1980 *Mathematics Subject Classification*. Primary 44A35; Secondary 42A85, 46E30.

0081-5438/84 $1.00 + $.25 per page

on some dense subset of L^p, with a constant C independent of f; here $\|\cdot\|_p$ is the norm in L^p. The second problem is connected with the fact that the operator in question is an integral operator on the whole class L^p, and is formulated as follows. Let L^0 be the space of all measurable a.e. finite functions on R^N.

PROBLEM B. *What are necessary and sufficient conditions for a function $k \in L^0$ to be the kernel of an integral convolution operator T_k acting from L^p to L^0, i.e., for the inequality* (2) *to hold for any $f \in L^p$?*

In our view, a solution of Problem B is useful because it permits us to distinguish convolution operators that are integral operators on all of L^p from operators that are integral only on dense subsets of L^p and are understood as linear extensions on the whole class L^p.

In this article we solve Problem B. Corollaries to this result are given; in particular, an example is constructed of an integral convolution operator that is bounded from $L^p(R^1)$ to $L^p(R^1)$ only when $p = 2$. We also obtain a criterion for a linear operator from L^p to L^0 to be representable as an integral convolution operator.

The author expresses deep gratitude to V. B. Korotkov for his constant attention and for a discussion of the results.

§1

THEOREM 1. *Suppose that $1 \leqslant p \leqslant \infty$ and $k \in L^0$. Then the inequality* (2) *holds for any function $f \in L^p$ if and only if*

$$g(y) \equiv \int_Q |k(x-y)|\,dx \in L^{p'} \cap L^\infty, \tag{4}$$

where $Q \subset R^N$ is a cube with edges of positive finite length and center at 0, *$p' = p/(p-1)$ for $1 < p < \infty$, $p' = \infty$ for $p = 1$, and $p' = 1$ for $p = \infty$.*

PROOF. *Sufficiency.* If $f \in L^p$, then Fubini's theorem, Hölder's inequality, and (4) give us that for any cube $P \subset R^N$ of finite size

$$\int_P \int_{R^N} |k(x-y)f(y)|\,dy\,dx = \int_{R^N} |f(y)| \int_P |k(x-y)|\,dx\,dy$$
$$\leqslant \frac{\operatorname{meas} P}{\operatorname{meas} Q} \|f\|_p \|g\|_{p'} < \infty.$$

From this and the arbitrariness of P we get (2).

Necessity. Step 1. Let us show that for any compact sets $V, \Omega \subset R^N$

$$g_V(y) \equiv \int_V |k(x-y)|\,dx \in L^\infty(\Omega). \tag{5}$$

Since V is compact, there exists a ball $S_r \subset R^N$ about 0 of radius r such that $V \subset S_r$ and $S_r - y \supset V$ for any $y \in R^N$ with $|y| < 1/2$, where $S_r - y$ is the translate of S_r by the vector $-y$. From condition (2),

$$\int_{S_r} |k(x-y)|\,dx < \infty \quad \text{for a.a. } y \in R^N.$$

Consequently, there exists a y_0 such that $|y_0| < 1/2$ and

$$\int_{S_r} |k(x-y_0)|\,dx = C_r < \infty.$$

We have that

$$\infty > C_r = \int_{S_r} |k(x - y_0)|\, dx = \int_{S_r - y_0} |k(u)|\, du \geqslant \int_V |k(u)|\, du. \qquad (6)$$

Since V is an arbitrary compact set, (6) implies that $k \in L^1_{\text{loc}}$. Since V and Ω are compact, there exists a ball S_k such that $V - \Omega \subset S_k$, where

$$V - \Omega = \{x \in R^N : x = z - y,\, z \in V,\, y \in \Omega\}.$$

For $y \in \Omega$

$$g_V(y) = \int_{V-y} |k(u)|\, du \leqslant \int_{S_R} |k(u)|\, du = C_R < \infty,$$

which implies (5).

Step 2. Let us show that if $S_{1/2}$ is the ball about 0 of radius 1/2, then

$$h(y) \equiv \int_{S_{1/2}} |k(x - y)|\, dx \in L^\infty. \qquad (7)$$

If $p = \infty$, then condition (2) with a suitable choice of $f(x)$ implies that $k \in L^1$, and (7) is trivial for this case. Suppose that $1 \leqslant p < \infty$ and $0 < \lambda < \infty$. Let $h = h_\lambda + h^\lambda$, where $h^\lambda(x) = h(x)$ if $h(x) \leqslant \lambda$, and $h^\lambda(x) = 0$ if $h(x) > \lambda$. We show that there exist λ and a compact set $\Omega_\lambda \subset R^N$ such that $\operatorname{supp} h_\lambda \subseteq \Omega_\lambda$; then it follows from the assertion in Step 1 that $h_\lambda \in L^\infty$, and together with the inequality $h^\lambda(x) \leqslant \lambda$ this gives us that $h \in L^\infty$. Assume not. Suppose that the supports of the functions h_{λ_n} are unbounded for any sequence $\{\lambda_n\}$ with $0 < \lambda_n < \infty$ and $\lim_{n\to\infty} \lambda_n = \infty$. This implies that there is a sequence $\{y_n\} \subset R^N$ such that

$$y_n \in \operatorname{supp} h_{\lambda_n}, \quad 2 + |y_n| < |y_{n+1}|, \quad \int_{S_{1/2}} |k(x - y_n)|\, dx > \lambda_n. \qquad (8)$$

For $|y| < 1/2$ we have that $S_{1/2} \subset S_1 - y$, where S_1 is the ball about 0 of radius 1. From this and (8) we get

$$\int_{S_1 - y} |k(x - y_n)|\, dx \geqslant \int_{S_{1/2}} |k(x - y_n)|\, dx > \lambda_n, \qquad |y| < 1/2.$$

It is not hard to show from this by a change of variable that

$$\int_{S_1 - y_n} |k(x - y)|\, dx > \lambda_n, \qquad |y| < 1/2.$$

Under condition (8) the sets $\{S_1 - y_n\}$ are pairwise disjoint. Let $\lambda_n = n^2$, and let $f(x)$ be given by

$$f(x) = \sum_{n=1}^{\infty} \frac{1}{n \log^2(n+1)} \chi_{S_1 - y_n}(x),$$

where $\chi_\Omega(x)$ is the characteristic function of a set $\Omega \subset R^N$.

It is easy to see that $f \in L^p$ for any $p \geqslant 1$; therefore, from (2) it is not hard to get the inequality

$$\int_{R^N} |k(x - y) f(x)|\, dx < \infty \quad \text{for a.a. } y \in R^N.$$

However, by construction,

$$\int_{R^N} |k(x-y)f(x)|\,dx = \sum_{n=1}^{\infty} \frac{1}{n\log^2(n+1)} \int_{S_1 - y_n} |k(x-y)|\,dx$$
$$\geqslant \sum_{n=1}^{\infty} \frac{n}{\log^2(n+1)} = \infty$$

for $|y| < 1/2$. This contradiction proves the assertion of Step 2, which easily implies the assertion that $g \in L^\infty$ in the theorem.

Step 3. It remains to show that $g \in L^{p'}$. If $p = 1$ or $p = \infty$, then what is required has already been obtained; therefore, consider the case $1 < p < \infty$. We need a lemma which is a special case of an assertion due to Korotkov [9].

FACTORIZATION LEMMA. *Suppose that $1 < p < \infty$ and that T_k is an integral convolution operator with kernel $k(x)$ and acting from L^p to L^0. Then T_k can be represented in the form $T_k = A \cdot B$, where $B\colon L^p \to L^p$ is a linear regular integral operator with kernel $b(x, y)$,*[1] *and A is the operator of multiplication by a particular positive function $a \in L^0$, $Af(x) = a(x)f(x)$, with $1/a \in L^1_{\mathrm{loc}}$.*

Suppose that $0 < \Phi(x) \in L^{p'}$ and is bounded. It follows from the factorization lemma that $|k(x-y)| = a(x)|b(x, y)|$; therefore, we write

$$\int_{R^N} |k(x-y)| \frac{\Phi(x)}{a(x)}\,dx = \int_{R^N} |b(x, y)|\Phi(x)\,dx = (|B|^*\Phi)(y). \tag{9}$$

By the factorization lemma, the operator $|B|$ with nonnegative kernel $|b(x, y)|$ acts from L^p to L^p. Hence, the adjoint operator $|B|^*$ acts from $L^{p'}$ to $L^{p'}$; therefore, if $F(x) = \Phi(x)/a(x)$, then it follows from (9) that

$$g_F(y) \equiv \int_{R^N} |k(x-y)|F(x)\,dx \in L^{p'}.$$

Since $F(x) > 0$ and is locally integrable, for $\varphi(x) = \chi_{S_1}(x)$ we have

$$L^{p'} \ni (g_F * \varphi)(z) = \int_{R^N} |k(x-z)|(F * \varphi)(x)\,dx. \tag{10}$$

The function $(F * \varphi)(x)$ is positive and continuous, so there exist a cube $Q \subset R^N$ with edges of positive finite length and a constant $c > 0$ such that $(F * \varphi)(x) \geqslant c\chi_Q(x)$ for $x \in Q$. By this and (10),

$$L^{p'} \ni (g_F * \varphi)(z) \geqslant c\int_Q |k(x-z)|\,dx,$$

and the conclusion of the theorem can now be obtained without difficulty.

REMARK. *It follows from the proof of Theorem* 1 *that the theorem remains valid if the cube $Q \subset R^N$ centered at* 0 *is replaced by a ball of positive radius or any other compact subset of R^N with nonempty interior.*

---

(1) An integral operator $B\colon L^p \to L^p$ with kernel $b(x, y)$ such that $Bf(x) = \int_{R^N} b(x, y)f(y)\,dy$ for a.a. $x \in R^N$ when $f \in L^p$ is said to be *regular* if the integral operator $|B|$ with $|B|f(x) = \int_{R^N} |b(x, y)|f(y)\,dy$ is defined on L^p and bounded from L^p to L^p.

DEFINITION. Let $1 \leqslant p \leqslant \infty$. Denote by K_p the collection of all measurable functions $k \in L^0$ that determine integral convolution operators from L^p to L^0. The statements $k \in K_p$ and T_k: $L^p \to L^0$ are regarded as equivalent. The statement T_k: $L^p \to L^q$ means that $k \in K_p$, and the operator T_k of the form (1) is bounded from L^p to L^q, i.e., the estimate (3) holds.

We formulate corollaries that are not hard to get from Theorem 1.

COROLLARY 1. *If* $1 < p < \infty$, *then* $L^1 = K_\infty \subset K_p \subset K_1 \subset L^1_{\text{loc}}$, *and if* $1 \leqslant p_1 < p_2 \leqslant \infty$, *then* $K_{p_1} \supset K_{p_2}$ *and the inclusion is proper.*

COROLLARY 2. *Let* $p \geqslant 2$. *If* T_k: $L^p \to L^p$, *then* T_k: $L^q \to L^q$ *for* $q \in [p', p]$.

From the last assertion it follows, in particular, that if $p \geqslant 2$ and T_k: $L^p \to L^p$, then the adjoint operator T_k^*: $L^{p'} \to L^{p'}$ is an integral operator on the whole of $L^{p'}$. This generalizes Theorem 2 in [9]. For $1 < p < 2$ the operator adjoint to T_k: $L^p \to L^p$ may not be an integral operator on $L^{p'}$ in general, as is clear from the following example.

EXAMPLE. Suppose that $N = 1$, $a > 0$, $a \neq 1$, $1 > \alpha \geqslant 1 - a/2$, $\alpha > 0$, and

$$k_{\alpha,a}(x) = |x|^{-\alpha} e^{i|x|^a}, \qquad T_{\alpha,a}f = k_{\alpha,a} * f.$$

It is known [10] that $T_{\alpha,a}$: $L^p \to L^p$ if

$$\frac{a}{a+\alpha-1} \leqslant p \leqslant \frac{a}{1-\alpha}.$$

It follows from Theorem 1 that $T_{\alpha,a}$ is an integral operator on L^p only when $1 \leqslant p < 1/(1-\alpha)$. Thus, for $0 < \alpha \leqslant 1/2$ it is possible to choose a and $p \in (1,2)$ such that $T_{\alpha,a}$: $L^p \to L^p$ is an integral operator on L^p but the adjoint operator $T^*_{\alpha,a}$ is not an integral operator on $L^{p'}$. For example, take $\alpha = 1/4$, $a = 9/2$, and $p = 6/5$.

We fix a cube $Q_0 \subset R^N$ centered at 0 with edges of unit length and with all the edges parallel to the corresponding coordinate axes. Let Λ be the additive group of points in R^N with integer coordinates, and $\{Q_0 + m\}$, $m \in \Lambda$, the collection of cubes centered at points $m \in \Lambda$ obtained by translating Q_0. If $k \in L^1_{\text{loc}}$, then for any positive number λ and $m \in \Lambda$ we let

$$E_m^\lambda = \{x \in Q_0 + m, |k(x)| > \lambda\}. \tag{11}$$

The following theorem makes more precise the admissible local sizes of the kernels of integral convolution operators in terms of the measure of portions of the form (11).

THEOREM 2. *If* $k \in K_2$, *then there exists a constant* $C > 0$, *depending only on k and the dimension N, such that*

$$\left(\sum_{m \in \Lambda} \left(\text{meas } E_m^\lambda\right)^2 \right)^{1/2} \leqslant \frac{C}{\lambda}, \qquad \lambda > 0; \tag{12}$$

moreover, the exponent 2 *on the left-hand side of the inequality cannot in general be replaced here by a smaller exponent, not even if it is assumed that the integral convolution operator with kernel k is bounded in* L^2.

For simplicity of the arguments the proof is carried out only for the one-dimensional case $N = 1$. The case $N > 1$ is treated similarly. By Theorem 1,

$$g(y) \equiv \int_{Q_0} |k(x+y)|\, dx \in L^2. \tag{13}$$

Here $Q_0 = [-1/2, 1/2]$. If $y = m - 1/4 + \eta$ and $|\eta| < 1/2$, then

$$g\left(m - \frac{1}{4} + \eta\right) = \int_{Q_0 + m - 1/4 + \eta} |k(u)|\, du. \tag{14}$$

By (13),

$$\|g\|_2^2 = \sum_{m=-\infty}^{\infty} \int_{-1/2}^{1/2} \left| g\left(m - \frac{1}{4} + \eta\right)\right|^2 d\eta \equiv C < \infty.$$

Hence,

$$\sum_{m=-\infty}^{\infty} \int_{-1/4}^{1/4} \left| g\left(m - \frac{1}{4} + \eta\right)\right|^2 d\eta \leqslant C < \infty.$$

In an analogous way we get that

$$\sum_{m=-\infty}^{\infty} \int_{-1/4}^{1/4} \left| g\left(m + \frac{1}{4} + \eta\right)\right|^2 d\eta \leqslant C < \infty.$$

For a fixed $m \in \Lambda$

$$Q_0 + m - 1/4 + \eta \supset \{u\colon 0 \leqslant m - u \leqslant 1/2\}, \qquad |\eta| < 1/4,$$
$$Q_0 + m + 1/4 + \eta \supset \{u\colon -1/2 \leqslant m - u \leqslant 0\}, \qquad |\eta| < 1/4.$$

Hence,

$$\{Q_0 + m - 1/4 + \eta\} \cup \{Q_0 + m + 1/4 + \eta\} \supset Q_0 + m, \qquad |\eta| < 1/4.$$

Further, for any η such that $|\eta| < 1/4$ it follows from (14) that

$$\begin{aligned} &g\left(m - \frac{1}{4} + \eta\right) + g\left(m + \frac{1}{4} + \eta\right) \\ &\quad = \int_{Q_0 + m - 1/4 + \eta} |k(u)|\, du + \int_{Q_0 + m + 1/4 + \eta} |k(u)|\, du \\ &\quad \geqslant \left[\int_{\{u\colon 0 \leqslant m - u \leqslant 1/2,\, |k(u)| > \lambda\}} du + \int_{\{u\colon -1/2 \leqslant m - u \leqslant 0,\, |k(u)| > \lambda\}} du\right] \\ &\quad \geqslant \lambda \int_{E_m^\lambda} du = \lambda \operatorname{meas} E_m^\lambda. \end{aligned}$$

Moreover,

$$\left(\operatorname{meas} E_m^\lambda\right)^2 \leqslant (2/\lambda^2)\left(|g(m - 1/4 + \eta)|^2 + |g(m + 1/4 + \eta)|^2\right), \quad |\eta| < 1/4.$$

Integration gives us that

$$\frac{1}{2}\left(\operatorname{meas} E_m^\lambda\right)^2 \leqslant \frac{2}{\lambda^2} \int_{-1/4}^{1/4} \left[\left|g\left(m - \frac{1}{4} + \eta\right)\right|^2 + \left|g\left(m + \frac{1}{4} + \eta\right)\right|^2\right] d\eta.$$

Therefore,

$$\sum_{m=-\infty}^{\infty} \left(\text{meas } E_m^\lambda\right)^2 \leqslant \frac{8\|g\|_2^2}{\lambda^2} = \frac{8C}{\lambda^2},$$

which is equivalent to (12), and the first part of the theorem is proved.

To prove the second part we consider the Fourier series

$$h(x) = \sum_{n=2}^{\infty} \frac{e^{in\log n}}{n^{1/2}\log^2 n} e^{2\pi inx} \tag{15}$$

and the integral convolution operator with kernel

$$H(x) = \sum_{|n|\geqslant 2} e^{i|n|\log|n|}\chi_{[-1/2|n|^{1/2}\log^2|n|,\,1/2|n|^{1/2}\log^2|n|]}(x-n) \tag{16}$$

connected with it, where $\chi_{[a,b]}$ is the characteristic function of the interval $[a, b] \subset R^1$.

By a theorem of Carleman ([11], Theorem 4.9), the series (15) converges uniformly; consequently, its sum $h(x)$ is in $L^\infty(R^1)$. Using this fact, we show below that the convolution operator T_H corresponding to the kernel $H(x)$ is bounded in L^2; however, for $0 \leqslant \lambda < 1$ and $p < 2$

$$\text{meas } E_m^\lambda = \frac{1}{|m|^{1/2}\log^2|m|}, \quad |m| \geqslant 2; \qquad \sum_{|m|\geqslant 2} \left(\text{meas } E_m^\lambda\right)^p = \infty.$$

The theorem is proved.

The next theorem is proved similarly.

THEOREM 3. *Suppose that $1 < p \leqslant \infty$ and $k \in K_p$. Then there exists a constant $C > 0$, depending only on k, p, and the dimension N, such that*

$$\left[\sum_{m\in\Lambda} \left(\text{meas } E_m^\lambda\right)^{p'}\right]^{1/p'} \leqslant \frac{C}{\lambda}, \qquad \lambda > 0,$$

where $p' = p/(p-1)$ for $1 < p < \infty$ and $p' = 1$ for $p = \infty$.

We now show that the integral convolution operator T_H: $L^p \to L^p$ with kernel $H(x)$ of the form (16) is bounded only when $p = 2$. From Theorem 1 it follows easily that $H \in K_p$ only when $1 \leqslant p \leqslant 2$. For $\tau > 0$ let $\chi_\tau(x) = \chi_{[-\tau/2,\tau/2]}(x)$. Consider the action of T_H on the function $\chi_\tau(x)$. For sufficiently small τ

$$|T_H\chi_\tau(x)|^p = \sum_{|n|\geqslant 2} \left|\chi_{[-1/2|n|^{1/2}\log^2|n|,\,1/2|n|^{1/2}\log^2|n|]} * \chi_\tau(x)\right|^p.$$

Therefore

$$\|T_H\chi_\tau\|_p^p \geqslant \tau \sum_{n^{1/2}\log^2 n > 2/\tau} \frac{1}{n^{p/2}\log^{2p} n};$$

consequently, $\|T_H\chi_\tau\|_p = \infty$ for $p < 2$, so T_H is unbounded from L^p to L^p when $p < 2$. It follows from duality that T_H is unbounded from L^p to L^p when $p > 2$.

The fact that T_H: $L^2 \to L^2$ is bounded follows from the condition $\hat{H} \in L^\infty$ and Parseval's equality, where $\hat{H}$ is the Fourier transform of H in the distribution sense, which is not hard to compute:

$$H^{\wedge}(\xi) = \frac{2}{\pi\xi}\sum_{n=2}^{\infty} e^{in\log n}\sin\frac{\pi\xi}{n^{1/2}\log^2 n}\cos 2\pi\xi n. \tag{17}$$

The condition $H^{\wedge} \in L^{\infty}$ follows from the convergence of the function series (17) and the uniform boundedness of its partial sums. Both these assertions can be shown by following the proof of the aforementioned theorem of Carleman. We prove the second one, with the letter A standing for an absolute constant that is possibly different in different places. For $|\xi| < 1$ the boundedness of the partial sums of (15) implies the estimate

$$\begin{aligned} |H_M^{\wedge}(\xi)| &= \frac{2}{\pi|\xi|}\left|\sum_{n=2}^{M} e^{in\log n}\sin\frac{\pi\xi}{n^{1/2}\log^2 n}\cos 2\pi\xi n\right| \\ &\leqslant 2\left|\sum_{n=2}^{M}\frac{e^{in\log n}}{n^{1/2}\log^2 n}\cos 2\pi\xi n\right| + 2\pi^2|\xi|^2\sum_{n=2}^{M}\frac{1}{n^{3/2}\log^6 n} \leqslant A. \end{aligned} \tag{18}$$

For $|\xi| > 1$ we use the Abel transformation. We have

$$\begin{aligned} H_M^{\wedge}(\xi) &= \frac{2}{\pi\xi}\sum_{n=2}^{M-1}\left(\sum_{k=2}^{n} e^{ik\log k}\cos 2\pi\xi k\right) \\ &\quad\times\left(\sin\frac{\pi\xi}{n^{1/2}\log^2 n} - \sin\frac{\pi\xi}{(n+1)^{1/2}\log^2(n+1)}\right) \\ &\quad + \frac{2}{\pi\xi}\left(\sum_{n=2}^{M} e^{in\log n}\cos 2\pi\xi n\right)\sin\frac{\pi\xi}{M^{1/2}\log^2 M}. \end{aligned} \tag{19}$$

It is known ([11], Theorem 4.7) that for any integer $j > 0$

$$\left|\sum_{n=2}^{j} e^{in\log n}\cos 2\pi\xi n\right| \leqslant Aj^{1/2}.$$

From this and (19) it follows that for $|\xi| > 1$

$$|H_M^{\wedge}(\xi)| \leqslant A\left[\frac{1}{|\xi|}\sum_{n=2}^{M-1}\frac{1}{n\log^2 n} + \frac{1}{|\xi|}M^{1/2}\left|\sin\frac{\pi\xi}{M^{1/2}\log^2 M}\right|\right],$$

from which, by (18), we get that $|H_M^{\wedge}(\xi)| \leqslant A$. Thus, $H^{\wedge} \in L^{\infty}$ and $T_H\colon L^2 \to L^2$.

§2

In this section we obtain a criterion for a linear operator from L^p to L^0 to be representable as an integral convolution operator; the criterion complements a theorem of Bukhvalov [12] on integral representation of a linear operator.

If $1 \leqslant p \leqslant \infty$ and $k \in K_p$, then $k * f \in L^1_{\text{loc}}$ for $f \in L^p$. Indeed, from Hölder's inequality and Theorem 1 it follows that for any cube $P \subset R^N$ of finite size

$$\begin{aligned} \int_P |(k*f)(x)|\,dx &\leqslant \int_{R^N}|f(y)|\int_P|k(x-y)|\,dx\,dy \\ &\leqslant \|f\|_p\left\|\int_P|k(x-y)|\,dx\right\|_{p'} < \infty. \end{aligned}$$

If the spaces L^0 and L^1_{loc} are equipped with the topology of convergence in measure and the topology of convergence in the L^1-norm on bounded sets, respectively, then the remarks made above give us that T_k: $L^p \to L^0$ if and only if T_k: $L^p \to L^1_{\text{loc}}$.

Let $D = C_0^\infty(R^N)$ be the space of infinitely differentiable compactly supported functions on R^N, and D' the space of distributions dual to D. The next assertion can be obtained by following the proof of Theorem 1.2 in [1].

HÖRMANDER'S THEOREM. *Suppose that* $1 \leqslant p < \infty$ *and* T: $L^p \to L^1_{\text{loc}}$ *is a translation-invariant linear operator, i.e.,* $\tau_h Tf = T\tau_h f$ *for any* $h \in R^N$, *where* $\tau_h f(x) = f(x-h)$. *Then there exists a unique distribution* $L \in D'$ *such that* $Tf = L * f$ *for* $f \in D$, *where the convolution is understood in the distribution sense.*

Suppose that $\psi \in D$, $\psi \geqslant 0$, $\|\psi\|_1 = 1$ and $\psi_n(x) = n^N\psi(nx)$ for $n = 1, 2, \ldots$. It is well known that

$$\lim_{n\to\infty} \|\psi_n * f - f\|_p = 0$$

for any $f \in L^p$ when $1 \leqslant p < \infty$. Let $\|\cdot\|_{p,\Omega}$ denote the norm in $L^p(\Omega)$.

THEOREM 4. *Suppose that* $1 \leqslant p < \infty$, *and* T: $L^p \to L^1_{\text{loc}}$ *is a linear operator. Then the following conditions are equivalent.*

I. *T is an integral convolution operator T_k with kernel $k \in K_p$.*

II. a) *T is translation-invariant;*

b) *the limit* $\lim_{n\to\infty} T\psi_n = k$ *exists in* L^1_{loc}, *i.e.,* $\lim_{n\to\infty}\|T\psi_n - k\|_{1,P} = 0$ *for any bounded cube* $P \subset R^N$; *and*

c) $s_Q|k(x-y)|\,dx \in L^{p'}$ *for some cube* $Q \subset R^N$ *with edges of positive finite length.*

PROOF. I $\to$ II by Theorem 1. Let us prove the converse. By IIa) and Hörmander's theorem, there exists a distribution $L \in D'$ such that $Tf = L * f$ for $f \in D$. Hence,

$$T\psi_n * f = L * \psi_n * f = \psi_n * Tf \quad \text{for } f \in D.$$

Fix $f \in D$. Let P_1 be a cube of R^N of bounded size, and P_2 a cube such that $P_2 \supset P_1 + \operatorname{supp} f$. It follows from IIb) that $T\psi_n(x)$ converges to $k(x)$ in measure on P_2 as $n \to \infty$, and it can be assumed without loss of generality that $|T\psi_n(\xi)| \leqslant F(\xi)$ for a.a. $\xi \in P_2$ and $n = 1, 2, \ldots$, where $F \in L^1(P_2)$. Then for a particular $x \in P_1$ the sequence $T\psi_n(x-y)$ converges in measure to $k(x-y)$ on $\operatorname{supp} f$ as $n \to \infty$. By Lebesgue's dominated convergence theorem, this implies that

$$\lim_{n\to\infty} (T\psi_n * f)(x) = (k * f)(x) \quad \text{for any } x \in P_1.$$

On the other hand, since $Tf \in L^1(P_1)$, there is a subsequence such that

$$\lim_{k\to\infty} (\psi_{n_k} * Tf)(x) = Tf(x) \quad \text{for a.a. } x \in P_1.$$

Thus, $Tf(x) = (k * f)(x)$ for a.a. $x \in P_1$ and a particular function $f \in D$. From this, by covering R^N by a countable number of cubes obtained by translating P_1, we get that for $f \in D$

$$Tf(x) = (k * f)(x) \quad \text{for a.a. } x \in R^N.$$

For any $f \in L^p$ there is a sequence $f_j \in D$ $(j = 1, 2, \ldots)$ such that

$$|f_j(x)| \leqslant F(x) \in L^p, \qquad \lim_{j\to\infty} f_j(x) = f(x)$$

for a.a. $x \in R^N$, and $\lim_{j\to\infty}\|f_j - f\|_p = 0$. By Lebesgue's theorem and IIc),

$$\lim_{j\to\infty} k * f_j(x) = k * f(x) \quad \text{for a.a. } x \in R^N.$$

Since $T\colon L^p \to L^1_{\mathrm{loc}}$,

$$\lim_{j\to\infty} \|Tf_j - Tf\|_{1,P} = 0$$

for any bounded cube $P \subset R^N$. Hence, there is a subsequence such that

$$\lim_{k\to\infty} Tf_{j_k}(x) = Tf(x)$$

for a.a. $x \in P$, and thus $Tf(x) = (k * f)(x)$ for a.a. $x \in P$. By translating the cube P by integer vectors and choosing a subsequence of $\{j_k\}$ if necessary, we get the required equality for the translated cubes. A diagonal process gives us that $Tf(x) = (k * f)(x)$ for a.a. $x \in R^n$, which is what we were required to prove.

Bibliography

1. Lars Hörmander, *Estimates for translation-invariant operators in L^p-spaces*, Acta Math. **104** (1960), 93–140.

2. O. V. Besov, V. P. Il'in and S. M. Nikol'skiĭ, *Integral representations of functions and imbedding theorems*, "Nauka", Moscow, 1975; English transl., Vols. 1, 2, Wiley, 1979.

3. Elias M. Stein, *Singular integrals and differentiability properties of functions*, Princeton Univ. Press, Princeton, N. J., 1970.

4. Elias M. Stein and Guido Weiss, *Introduction to Fourier analysis on Euclidean spaces*, Princeton Univ. Press, Princeton, N. J., 1971.

5. Michael Münster, *Opérateurs intégraux et multiplicateurs de convolution*, Bull. Soc. Roy. Sci. Liège **39** (1970), 96–100.

6. G. Sampson, A. Naparstek and V. Drobot, *(L_p, L_q) mapping properties of convolution transforms*, Studia Math. **55** (1975/76), 41–70.

7. V. D. Stepanov, *Operators in the spaces $L_p(R_n)$ that commute with shifts*, Sibirsk. Mat. Zh. **15** (1974), 693–699; English transl. in Siberian Math. J. **15** (1974).

8. ______, *On integral convolution operators*, Dokl. Akad. Nauk SSSR **243** (1978), 45–48; English transl. in Soviet Math. Dokl. **19** (1978).

9. V. B. Korotkov and V. D. Stepanov, *On certain properties of integral convolution operators*, Application of the Methods of Functional Analysis to Problems in Mathematical Physics and Numerical Analysis (School-Sem., 1978; V. N. Vragov, editor), Inst. Mat. Sibirsk. Otdel. Akad. Nauk SSSR, Novosibirsk, 1979, pp. 64–68. (Russian)

10. Per Sjölin, *Convolution with oscillating kernels*, Indiana Univ. Math. J. **30** (1981), 47–55.*

11. A. Zygmund, *Trigonometric series*, 2nd rev. ed., Vols. I, II, Cambridge Univ. Press, 1959.

12. A. V. Bukhvalov, *A criterion for integral representability of linear operators*, Funktsional. Anal. i Prilozhen. **9** (1975), no. 1, 51; English transl. in Functional Anal. Appl. **9** (1975).

Translated by H. H. McFADEN

**Editor's note.* The Russian original cites a preprint of this article.

Труды Матем. Инст. Стеклов.
161(1983)

Proceedings of the Steklov
Institute of Mathematics
1984, Issue 4

CORRECTION TO THE PAPER "DESCRIPTIONS OF FUNCTIONS FROM CLASSES OF BESOV-LIZORKIN-TRIEBEL TYPE"

UDC 517.518.22

G. A. KALYABIN

ABSTRACT. An inaccuracy is removed in the formulation of the majorant property of the Hardy-Littlewood maximal function. The corrected error does not affect the validity of the results in the paper.

In §3 of the paper [1] there is an error in the formulation of the majorant property of the Hardy-Littlewood maximum operator.

In the right sides of inequalities (3.16) and (3.24), in place of the L_1-norm of the function $Q(t)$ should stand the norm in L_1 of its decreasing majorant

$$Q^*(t) = \operatorname{ess\,supp}\{|Q(v): |v| > |t|\}.$$

A simple proof of the strengthened inequalities may be found, for example, in the book [2], Russian pp. 71–72.

Our correction does not affect the validity of the results obtained in [1], since in all cases where inequalities (3.16) and (3.24) are used (namely to derive the relations (4.9), (5.26), (6.9), (7.11), (7.41), (8.11), (8.16), (8.21) and (8.28)) lower bounds are established for the functions $Q^*(t)$.

As for inequality (3.23), which is false in general, it is not used in [1].

BIBLIOGRAPHY

1. G. A. Kalyabin, *Descriptions of functions from classes of Besov-Lizorkin-Triebel type*, Trudy Mat. Inst. Steklov. **156** (1980), 82–109 = Proc. Steklov Inst. Math. **1983**, no. 2, 89–118.

2. Elias M. Stein and Guido Weiss, *Introduction to Fourier analysis on Euclidean spaces*, Princeton Univ. Press, Princeton, N.J., 1971; Russian transl., Mir, Moscow, 1974.

0081-5438/84 $1.00 + $.25 per page

BCDEFGHIJ–89876